Zebrafish

Zebrafish
Methods for Assessing Drug Safety and Toxicity

Edited by

Patricia McGrath
Phylonix, Cambridge, MA, USA

A JOHN WILEY & SONS, INC., PUBLICATION

Library of Congress Cataloging-in-Publication Data:

Zebrafish : methods for assessing drug safety and toxicity / edited by Patricia McGrath.
    p. cm.
  ISBN 978-0-470-42513-8 (cloth)
  1. Logperch–Genetics. 2. Drugs–Safety measures. 3. Toxicology–Animal models. 4. Fish as laboratory animals. 5. Animal models in research. I. McGrath, Patricia, 1949–
  QL638.P4Z43 2011
  597′.482–dc22

                                                            2011009822

10 9 8 7 6 5 4 3 2 1

Contents

Preface

The zebrafish model organism is increasingly used for assessing compound toxicity, safety, and efficacy and numerous studies confirm that mammalian and zebrafish toxicity profiles are strikingly similar. This convenient, predictive animal model can be used at an intermediate stage between performing cell-based assays and conventional animal testing. Although *in vitro* assays using cultured cells are commonly used to evaluate potential drug effects, they are frequently not predictive of the complex metabolism that affects drug efficacy and causes toxicity in animals. Therefore, many compounds that appear effective *in vitro* fail during costly animal trials.

Currently, there is no single reference source for toxicity testing using this emerging model organism. Investigators seeking general information on toxicity methods and results currently refer to toxicology textbooks that focus on mammalian models. The target readership of this timely book includes students (undergraduates and graduate level) and professionals in all biomedical sciences, including drug research and development, environmental testing, and product safety assessment.

This initial volume describes methods for assessing compound-induced toxicity in all major organs, including heart (Chapters 4, 5, 6, and 11), liver (Chapters 8, 9, and 11), kidney (Chapter 11), central nervous system (Chapters 10, 11, 12, 13, and 14), eye (Chapters 15 and 16), ear (Chapter 19), hematopoietic system (Chapter 7), and overall development (Chapters 2 and 3).

This vertebrate model offers several compelling experimental advantages including drug delivery directly in the fish water, small amount of drug required per experiment, statistically significant number of animals per test, and low cost. Animal transparency makes it possible to visually assess compound-induced effects on morphology and fluorescently labeled probes and antibodies can be used to localize and quantitate compound effects in physiologically intact animals. Compounds can be assessed using wild-type, mutant, transgenic, knockdown, and knock-in animals. In addition, several chemical-induced disease models, phenocopies, designed to identify potential drug candidates, are described (Chapters 14, 16, 17, 18, 19, and 21). Assays used to develop disease models can also be used to assess compound-induced toxicity on specific end points. Several widely used cell-based assay techniques have been adapted for use with this small model organism and quantitative morphometric image analysis (Chapters 10, 14, and 18) and microplate formats (9, 16, and 17) offer unprecedented throughput for assessing compound effects in whole animals. Additional analytical tools adapted for use with zebrafish, including ECG (Chapter 6) and motion detectors (Chapters 10, 12, 13, 15, and 18), are described.

Improvements in breeding and spawning, which address requirements of industrial scale screening, are discussed (Chapter 1). As a reference source to be used as a companion document for assessing data presented in individual chapters, we have reprinted a description of zebrafish stages during organogenesis. An interesting recent development that successfully pairs this emerging model with an emerging market need is the use of zebrafish for assessing safety of nanoparticles (Chapter 20), which are now incorporated in virtually all product categories. In addition, the unique ability of this animal to regenerate tissue and organs offers potential for compound screening for cell-based therapies (Chapter 22).

An important recent development impacting wider use of zebrafish for toxicity testing is that the Organization for Economic Cooperation and Development (OECD), an international organization helping governments tackle the economic social and governance challenges of the globalized economy, is developing standards for using zebrafish to assess chemical toxicity.

Further supporting wider use of this emerging model organism, the European Union recently enacted Registration, Evaluation, Authorisation and Registration of Chemicals (REACH) legislation that requires toxicity assessment for any chemical imported or manufactured in the region and is expected to have far-reaching impact on new product introductions and animal testing, including zebrafish.

Confounding interpretation of drug-induced toxicity and limiting wider acceptance of this model organism, reported results show that inter- and intralaboratory standards vary widely, although cooperation among academic and industry laboratories to develop standard operating procedures for performing compound assessment in zebrafish is increasing. Understanding all aspects of current toxicology testing will facilitate more uniform approaches across industries and enhance acceptance from regulatory authorities around the world. Full validation of this model organism will require assessment of large numbers of compounds from diverse classes in a wide variety of assays and disease models. I hope that methods and data reported here will facilitate standardization and support increased use of zebrafish for compound screening.

PATRICIA McGRATH

Cambridge, MA

Contributors

Wendy Alderton, CB1 Bio Ltd, Cambridge, UK

Jessica Awerman, Phylonix, Cambridge, MA, USA

Florian Beuerle, The Institute für Organische Chemie, Universität Erlangen-Nürnberg, Erlangen, Germany

Louis D'Amico, Phylonix, Cambridge, MA, USA

Myrtle Davis, NCI, NIH, Bethesda, MD, USA

Anthony DeLise, Sanofi-Aventis, Bridgewater, NJ, USA

Adam P. Dicker, Department of Radiation Oncology, Thomas Jefferson University, Philadelphia, PA, USA

Elizabeth Glaze, NCI, NIH, Bethesda, MD, USA

Maryann Haldi, Phylonix, Cambridge, MA, USA

Maegan Harden, Phylonix, Cambridge, MA, USA

Uwe Hartnagel, The Institute für Organische Chemie, Universität Erlangen-Nürnberg, Erlangen, Germany

Adrian Hill, Evotec (UK) Ltd, Abingdon, Oxfordshire, UK

Andreas Hirsch, The Institute für Organische Chemie, Universität Erlangen-Nürnberg, Erlangen, Germany; and C-Sixty Inc., Houston, TX, USA

Deborah L. Hunter, Integrated Systems Toxicology Division, National Health and Environmental Effects Research Laboratory, Office of Research and Development, U.S. Environmental Protection Agency, Research Triangle Park, NC, USA

Terra D. Irons, Curriculum in Toxicology, University of North Carolina, Chapel Hill, NC, USA

Gabor Kari, Department of Radiation Oncology, Thomas Jefferson University, Philadelphia, PA, USA

Christian Lawrence, Aquatic Resources Program, Children's Hospital Boston, Boston, MA, USA

Russell Lebovitz, C-Sixty Inc., Houston, TX, USA

Chunqi Li, Phylonix, Cambridge, MA, USA

Yingxin Lin, Phylonix, Cambridge, MA, USA

Liqing Luo, Phylonix, Cambridge, MA, USA

Robert C. MacPhail, Toxicity Assessment Division, National Health and Environmental Effects Research Laboratory, Office of Research and Development, U.S. Environmental Protection Agency, Research Triangle Park, NC, USA

Calum A. MacRae, Cardiovascular Division, Brigham and Women's Hospital, Boston, MA, USA

Patricia McGrath, Phylonix, Cambridge, MA, USA

Joshua Meidenbauer, Phylonix, Cambridge, MA, USA

David J. Milan, Cardiovascular Research Center and Cardiology Division, Massachusetts General Hospital, Boston, MA, USA

Stephanie Padilla, Integrated Systems Toxicology Division, National Health and Environmental Effects Research Laboratory, Office of Research and Development, U.S. Environmental Protection Agency, Research Triangle Park, NC, USA

Demian Park, Phylonix, Cambridge, MA, USA

Chuenlei Parng, Phylonix, Cambridge, MA, USA

Ulrich Rodeck, Department of Radiation Oncology, Thomas Jefferson University, Philadelphia, PA, USA; and Department of Dermatology, Thomas Jefferson University, Philadelphia, PA, USA

Katerine S. Saili, Department of Environmental and Molecular Toxicology, Environmental Health Sciences Center, Oregon State University, Corvallis, OR, USA

Wen Lin Seng, Phylonix, Cambridge, MA, USA

Sumitra Sengupta, Department of Environmental and Molecular Toxicology, Environmental Health Sciences Center, Oregon State University, Corvallis, OR, USA

Michael T. Simonich, Department of Environmental and Molecular Toxicology, Environmental Health Sciences Center, Oregon State University, Corvallis, OR, USA

Breanne Sparta, Phylonix, Cambridge, MA, USA

Willi Suter, Novartis Pharmaceuticals, East Hanover, NJ, USA

Tamara Tal, Department of Environmental and Molecular Toxicology, Environmental Health Sciences Center, Oregon State University, Corvallis, OR, USA

Jian Tang, Phylonix, Cambridge, MA, USA

Susie Tang, Phylonix, Cambridge, MA, USA

Robert L. Tanguay, Department of Environmental and Molecular Toxicology, Environmental Health Sciences Center, Oregon State University, Corvallis, OR, USA

Alison M. Taylor, Stem Cell Program and Division of Hematology/Oncology, Children's Hospital Boston and Dona Farber Cancer Institute, Harvard Medical School

Lisa Truong, Department of Environmental and Molecular Toxicology, Environmental Health Sciences Center, Oregon State University, Corvallis, OR, USA

Patrick Witte, The Institute für Organische Chemie, Universität Erlangen-Nürnberg, Erlangen, Germany

Yi Yang, Novartis Pharmaceuticals, East Hanover, NJ, USA

Lisa Zhong, Phylonix, Cambridge, MA, USA

Leonard I. Zon, Stem Cell Program and Division of Hematology/Oncology, Children's Hospital Boston and Dona Farber Cancer Institute, Harvard Medical School

Acknowledgments

Special thanks to the contributing authors who are at the forefront of developing methods for compound screening in zebrafish. Thanks also to the Phylonix team for their patience and for taking a backseat while this book took shape; Yingli Duan, Kristine Karklins, Demian Park, and Wen Lin Seng doggedly edited all chapters and generated data describing state-of-the-art assays for compound assessment in zebrafish.

Chapter 1

The Reproductive Biology and Spawning of Zebrafish in Laboratory Settings*

Christian Lawrence

Aquatic Resources Program, Children's Hospital Boston, Boston, MA, USA

1.1 INTRODUCTION

There is growing demand for new, robust, and cost-effective ways to assess chemicals for their effect on human health, particularly during early development. Traditional mammalian models for toxicology are both expensive and difficult to work with during embryonic stages. The zebrafish (*Danio rerio*) has a number of features that make it an excellent alternative model for toxicology studies, including its small size, rapid external development, optical transparency during early development, permeability to small molecules, amenability to high-throughput screening, and genetic similarity to humans (Lieschke and Currie, 2007; Peterson et al., 2008).

A major underpinning of the use of zebrafish in this arena is their great fecundity, which supports high-throughput analysis and increases the statistical power of experiments. Adult female zebrafish can spawn on a daily basis, and individual clutch sizes can exceed 1000 embryos (Spence and Smith, 2005; Castranova et al., 2011). However, consistent production at these high levels is greatly dependent upon sound management of laboratory breeding stocks, which must be grounded in a thorough understanding of the reproductive biology and behavior of the animal. Management practices must also address key elements of husbandry, most notably water quality, nutrition, and behavioral and genetic management.

---

*Some information in this chapter was originally published in Harper and Lawrence, The Laboratory Zebrafish, CRC Press/Taylor and Francis Group, 2011. Used with permission.

Zebrafish: Methods for Assessing Drug Safety and Toxicity, Edited by Patricia McGrath.

1.2 OVERVIEW OF ZEBRAFISH REPRODUCTIVE BIOLOGY AND BEHAVIOR

1.2.1 Natural History

Zebrafish are native to South Asia, and are distributed primarily throughout the lower reaches of many of the major river drainages of India, Bangladesh, and Nepal (Spence et al., 2008). This geographic region is characterized by its monsoonal climate, with pronounced rainy and dry seasons. Such seasonality in rainfall profoundly affects both the physicochemical conditions in zebrafish habitats and resource availability. These factors also shape reproductive biology and behavior.

Data gathered from the relatively small number of field studies suggest that zebrafish are primarily a floodplain species, most commonly found in shallow, standing, or slow-moving bodies of water with submerged aquatic vegetation and a silt-covered substratum (Spence et al., 2008). Environmental conditions in these habitats are highly variable in both space and time. For example, pooled environmental data from zebrafish collection sites in India in the summer rainy season (Engeszer et al., 2007) and Bangladesh in the winter dry season (Spence et al., 2006) show that pH ranges from 5.9 to 8.5, conductivity from 10 to 2000 μS, and temperature from 16 to 38°C. These differences, which reflect changes in seasonality and geography, provide strong evidence that zebrafish are adapted to wide swings in environmental conditions. Results of laboratory experiments demonstrating their tolerance to both thermal (Cortemeglia and Beitinger, 2005) and ionic (Boisen et al., 2003) fluctuations support this hypothesis.

Zebrafish feed mainly on a wide variety of zooplankton and insects (both aquatic and terrestrial), and to a lesser extent, algae, detritus, and various other organic materials (McClure et al., 2006; Spence et al., 2007a). Gut content analyses of wild collected animals indicate that they feed primarily in the water column, but also take items off the surface and the benthos (Spence et al., 2007a).

Zebrafish are a shoaling species, most often occurring in small schools of 5–20 individuals (Pritchard et al., 2001), although shoals of much larger numbers have been observed (Engeszer et al., 2007). Reproduction takes place primarily during the monsoons, a period of resource abundance (Talwar and Jhingran, 1991). Fish spawn in small groups during the early morning, along the margins of flooded water bodies, often in shallow, still, and heavily vegetated areas (Laale, 1977). There has also been at least one report of fish spawning during periods of heavy rain later on in the day (Spence et al., 2008). Females scatter clutches of eggs over the substratum, and there is no parental care. The eggs, which are demersal and nonadhesive, develop and hatch within 48–72 h at 28.5°C. After hatching, larvae adhere to available submerged surfaces by means of specialized cells on the head (Eaton and Farley, 1974). Within 24–48 h post hatch, they inflate their gas bladders and begin to actively feed on small zooplankton. Larval fish remain in these nursery areas as they develop, and move into deeper, open water as they mature and floodwaters recede (Engeszer et al., 2007).

1.2.2 Reproductive Cycle and Controlling Factors

Zebrafish typically attain sexual maturity within 3–6 months post fertilization in laboratory settings, although this may vary considerably with environmental conditions, most importantly rearing densities, temperature, and food availability (Spence et al., 2008). Consequently, it may be more appropriate to relate reproductive maturity to size rather than age. Data from a number of studies indicate that a standard length of approximately 23 mm corresponds with attainment of reproductive maturity in this species (Eaton and Farley, 1974; Spence et al., 2008).

Under favorable conditions, zebrafish spawn continuously upon attainment of sexual maturation (Breder and Rosen, 1966). Females are capable of spawning on a daily basis. Eaton and Farley (1974) found that females would spawn once every 1.9 days if continuously housed with a male, and Spence and Smith (2006) reported that females were capable of producing clutches every day over a period of at least 12 days, though variance in egg production was substantial. This interval is likely to be greater when the environment (water chemistry, nutrition, behavioral setting, etc.) is suboptimal or if the fish are used for production frequently (Lawrence, 2007).

Olfactory cues play a determining role in zebrafish reproduction and spawning behavior (Fig. 1.1). The release of steroid glucuronides into the water by males induces ovulation in females (Chen and Nartinich, 1975; Hurk and Lambesrt, 1983). Gerlach (2006) reported that females exposed to male pheromones showed significant increases in spawning frequencies, clutch size, and egg viability when compared with females held in isolation. Upon ovulation, females release pheromones that in turn

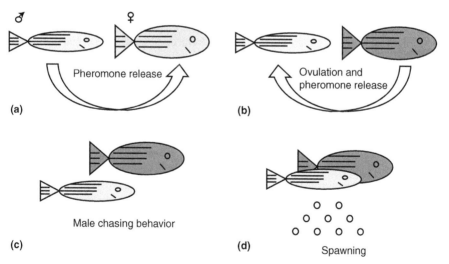

Figure 1.1 Simplified model of olfactory control of zebrafish reproduction. (a) Male (yellow) releases pheromone into water in vicinity of female (blue). (b) Female smells pheromone, which triggers ovulation (indicated by female color change to green), which is then followed by female release of postovulatory pheromones. (c) Male senses pheromones, which trigger mating and chasing behavior. (d) Spawning. (See the color version of this figure in Color Plates section.)

prompt male mating behavior that immediately precedes and elicits oviposition and spawning (Hurk and Lambesrt, 1983). Pheromonal release in some cases also appears to suppress reproduction, as holding water from "dominant" female zebrafish has been shown to inhibit spawning of subordinate females (Gerlach, 2006).

Reproduction in zebrafish is also influenced by photoperiod. Ovulation most typically occurs just prior to dawn (Selman et al., 2005) and spawning commences within the first few hours of daylight (Spence et al., 2006; Engeszer et al., 2007). However, spawning is not strictly limited to this time period. Zebrafish will breed in the laboratory throughout the day, particularly during the evenings, although spawning is most reliable and intense in the early morning (personal observation). In the wild, zebrafish have also been observed spawning during the afternoon following the onset of heavy rain (Spence et al., 2008).

1.2.3   Reproductive Behavior

Zebrafish display ritualized courtship behaviors prior to and during spawning. During courtship, males swim in tight circles or hover, with fins raised, above a spawning site in clear view of nearby females. If females do not approach, males will chase them to the site, snout to flank. When spawning, a male swims parallel to a female and wraps his body around hers, triggering oviposition and releasing sperm simultaneously (Spence et al., 2008). This ritualized mating behavior and the fact that males are known to establish and defend territories indicate that females are selective (Darrow and Harris, 2004; Spence and Smith, 2005). This is supported by the fact that females will produce larger clutches and spawn more frequently when paired with certain males (Spence and Smith, 2006).

Females may exert choice on the basis of several combined factors. The quality of a spawning site is clearly important, as both male and female zebrafish show a strong preference for oviposition site, selecting and preferentially spawning over gravel versus silt in both laboratory and field-based experiments (Spence et al., 2007b). If given the choice, fish will also spawn preferentially in vegetated versus nonvegetated sites (Spence et al., 2007b) and in shallow versus deep water (Sessa et al., 2008; Adatto et al., 2011).

Male defense of territories may be one cue that females use to select males. Spence and Smith (2005, 2006) found that territorial males had a marginally higher reproductive success than nonterritorial males at low densities, though there was no difference at higher fish densities, and that male dominance rank did not correlate with female egg production. This fact, coupled with female preferences for substrate, depth, and structure for spawning, suggests that male defense of desirable spawning locations over which females are choosy may be the basis to the zebrafish mating system.

Females appear to select males based on their genotype. Many fish, including zebrafish, use olfactory cues to differentiate between kin and nonkin, and this mechanism may be utilized during breeding to avoid inbreeding. Zebrafish also appear to use olfactory cues to make social and reproductive decisions. Using odor plume tests, Gerlach and Lysiak (2006) showed that adult female zebrafish chose the odors of nonrelated, unfamiliar (reared and maintained separately) males over those of unfamiliar brothers for mating. The underlying genetic basis of this preference is unknown, but may be the major histocompatibility complex (MHC) genes that are important in kin recognition in other fish species (Apanius et al., 1997).

1.3 SPAWNING TECHNIQUES AND TECHNOLOGY

1.3.1 In-Tank Strategies

One general approach to breeding zebrafish in the laboratory is to simply provide a spawning site or substrate directly in holding tanks, while fish remain "on system" or in flow. This type of technique relies on the "natural" production of fish kept in mixed sex groups with minimal manipulation of individuals. Another important feature of this basic approach is that because fish remain on flow, water quality is regulated and maintained throughout breeding events. Finally, it also largely minimizes the handling of fish, which can be a stressful event (Davis et al., 2002).

The first formally described technique for breeding laboratory zebrafish is the most basic example of an in-tank breeding method. In this approach, glass marbles are placed at the bottom of holding tanks to provide a spawning substrate for the animals. Fish spawn over the marbles, and the eggs drop into the spaces in between, preventing egg cannibalism and facilitating their subsequent collection by siphoning (Westfield, 1995; Brand et al., 2002). While this method may be effective to some extent, it is generally impractical for use in large culturing facilities with hundreds or thousands of tanks. Despite its shortcomings, it is still frequently cited in the methods sections of zebrafish papers, and is often used by investigators breeding zebrafish for the first time.

A slightly more advanced in-tank approach involves placing a breeding box or container in holding tanks that fish will spawn over during breeding events (Fig. 1.2a). A common feature of this method is that the box/container will have a mesh-type top through which spawned eggs drop and are subsequently protected from cannibalism. The box will also typically have some plastic plants affixed to it to make it more attractive as a spawning site. This type of method is more facile than the marbles technique, as boxes can be moved freely in and out of holding tanks as desired. It also better facilitates the collection of staged embryos from groups of fish, and can also be used for breeding pairs. This method is utilized relatively infrequently, and thus no commercially fabricated equipment of this type is available. When this method is chosen, the box/container must be custom-made to fit with the needs of the particular facility in which it is being utilized.

Another form of in-tank breeding involves the use of a specially manufactured crossing cage that is designed to fit inside holding tanks. The fish to be crossed are netted out of holding tanks and transferred to the crossing cage. Eggs are collected after breeding takes place by siphoning or after removal of the fish from the tank. This method allows for production of time-staged embryos because it can include a divider to separate males and females until eggs are needed for experiments. This technology

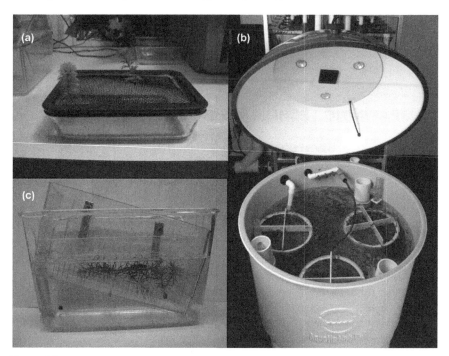

Figure 1.2 Representative examples of zebrafish spawning technology. (a) In-tank breeding container. (b) MEPS™. (c) Typical static tank mating tank with insert.

has a number of drawbacks, including the fact that all fish in the housing tanks where breeding is taking place must be either in the crossing cage or transferred to other tanks so that eggs are not cannibalized. This requires extensive handling of animals, offsetting one inherent advantage of the in-tank breeding methodology. Second, in most cases, flow of clean water into tanks must be either shut off or reduced to prevent spawned eggs from being flushed out of the tanks. There may be means by which to collect these eggs when flow remains on, but if not, another strength of the in-tank system is taken away when using this method. Finally, although various cages of this type are commercially available, they are often system vendor specific, which limits their applicability to the users of the associated system.

The most recent development in in-tank breeding technology is the Mass Embryo Production System (MEPS™), designed by Aquatic Habitats, an aquatic animal housing system manufacturer. The MEPS™ is a large spawning vessel, with a holding capacity of 80 or 250 L, which can be plumbed directly into any existing recirculating or flow-through system. The MEPS™, which can house large populations (up to 1000 or more) of breeding fish, contains one or more spawning platforms, which are specially fabricated funnels capped with plastic mesh screens that can be located at various depths inside the vessel (Fig. 1.2b). When the spawning platforms are placed inside the vessel, fish breed over and on the platforms, and spawned eggs fall through the mesh into the associated funnels. The eggs are then pumped through an attached tube into separate collection screens by means of pressurized air directed into the funnels, allowing eggs to be collected without disturbing the fish. The units also have the capability to be run on altered photoperiods via the use of an attached light cycle dome with a programmable light cycle dimmer.

The MEPS™ system capitalizes upon several attributes of the general in-tank breeding approach, including consistent water quality and minimal handling of animals, with the added benefits of reduced labor input and increased space efficiency. When used properly, this technology is capable of supporting high-level egg production on the order of tens of thousands of embryos per event, and is therefore well suited for experimental applications requiring large numbers of time-staged eggs. However, this approach is not without its limitations and specific challenges. For example, its use is limited to experiments where the individual identity of parents is not necessary, which excludes it from being used for certain types of genetic screens, which are an important component of the zebrafish model system. The performance of fish in this type of breeding unit is also very dependent upon management. Detailed understanding of reproductive behavior and biology of the fish is imperative to maximize efficiency, and therefore the MEPS™ may be less suitable for newly established zebrafish laboratories where such expertise is not available.

1.3.2 Static Tank Strategies

The alternative to in-tank breeding strategies is to remove fish from holding tanks and to spawn them in off-system or "static water" breeding chambers. This general approach, which is utilized in the great majority of zebrafish breeding facilities, adheres to the following general principles: a small (typically <1 L) plastic mating cage or insert with a mesh or grill bottom is placed inside a slightly larger container that is filled with water. Fish (pairs or small groups) are then added to the insert in the evening. When the fish spawn, the fertilized eggs fall through the "floor" of the insert and are thereby protected from cannibalism by adults (Mullins et al., 1994).

This technique has proven to be generally effective and, consequently, derivations of the static tank design are manufactured by a number of aquaculture and laboratory product supply companies. Available products vary slightly in size, shape, depth, and total volume, as well as adjustability of inserts in the static spawning chamber (Fig. 1.2c). A very small number of studies have explored the effects of variations of these parameters on reproductive success and spawning efficiency. Sessa et al. (2008) showed that fish set up in crossing cages in which spawning inserts were titled to provide a deep to shallow water gradient showed statistically significant increases in egg production when compared with fish set up in cages in which the inserts were not titled (no gradient). Fish that were set up in chambers with titled inserts displayed both a preference to spawn in shallow water and specific breeding behaviors that were limited to the tilted physical configuration. Indeed, this behavior is the basis of a newly developed approach for collecting large numbers of developmentally synchronized embryos from groups of fish in a static breeding vessel (Adatto et al., 2011).

Little else has been published in this area, although a study of the effects of varying the size of the breeding insert itself on spawning success and egg production showed no difference in spawning success between control cage of 3.5 L and test cages of 500, 400, 300, 200, and 100 mL, and reduced production in 200 and 100 mL sizes (Goolish et al., 1998). However, since this particular study was conducted in recirculating water (test chambers were placed inside large on-system tanks), it does not present a clear picture of the effect of chamber size on breeding efficiency in static tanks.

There are a number of strengths to the static tank approach. Virtually any type of experiment can be supported using this technique, as fish of any desired genotype can be set up in pairs or smaller groups in a varying number of crosses. Because fish are removed from holding tanks, the effects of behavioral hierarchies established in holding tanks that can be counterproductive to breeding are negated. Static tank technologies also allow for direct manipulation of water quality parameters; changes in water chemistry, such as decreases in salinity, pH, and temperature, are thought to promote spawning in fish adapted to monsoonal climate regimes (Murno, 1990). These factors may also affect reproduction in zebrafish (Breder and Rosen, 1966).

There are drawbacks to static tank breeding strategies. Because the chambers are off-flow, water quality conditions in the spawning setups deteriorate over time. Although this has not been formally investigated, metabolites such as total ammonia nitrogen and carbon dioxide accumulate in the water and are likely to have a negative effect on spawning. Tanks may be flushed with fresh water to offset these potential problems, but this represents added labor. Using static setups also necessitates that fish are handled constantly, which may be a source of long-term stress for breeding populations.

1.4 DETERMINING FACTORS FOR REPRODUCTION IN LABORATORY STOCKS OF ZEBRAFISH

That zebrafish will readily spawn under a wide range of conditions in captivity has undoubtedly played an important role in their rapid rise to prominence as a model organism. This flexibility also suggests that there is considerable spread in reproductive performance of laboratory stocks. Indeed, there are a number of key husbandry factors that impact breeding efficiency, including water quality, nutrition, behavior, and genetic management. A detailed understanding of how each of these factors contributes to reproductive success is vital to maximizing production of zebrafish in controlled settings. These concepts are touched upon briefly below. For a more in-depth treatment of these subjects, see reviews by Lawrence (2007) and Spence et al. (2008).

1.4.1 Water Quality

Zebrafish tolerate a wide range of environmental conditions in captivity. This flexibility is a reflection of their distribution in the wild, as they are found across a range of habitat types that vary considerably in their physicochemical properties as a result of local geology and pronounced seasonal fluctuations in rainfall patterns (Talwar and Jhingran, 1991). However, it should be recognized that there is an energetic cost to fish in operating outside their optimum range of environmental parameters. Animals maintained under suboptimal conditions must devote an increasing proportion of energy toward maintaining homeostasis, rather than on growth, reproduction, and immune function (Wooton, 1998). Consequently, one major consequence of fish being held under suboptimal conditions is a decrease in the number and quality of offspring (Haywood, 1993). Thus, it is vital to manage water chemistry as close to optimal as possible to ensure that fish allocate resources to reproductive function.

Stability within a given range of each parameter is also crucial, and may be more important than maintaining at optimum, especially for a generalist species like zebrafish. Adapting to constantly fluctuating environmental conditions is energy intensive, and can be a source of chronic stress that manifests itself in decreases in number and quality of offspring (Wooton, 1998; Conte, 2004).

While managing water quality for stability within optimum ranges is straightforward conceptually, it is a bit more challenging to achieve in practice, primarily because optimum environmental conditions for zebrafish for the most part have yet to be demonstrated experimentally. Until such data are available, the soundest practice is to base management on the best available scientific information. Observational data from years of experimental use along with concepts gleaned from biological studies of zebrafish allow for a reasonable place to start, however. A detailed treatment of each of these factors relative to the management of zebrafish is given in the review by Lawrence (2007).

1.4.2   Nutrition and Feeding

Nutrition and feeding are among the most important determinants of reproductive success—or failure—in zebrafish facilities. Therefore, to ensure efficient and scientifically sound management of breeding stocks, it is essential that managers and technicians possess a thorough understanding of fish nutrition and the different types of feeds available, as well as the techniques to deliver them.

While the specific nutritional requirements of zebrafish are yet to be determined, it possible to apply scientific principles of finfish nutrition, along with what zebrafish specific data does exist in the design of diets and feeding regimens that will support high levels of production. At the most general level, stocks should be fed balanced diets with adequate levels of essential nutrients: proteins, lipids, carbohydrates, vitamins, and minerals. Deficiencies in essential nutrients will result in reduced production, low growth, and decreased immune function, among other problems.

At minimum, it is also crucial to ensure that diets used for breeding populations of zebrafish contain adequate levels of specific nutrients known to support reproductive function in fish. Most notably, these include the highly unsaturated fatty acids (HUFAs) eicosapentaenoic acid (20:5n-3; EPA), docosahexaenoic acid (22:6n-3;

DHA), and arachidonic acid (20:4n-6; AA), all of which are of pivotal importance for the production of high-quality gametes and offspring (Watanabe, 1982), and have been specifically shown to enhance reproduction in zebrafish (Jaya-Ram et al., 2008). Certain vitamins, including retinoids and ascorbic acid in particular, are also known to be extremely important for long-term reproductive quality and health, and should be considered in diet selection (Dabrowski and Ciereszgo, 2001; Alsop et al., 2008).

The type of feed is also of critical importance. Zebrafish may be fed live prey items, processed diets, or some mixture of the two. Since the specific nutritional requirements of zebrafish have yet to be determined, and may be fundamentally different from even closely related species, it may be unwise to feed an exclusively processed diet, especially since systematic studies of adult zebrafish performance on these diets are not available. Live prey items such as *Artemia* typically possess relatively balanced nutritional profiles (Watanabe, 1982) and therefore are most likely to meet much of the requirements of zebrafish. Processed diets may be included to the diet as a supplement to *Artemia*, as they can be used to deliver specific nutrients that may not be present in sufficient levels in *Artemia* or other live prey items. For example, *Artemia* are deficient in DHA and in stabilized vitamin C (Lavens and Sorgeloos, 1996). One way to address these inadequacies is to incorporate a prepared feed containing known levels of these nutrients into the diet to help ensure that these dietary requirements are adequately met and reproductive function is supported.

Finally, it is essential that feeds be stored and administered properly. This is particularly critical for processed feeds. The typical maximal shelf life of a processed feed does not exceed 3 months, when maintained in cool, dry conditions (Craig and Helfrich, 2002). Oxidation of feed components, particularly fatty acids, increases with temperature. Thus, feeds should be kept in airtight containers, refrigerated, and discarded after 3 months to ensure that fish stocks derive maximal nutritional benefit from their application. In terms of delivery, processed feeds should be fed dry to minimize leaching of water-soluble amino acids and vitamins upon administration (Pannevis and Earle, 1994; Kvåle et al., 2007).

1.4.3 Genetic Management

Small, closed populations of laboratory strains of animals such as zebrafish are subject to a continuous loss of genetic diversity stemming from founder effects, genetic drift, and population bottlenecking (Stohler et al., 2004). This loss of genetic diversity can cause a number of problems relative to reproductive potential of zebrafish breeding stocks. Continued breeding between close relatives will lead to accumulation of deleterious alleles in breeding populations. These alleles may directly affect a number of factors related to reproduction, including reduced quantity and quality of embryos. Reduced genetic diversity may also manifest itself in reduced spawning rates, as zebrafish show preference to associate with nonrelatives over siblings or closely related individuals (Gerlach and Lysiak, 2006). This mode of kin recognition, which is thought to help avoid inbreeding in natural populations, may result in decreased spawning rates when fish in a breeding population are closely related.

These and other problems related to the loss of genetic diversity may be alleviated to a certain extent by careful genetic breeding programs that (1) maximize effective population size and (2) minimize breeding between siblings or close relatives. Genetic diversity may also be maintained or enhanced by periodically importing fish from outside populations and breeding them programmatically with existing stocks.

1.4.4 Behavioral Management

Zebrafish reproductive behavior is complex and undoubtedly exerts myriad effects on reproductive potential of breeding stocks. The most notable instance of this type of dynamic involves social interactions between fish in holding tanks. Dominant females have been shown to suppress egg production in subordinate females via the release of pheromones (Gerlach, 2006). Further, aggression arising during formation of dominance hierarchies and territory establishment by both males and females is a source of both acute and chronic stress that may also decrease reproductive output (Pottinger and Pickering, 1992; Fox et al., 1997).

Employing various strategies of behavioral management may help to minimize the potentially negative effects of such interactions on the reproductive capacity of breeding stocks. For example, the establishment of dominance hierarchies detrimental to breeding may be prevented to some extent by regularly mixing fish from different tanks and periodically flushing tanks and systems with fresh water to reduce concentrations of repressive pheromones circulating in the water (Adatto et al., 2011). In addition, maintaining fish at intermediate densities in holding tanks may also reduce the frequency and intensity of antagonistic interactions, which are highest when densities are low and territories are easiest to defend (Pickering and Pottinger, 1987). While an "optimal" density value has not yet been determined for zebrafish, the results of one recent study suggest that using stocking densities as high as 12 fish per liter has no negative impact on reproductive performance (Castranova et al., 2011).

1.5 CONCLUSIONS

The impressive reproductive capacity of zebrafish makes them an attractive experimental model for toxicology studies. There are a number of existing tools that are currently available to breed zebrafish in laboratory settings, and the application of emerging data on the reproductive biology and behavior of this species will lead to improved technologies. The efficiency of zebrafish production is dependent upon sound stock management grounded in key principles of fish husbandry.

REFERENCES

ADATTO I, LAWRENCE C, THOMPSON M, and ZON LI (2011). A new system for the rapid collection of large numbers of developmentally synchronized zebrafish embryos. PLoS One 6(6): e21715.
ALSOP D, MATSUMOTO J, BROWN S, and VAN DER KRAAK G (2008). Retinoid requirements in the reproduction of zebrafish. Gen Comp Endocrinol 156(1): 51–62.

APANIUS V, PENN D, SLEV PR, RUFF LR, and POTTS WK (1997). The nature of selection on the major histocompatibility complex. Crit Rev Immunol 17(2): 179–224.

BOISEN AM, AMSTRUP J, NOVAK I, and GROSELL M (2003). Sodium and chloride transport in soft water and hard water acclimated zebrafish (*Danio rerio*). Biochim Biophys Acta 1618(2): 207–218.

BRAND M, GRANATO M, and NÜSSLEIN-VOLHARD C (2002). Keeping and raising zebrafish. In: Nüsslein-Volhard C and Dahm R (Eds.), Zebrafish: A Practical Approach. Oxford University Press, Oxford, pp. 7–37.

BREDER CM and ROSEN DE (1966). Modes of Reproduction in Fishes. The Natural History Press, New York, 941 pp.

CASTRANOVA D, LAWTON A, LAWRENCE C, BAUMANN DP, BEST J, COSCOLLA J, DOHERTY A, RAMOS J, HAKKESTEEG J, WANG C, WILSON C, MALLEY J, and WEINSTEIN, BM (2011). The effect of stocking densities on reproductive performance in laboratory zebrafish (Danio rerio). Zebrafish epub ahead of print.

CHEN L and NARTINICH R (1975). Pheromonal stimulation and metabolite inhibition of ovulation in the zebrafish, *Brachydanio rerio*. Fish Bull 73: 889–893.

CONTE FS (2004). Stress and the welfare of cultured fish. Appl Anim Behav Sci 86: 205–223.

CORTEMEGLIA C and BEITINGER TL (2005). Temperature tolerances of wild-type and red transgenic zebra danios. Trans Am Fish Soc 134: 1431–1437.

CRAIG S and HELFRICH L (2002). Understanding fish nutrition, feeds, and feeding. Virginia Cooperative Extension, Publication No. 420-256.

DABROWSKI KR and CIERESZGO A (2001). Ascorbic acid and reproduction in fish: endocrine regulation and gamete quality. Aquacult Res 32: 623–638.

DARROW KO and HARRIS WA (2004). Characterization and development of courtship in zebrafish, *Danio rerio*. Zebrafish 1(1): 40–45.

DAVIS KB, GRIFFIN BR, and GRAY WL (2002). Effect of handling stress on susceptibility of channel catfish *Ictalurus punctatus* to *Ichthyophthirius multifiliis* and channel catfish virus infection. Aquaculture 214: 55–66.

EATON R and FARLEY RD (1974). Spawning cycle and egg production in zebrafish, *Brachydanio rerio,* reared in the laboratory. Copeia (1): 195–204.

ENGESZER RE, PATTERSON LB, RAO AA, and PARICHY DM (2007). Zebrafish in the wild: a review of natural history and new notes from the field. Zebrafish 4(1): 21–40.

FOX HE, WHITE SA, KAO MH, and FERNALD RD (1997). Stress and dominance in a social fish. J Neurosci 17(16): 6463–6469.

GERLACH G (2006). Pheromonal regulation of reproductive success in female zebrafish: female suppression and male enhancement. Anim Behav 72: 1119–1124.

GERLACH G and LYSIAK N (2006). Kin recognition in zebrafish based on phenotype matching. Anim Behav 71: 1371–1377.

GOOLISH EM, EVANS R, and MAX R (1998). Chamber volume requirements for reproduction of the zebrafish *Danio rerio*. Prog Fish Cult 60: 127–132.

HAYWOOD GA (1993). Ammonia toxicity in teleost fish: a review. Can Tech Rep Aquat Sci 1177: 1–35.

HURK VR and LAMBESRT JGD (1983). Ovarian steroid glucuronides function as sex pheromones for male zebrafish, *Brachydanio rerio*. Can J Zool 61: 2381–2387.

JAYA-RAM A, KUAH MK, LIM PS, KOLKOVSKI S, and SHU-CHIEN AC (2008). Influence of dietary HUFA levels on reproductive performance, tissue fatty acid profile and desaturase and elongase mRNAs expression in female zebrafish *Danio rerio*. Aquaculture 277: 275–281.

KVÅLE A, NORDGREEN A, TONHEIM S, and HAMRE K (2007). The problem of meeting dietary protein requirements in intensive aquaculture of marine fish larvae, with emphasis on Atlantic halibut (*Hippoglossus hippoglossus* L.). Aquacult Nutr 13: 170–185.

LAALE HW (1977). The biology and use of zebrafish *Brachydanio rerio* in fisheries research: a literature review. J Fish Biol 10: 121–173.

LAVENS L and SORGELOOS P (1996). Manual on the production and use of live food for aquaculture. FAO Fisheries Technical Paper 361, 295 pp.

LAWRENCE C (2007). The husbandry of zebrafish (*Danio rerio*): a review. Aquaculture 269: 1–20.

Lieschke GJ and Currie PD (2007). Animal models of human disease: zebrafish swim into view. Nat Rev Genet 8(5): 353–367.

McClure MM, McIntyre PB, and McCune AR (2006). Notes on the natural diet and habitat of eight danionin fishes, including the zebrafish *Danio rerio*. J Fish Biol 69: 553–570.

Mullins MC, Hammerschmidt M, Haffter P, and Nusslein-Volhard C (1994). Large-scale mutagenesis in the zebrafish: in search of genes controlling development in a vertebrate. Curr Biol 4(3): 189–202.

Murno AD (1990). Tropical freshwater fishes. In: Munro AD, Scott P, and Lam TJ (Eds.), Reproductive Seasonality in Teleosts: Environmental Influences. CRC Press, Boca Raton, FL, pp. 145–239.

Pannevis MC and Earle KE (1994). Nutrition of ornamental fish: water soluble vitamin leaching and growth of *Paracheirodon innesi*. J Nutr 124(12 Suppl): 2633S–2635S.

Peterson RT, Nass R, Boyd WA, Freedman JH, Dong K, and Narahashi T (2008). Use of non-mammalian alternative models for neurotoxicological study. Neurotoxicology 29(3): 546–555.

Pickering AD and Pottinger TG (1987). Crowding causes prolonged leucopenia in salmonid fish, despite interrenal acclimation. J Fish Biol 32: 701–712.

Pottinger TG and Pickering AD (1992). The influence of social interaction on the acclimation of rainbow trout, Oncorhynchus mykiss (Walbaum) to chronic stress. J Fish Biol 41: 435–447.

Pritchard VL, Lawrence J, Butlin RK, and Krause J (2001). Shoal choice in zebrafish, *Danio rerio*: the influence of shoal size and activity. Anim Behav 62: 1085–1088.

Selman K, Wallace RA, Sarka A, and Qi X (2005). Stages of oocyte development in the zebrafish, *Brachydanio rerio*. J Morphol 218(2): 203–224.

Sessa AK, White R, Houvras Y, Burke C, Pugach E, Baker B, Gilbert R, Thomas Look A, and Zon LI (2008). The effect of a depth gradient on the mating behavior, oviposition site preference, and embryo production in the zebrafish, *Danio rerio*. Zebrafish 5(4): 335–339.

Spence R, Fatema MK, Ellis S, Ahmed ZF, and Smith C (2007a). Diet, growth and recruitment of wild zebrafish in Bangladesh. J Fish Biol 71: 304–309.

Spence R, Ashton R, and Smith C (2007b). Adaptive oviposition choice in the zebrafish, *Danio rerio*. Behaviour 144: 953–966.

Spence R, Fatema MK, Reichard M, Huq KA, Wahab MA, Ahmed ZF, and Smith C (2006). The distribution and habitat preferences of the zebrafish in Bangladesh. J Fish Biol 69(5): 1435–1448.

Spence R, Gerlach G, Lawrence C, and Smith C (2008). The behaviour and ecology of the zebrafish, *Danio rerio*. Biol Rev Camb Philos Soc 83(1): 13–34.

Spence R and Smith C (2005). Male territoriality mediates density and sex ratio effects on oviposition in the zebrafish, *Danio rerio*. Anim Behav 69(6): 1317–1323.

Spence R and Smith C (2006). Mating preference of female zebrafish, *Danio rerio*, in relation to male dominance. Behav Ecol 17: 779–783.

Stohler RA, Curtis J, and Minchella DJ (2004). A comparison of microsatellite polymorphism and heterozygosity among field and laboratory populations of *Schistosoma mansoni*. Int J Parasitol 34(5): 595–601.

Talwar PK and Jhingran AG (1991). Inland Fishes of India and Adjacent Countries. A.A. Balkema, Rotterdam, 1158 pp.

Watanabe T (1982). Lipid nutrition in fish. Biochem Physiol 73B: 3–15.

Westfield M (1995). The Zebrafish Book: A Guide for the Laboratory Use of Zebrafish (*Danio rerio*), 3rd edition. University of Oregon Press, Eugene, OR, 385 pp.

Wooton RJ (1998). The Ecology of Teleost Fishes, 2nd edition. Chapman and Hall, London, 404 pp.

Chapter 2

Developmental Toxicity Assessment in Zebrafish

Maryann Haldi[1], Maegan Harden[1], Louis D'Amico[1], Anthony DeLise[2], and Wen Lin Seng[1]

[1]*Phylonix, Cambridge, MA, USA*
[2]*Sanofi-Aventis, Bridgewater, NJ, USA*

2.1 INTRODUCTION

After the developmental effects of thalidomide were recognized in 1966, the FDA established protocols to be used for assessing drug effects on reproduction and development prior to approval for human use (Marathe and Thomas, 1990). In addition, due to concerns about chemicals in the human food supply, the EPA issued similar guidelines for pesticides in 1982 and industrial chemicals in 1985. Current methods for studying developmental toxicity and teratogenicity include *in vivo* mammal segment studies, *in vitro* rodent whole embryo culture test, *in vitro* cell culture test, *in vitro* embryonic stem cell test, and FETAX (frog embryo teratogenesis assay—*Xenopus*). The *in vitro* culture tests have been shown to have limited value in predicting the effect of drugs on human embryonic and fetal development (Oberemm, 2000) and the FETAX assay is limited by the animal's partial transparency. In addition, frogs are insensitive to halogenated aromatic hydrocarbons (Fort et al., 1988), and responses to toxicant exposure in *Xenopus* differ from response in mammals. Although mammalian models remain the gold standard for assessing developmental toxicity, acceptance of zebrafish as a model for toxicological analysis is increasing in the US, mirroring widespread use in Europe for environmental assessment (Parng, 2005; Spitsbergen and Kent, 2003). There is strong rationale for performing developmental toxicity studies in zebrafish including the following (Ton et al., 2006): (1) zebrafish is a distinct species, potentially enhancing sensitivity for identifying compounds that exhibit teratogenicity *in vivo*, (2) developmental processes in zebrafish are highly conserved, (3) zebrafish embryos can be cultured

Zebrafish: Methods for Assessing Drug Safety and Toxicity, Edited by Patricia McGrath.
© 2012 John Wiley & Sons, Inc. Published 2012 by John Wiley & Sons, Inc.

throughout development—blastula through advanced organogenesis—compared to rodent embryo culture that is limited to early organogenesis, and (4) the zebrafish genome is well characterized and dysmorphology phenotypes linked to genomic targets can potentially enable rapid evaluation of mechanisms of action for compound-induced teratogenicity. Here, we describe methods for assessing developmental toxicity in zebrafish and recent results.

2.2 METHODS

2.2.1 Zebrafish Developmental Toxicity Screen

Compounds that induce defects in embryo morphology or organ function before causing death are defined as teratogens and the zebrafish toxicity assay is designed to assess lethality and toxicity curves. Multiple concentrations of each compound are tested for the duration of embryogenesis and 10–30 animals are typically used for each concentration.

Compared with other animal models, rapid development is a significant advantage for assessing compound developmental toxicity in zebrafish (Table 2.1); organogenesis is essentially complete within the first 5 days of development (Westerfield, 1993). Due to transparency, organs can be assessed *in vivo* directly by microscopy without use of complicated dissecting procedures.

2.2.2 Embryo Selection and Drug Treatment

The stages for initiating and terminating compound treatment are critical parameters for assessing developmental toxicity. Since zebrafish organ development is completed by 5dpf, this stage is usually used for terminating compound treatment. To evaluate compound effects throughout gastrulation, it is important to initiate compound

Table 2.1 Comparison of Embryo Development in Zebrafish, Mouse, and Humans

Stage of embryonic development	Zebrafish	Mouse	Humans
Cleavage period	0–2hpf	Days 1–2	Days 1–5
8-cell stage	$1\,^1/_4$hpf		Day 4
Gastrulation	$5\,^1/_4$–10hpf	Days 6–8	Weeks 2–4
Initial neurulation	9hpf	Day 8	Day 22
Formation of five brain ventricles	30hpf	Day 10	Weeks 5–9
Heart formation	22hpf	Day 8	Day 16
Heart beat	30hpf	Day 8	Day 22
Organogenesis	24–48hpf	Days 8–14	Weeks 4–8
Initial craniofacial development	24hpf	Day 8	Day 22

treatment as early as possible. Due to the presence of unfertilized eggs and the time required to select healthy, synchronized eggs prior to compound treatment, the stage for compound treatment is ordinarily 6hpf and not later than 24hpf.

Since spontaneous mutation and death can take place between 6 and 24hpf, selecting healthy, synchronized embryos is crucial for establishing a valid developmental toxicity screen. The earliest stage for distinguishing healthy, synchronized embryos is 4hpf, leaving a 2 h window for selecting embryos prior to compound treatment. Figure 2.1 illustrates typical phenotypes of 4hpf embryos observed in a pool of several clutches. Normal 4hpf embryos that exhibit intact chorion membranes (Fig. 2.1a and b) are selected and reexamined 2 h later; normal chorionated 6hpf phenotypes (Fig. 2.1c) are then used for compound treatment. Normal fertilized embryos are pooled for random selection for subsequent compound treatment. Six-well microplates are typically used for treating 30 embryos simultaneously with 4 mL compound solution in which embryos are incubated at 28°C in the dark for 114 h.

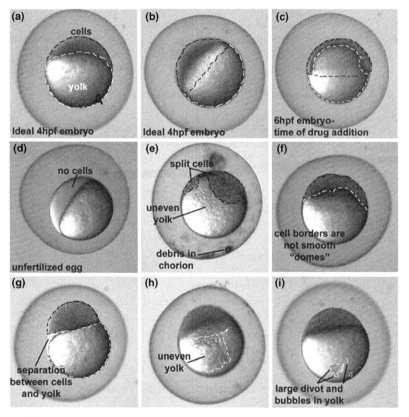

Figure 2.1 Phenotypes of normal and abnormal 4 and 6hpf zebrafish embryos. (a and b) Normal 4hpf embryos. (c) Normal 6hpf embryo. (d–i) Abnormal 4hpf embryos.

2.2.3 Lethality Concentration Determination (LC$_{50}$)

Since the concentration range that causes lethality or toxicity varies by compound, initially an exploratory experiment using 10 concentrations, 0.01, 0.05, 0.1, 0.5, 1, 5, 10, 50, 100, and 500 μM (or up to the highest concentration permitted by compound solubility), is performed. Additional higher or lower concentrations are tested, if necessary. Ideally, in these experiments, concentrations exhibiting 0–100% lethality are determined permitting generation of concentration–response curves; LC$_{50}$ can be accurately estimated from the lethality curves. Untreated and DMSO-treated zebrafish are used as controls; DMSO concentration usually ranges from 0.1% to 0.4%. Dead zebrafish are removed from each well daily and number of dead zebrafish is recorded. At the end of the test period, number of dead zebrafish and percent lethality are obtained for each concentration.

The overall objective of this initial 10-dose range-finding study is to select concentrations for subsequent studies by determining (a) the concentration range at which lethality occurs and (b) the concentration range at which any developmental malformation is observed. Once the dose range for LC$_{50}$ is established, LC$_{50}$ experiments are performed three times, using a narrower concentration range. To obtain a LC$_{50}$ $\pm$ SD (standard deviation) for each compound, best-fit concentration–response curves are then generated using the logistic regression function of statistical software (i.e., JMP, The SAS Institute, Cary, NC).

2.2.4 Toxicity Concentration Determination (EC$_{50}$)

During these exploratory experiments, the concentration range that causes malformation in zebrafish development is also identified. Based on initial observations, a narrower concentration range (six concentrations) is then established that allows accurate estimation of EC$_{50}$, which ideally includes a low concentration that exhibits few or no malformations, a high concentration that exhibits several malformations (but not death), and two concentrations in the middle of the range; a concentration–response curve can then be generated. Untreated and DMSO alone (usually between 0.1% and 0.4% maximum concentration) are used as controls. In addition, a positive control drug (such as 0.1 μM 9-*cis*-retinoic acid) is used to validate the assay. A negative control drug, such as penicillin, can also be included. After compound treatment, 10 randomly selected live zebrafish are examined by microscopy (Zeiss Stemi 2000C stereomicroscope equipped with a SPOT digital camera). For image capture, embryos are anesthetized using tricaine a.k.a. MESAB (0.5 mM 3-aminobenzoic acid ethyl ester, 2 mM Na$_2$HPO$_4$) and/or mounted in methylcellulose.

2.2.5 Quantitation of Developmental Toxicity

Table 2.2 describes common developmental defects scored in the developmental toxicity screen. Because circulation and motility are sensitive to temperature and

Table 2.2 Parameters for Scoring Developmental Toxicity at 5dpf

Common defects	Criteria
Heart morphology	Missing chambers, abnormal blood flow, asynchrony
Circulation	Absent, slow, or fast blood flow; defective vessel pattern
Edema	
Pericardial	Accumulation of fluid surrounding the heart
Trunk	Accumulation of fluid in the interstitial space surrounding an organ or tissue in the trunk
Eye/head	Accumulation of fluid around the eye or brain
Hemorrhage	Presence of pooled blood outside of vascular network
Brain morphology	Small or misshapen
Brain tissue	Brown/opaque tissue
Jaw morphology	Short or misshapen jaw
Tail morphology	Short, curly or bent tail
Eye morphology	Small or misshapen eye(s)
Eye pigmentation	Absent or patchy retinal pigmented epithelium
Notochord morphology	Notochord that is not straight (e.g., wavy or kinked)
Body pigmentation	Significantly higher or lower amount of black or yellow pigment
Tail/trunk muscle morphology	Presence of shredded, degenerating muscle within the somites
Motility	No movement or recoil after touch by a needle
Fin malformation	Absent or small pectoral fins
Liver morphology	Absent or abnormal size
Liver tissue	Presence of dark brown, opaque tissue with no passage of blood
Intestine malformation	Absent, no lumen or no infolded epithelium, does not extend to anal pore
Intestinal tissue	Brown/opaque tissue

anesthetic, to reduce experimental artifacts, these parameters are measured prior to tricaine addition.

For each concentration, number of zebrafish exhibiting defects is recorded to obtain percent incidence using formula (2.1):

$$\text{percent incidence} = \frac{\text{no. of zebrafish with defects}}{10} \times 100\%. \qquad (2.1)$$

A concentration–response curve (percent incidence versus concentration) is then generated for each compound using the nonlinear regression function available in statistical software (JMP, The SAS Institute).

2.2.6 Determination of Teratogenicity and Calculation of Teratogenic Index

Teratogenicity is evaluated by comparing compound concentrations that cause lethality and concentrations that cause developmental defects. Following the definition used for performing FETAX, the teratogenic index (TI) is defined as the LC_{50}/EC_{50} ratio for each compound. A compound is considered teratogenic if the TI is >1, indicating that the concentration at which 50% of the zebrafish exhibit developmental abnormalities is lower than the concentration at which 50% lethality is observed. A TI < 1 is considered nonteratogenic. If the TI = 1, the toxicity assessment must be performed at the exact LC_{50} concentration to determine if zebrafish exhibit malformations at that concentration. Obtaining a TI = 1 is unlikely.

2.3 RESULTS

2.3.1 Results Based on TI = LC_{50}/EC_{50}

To demonstrate different outcomes in zebrafish, we present results for four test compounds: (1) SSR101010 (FAAH inhibitor, Sanofi-Aventis), a nonteratogenic drug (rat and rabbit), (2) M100907 (5HT antagonist, Sanofi-Aventis), a teratogenic drug (rat and rabbit), (3) valproic acid (an anticonvulsant, Bristol-Myers Squibb), a weak teratogenic drug (mammalian models), and (4) 9-*cis*-retinoic acid (a cytostatic agent, Bristol-Myers Squibb), a potent teratogenic drug (mammalian models).

Table 2.3 summarizes the LC_{50} and EC_{50} for these four drugs. SSR101010 was not very soluble and precipitation was observed at concentrations $\geq 5\,\mu M$; however, since no lethality was observed at any test concentration (up to $500\,\mu M$), LC_{50} was not determined. M100907 precipitated at concentrations $\geq 200\,\mu M$, which exhibited

Table 2.3 LC_{50} and EC_{50} of Test Drugs

Drug	LC_{50}	EC_{50}	TI (LC_{50}/EC_{50})	Teratogenic
SSR101010[a]	ND[b]	ND[b]	ND[b]	No
M100907[a]	ND	95.8	>1.04[c]	Yes
Valproic acid[d]	59.8	28.7	2.08	Yes
9-*cis*-Retinoic acid[d]	ND[e]	ND[e]	ND[e]	Yes

ND: not determined, since 50% lethality was not observed at soluble concentrations. At $400\,\mu M$, precipitation was observed; however, 100% lethality was observed.
[a] Sanofi-Aventis, proprietary compounds.
[b] Not determined, since no lethality and no abnormality were observed.
[c] TI based on the highest soluble concentration divided by the EC_{50}.
[d] Bristol-Myers Squibb, validation compounds.
[e] Not determined, since no lethality was observed at the lowest testing concentration, $0.1\,\mu M$; however, abnormality was observed in 100% treated animals.

8.9% lethality. One hundred percent lethality was observed at 400 µM; however, precipitation was observed in the solution. Therefore, the actual LC_{50} concentration was not determined. Valproic acid has an estimated LC_{50} of 59.8 µM. 9-*cis*-Retinoic acid did not cause lethality at the lowest testing concentration, 0.1 µM; therefore, LC_{50} was not determined.

Figure 2.2a shows the organ structure of treated zebrafish at 8× magnification. Figure 2.2b shows the same images with organs outlined to highlight dysmorphogenesis: heart (blue), liver (red), and intestine (green). SSR101010-treated zebrafish exhibited similar morphology as untreated and DMSO controls at all test concentrations, although precipitation was observed at concentrations ≥5 µM. M100907-treated zebrafish exhibited abnormal jaw (blue arrow in the left panel), and dark opaque color in liver (black arrow in the left panel) and in intestine (yellow arrow in the left panel), although organ morphology was not affected (outlined in the right panel). In addition, arrhythmia was observed (not shown). Valproic acid-treated zebrafish exhibited pericardial edema (red arrow in the left panel); therefore, the heart chambers were elongated to form a thin tube (blue area in the right panel). In addition, small liver and intestine were observed (red and green areas, respectively, in the right panel). 9-*cis*-Retinoic acid-treated zebrafish exhibited pericardial edema with a small, elongated heart chamber (blue area in the right panel); small intestine (green area in the right panel) and absence of liver tissue were also observed (lack of red area in the right panel).

Using the TI definition, the values for SSR101010 could not be determined. However, since no abnormality was observed at any test concentration, we determined that SSR101010 was not teratogenic. Although LC_{50} was not determined for M100907, based on low lethality observed at the highest soluble concentration (100 µM), estimated TI was >1.04; therefore, M100907 was considered teratogenic. Valproic acid exhibited a TI value of 2.08, and was therefore teratogenic. Although TI was not determined for 9-*cis*-retinoic acid, 100% of treated zebrafish exhibited abnormalities and severe body deformation at the lowest tested nonlethal concentration (0.1 µM), indicating that this drug is a potent teratogen. Results are summarized in Table 2.3.

2.3.2 Results Based on Alternative TI Definition

In a recent comprehensive study, 12 blinded compounds, including valproic acid and 9-*cis*-retinoic acid (Bristol-Myers Squibb), were evaluated for developmental toxicity in zebrafish embryos using TI value defined as visual assessment at maximal nonlethal concentration. During the first phase of the study (LC_{50} determination), compounds were ranked according to LC_{50} values prior to performing the time-consuming developmental toxicity screen (Fig. 2.3). Teratogenicity properties of these compounds were further evaluated by visual assessment *in vivo* (Table 2.4).

The data from this study were evaluated for strength in predictivity and accuracy based on teratogenicity properties identified *in vivo* (Table 2.4). The assay was good (>70%, but <80%) for specificity and excellent (>80%) for sensitivity. That is, it

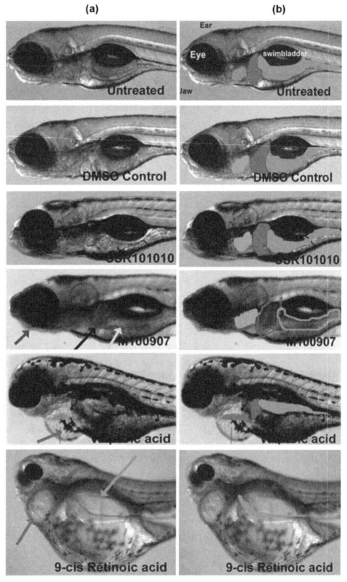

Figure 2.2 Organ structure of treated zebrafish at high magnification (8×). Panels (a) and (b) are the same images; however, organs are outlined in panel (b) to highlight dysmorphogenesis: heart (blue), liver (red), and intestine (green). SSR101010-treated zebrafish exhibited similar morphology as controls in all tested concentrations, although precipitation was observed at concentrations ≥5 µM. M100907-treated zebrafish exhibited abnormal jaw (blue arrow (a)), and dark opaque liver (black arrow, (a)) and intestine (yellow arrow, (a)), although organ morphology was not affected (outlined in (b)). In addition, arrhythmia was observed (data not shown). Valproic acid-treated zebrafish exhibited pericardial edema (red arrow in (a)); therefore, the heart chambers were stretched to form a thin tube (b). In addition, small liver and intestine were observed (b). 9-*cis*-Retinoic acid-treated zebrafish exhibited pericardial edema and a stretched tiny heart chamber (b); small intestine and absence of liver tissue were also observed (b). (See the color version of this figure in Color Plates section.)

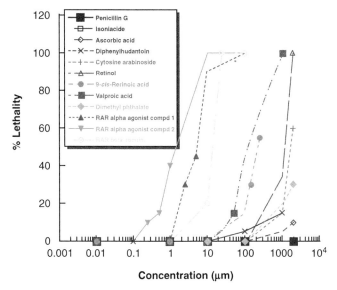

Figure 2.3 Lethality curves in zebrafish. (See the color version of this figure in Color Plates section.)

presented a 75% success rate in identifying nonteratogenic compounds and a 100% success rate in identifying teratogens. The accuracy for predicting relative teratogenic potency was excellent with a 92% success rate in predicting accuracy.

2.4 DISCUSSION

Here we describe 6hpf to 5dpf as the time frame for compound treatment; however, different time points including 24hpf for compound treatment initiation to 2dpf for compound termination have been used by other researchers (Henry et al., 1997; Prasch et al., 2003). Shorter treatment time increases throughput and is generally adequate for assessing effects in both the cardiovascular and nervous systems; however, it is not considered adequate for assessing compound effects on the developing digestive system, which develops at a later time point. Another consideration for these studies is the effect of the chorion membrane on compound availability.

Although the typical zebrafish developmental toxicity screen includes 14–21 parameters, animal transparency offers the unique opportunity to design screens to address specific interests. The parameters and TI definition used by the pharmaceutical industry and government agencies charged with responsibility for protecting citizens and the environment from effects of hazardous chemicals may differ. For example, since the retinoid class of drugs is known to induce neurological defects in humans, to determine compound toxicity on the developing nervous system, specific compartments in the brain, such as brain ventricle, can be assessed. For agrichemicals and pesticides, environmental concerns may warrant evaluation of all discernable

Table 2.4 Assessment of Compound Teratogenicity in Zebrafish

Compound	LC$_{50}$ (μM)	Zebrafish				Zebrafish teratogen	Mammalian teratogen	Correct prediction
		Body	Heart	Liver	GI			
Dimethyl phthalate	ND	No	No	No	No	None	None	Yes
Retinol	1146	Yes	Yes	Yes	Yes	Weak	Weak	Yes
Valproic acid	110.9	Yes	Yes	No	No	Weak	Weak	Yes
BMS*-A (RAR alpha agonist 1)	4.7	Yes	Yes	Yes	Yes	Potent	Potent	Yes
BMS*-B (RAR alpha agonist 2)	1.3	Yes	Yes	Yes	Yes	Potent	Potent	Yes
BMS*-C (RAR beta agonist)	15	Yes	Yes	Yes	No	Potent	Potent	Yes
Penicillin G	ND	No	No	No	No	None	None	Yes
Isoniacide	ND	No	No	No	No	None	None	Yes
Ascorbic acid	ND	No	No	No	No	Can't determine	None	Can't determine
Diphenylhydantoin	ND	No	No	No	No	None	Weak	No
Cytosine arabinoside	1478	Yes	Yes	Yes	Yes	Potent	Potent	Yes
9-cis-Retinoic acid	258.4	Yes	Yes	Yes	Yes	Potent	Potent	Yes

parameters, including swim bladder inflation and hatching time. The LC_{50}/EC_{50} definition of TI can be used for evaluating compound effects on environment. In contrast, an alternative TI definition based on the presence of any dysmorphogenesis at the maximal nonlethal concentration is often used to evaluate therapeutic compounds.

Although the developmental toxicity screen focuses on compound effects during organogenesis, the same visual methods are widely used to perform nondevelopmental toxicity screens. In these assays, since compounds are added after the organs are formed and functioning, it is unlikely that arrested development will be observed; however, morphological defects, including enlarged or small organs, can be scored.

Subsequent to visual assessment used in developmental toxicity screens, complementary quantitative assays are typically performed. Immunostaining with cell-specific antibodies can be used to confirm visual observations and quantitative assay formats include morphometric image analysis and microplate assessment in 96 and 384 wells.

REFERENCES

FORT DJ, DAWSON DA, and BANTLE JA (1988). Development of a metabolic activation system for the frog embryo teratogenesis assay: *Xenopus* (FETAX). Teratog Carcinog Mutagen 8(5): 251–263.

HENRY TR, SPITSBERGEN JM, HORNUNG MW, ABNET CC, and PETERSON RE (1997). Early life stage toxicity of 2,3,7,8-tetrachlorodibenzo-*p*-dioxin in zebrafish (*Danio rerio*). Toxicol Appl Pharmacol 142(1): 56–68.

MARATHE M and THOMAS G (1990). Current status of animal testing in reproductive toxicology. Indian J Pharmacol 22: 192–201.

OBEREMM A (2000). The use of a refined zebrafish embryo bioassay for the assessment of aquatic toxicity. Lab Anim 29: 32–40.

PARNG C. (2005). *In vivo* zebrafish assays for toxicity testing. Curr Opin Drug Discov Dev 8(1): 100–106.

PRASCH AL, TERAOKA H, CARNEY SA, DONG W, HIRAGA T, STEGEMAN JJ, HEIDEMAN W, and PETERSON RE (2003). Aryl hydrocarbon receptor 2 mediates 2,3,7,8-tetrachlorodibenzo-*p*-dioxin developmental toxicity in zebrafish. Toxicol Sci 76(1): 138–150.

SPITSBERGEN JM and KENT ML (2003). The state of the art of the zebrafish model for toxicology and toxicologic pathology research: advantages and current limitations. Toxicol Pathol 31(Suppl): 62–87.

TON C, LIN Y, and WILLETT C (2006). Zebrafish as a model for developmental neurotoxicity testing. Birth Defects Res A Clin Mol Teratol 76(7): 553–567.

WESTERFIELD M (1993). The Zebrafish Book: A Guide for the Laboratory Use of Zebrafish (*Danio rerio*). University of Oregon Press, Eugene, OR.

Chapter 3

Use of Emerging Models for Developmental Toxicity Testing

Patricia McGrath

Phylonix, Cambridge, MA, USA

3.1 IMPORTANCE OF ASSESSING DEVELOPMENTAL TOXICITY

Approximately 50% of all pregnancies in the United States result in a sick baby or prenatal or post natal death. Furthermore, major developmental defects, including neural tube and heart deformities, occur in approximately 120,000 of the 4 million infants born in the United States each year. Developmental toxicity profiles of most chemicals and approved drugs have not been extensively evaluated and conventional mammalian assays using fetuses are laborious and expensive.

3.2 CURRENT METHODS FOR ASSESSING DEVELOPMENTAL TOXICITY

In vivo mammal segment studies using female rats or rabbits are standardized screens for assessing developmental and reproductive toxicity (DART). The entire reproductive cycle is divided into three segments: Segment I is designated for reproductive toxicity and does not include fetal examination, Segment II encompasses the period from implantation through major organogenesis, and Segment III focuses on late pregnancy through post natal development. Segment II involves exposing pregnant animals during the period of major organogenesis and structural development. Dosing for Segment II studies consists of a control group and three or four compound concentrations; each group is comprised of 20 pregnant dams. The dams are sacrificed

Zebrafish: Methods for Assessing Drug Safety and Toxicity, Edited by Patricia McGrath.

just prior to normal delivery. Then, death and fetal defects, including external, skeletal, and visceral malformations, are assessed visually or by X-ray. External malformations including missing limbs and missing or malformed bones and visceral malformations including development of internal organs such as heart, brain, lungs, and so on are examined. Each specific malformation is typically recorded as a dichotomous variable (present/absent) and each fetus may exhibit several types of malformations. These conventional developmental toxicity tests are laborious, cost ineffective, and time consuming. As a result, developmental toxicity is currently assessed for only a small number of drugs during preclinical development.

3.3 USE OF EMERGING MODELS FOR DEVELOPMENTAL TOXICITY TESTING

Although whole embryo testing has previously been performed on invertebrates, including fruit flies and nematodes (Eisses, 1989; Hitchcock et al., 1997), these species are not closely related to humans and they lack many mammalian organs and enzymes. Therefore, use of *Drosophila* and *Caenorhabditis elegans* as comparative models for assessing toxicity, including developmental toxicity, has been limited.

3.3.1 Frog Embryo Teratogenesis Assay—*Xenopus*

The frog embryo teratogenesis assay—*Xenopus* (FETAX) has recently been developed as a screening assay for assessing developmental toxicity. This test uses early-stage South African clawed frog *Xenopus* embryos to assess chemical effects on (1) mortality, (2) malformation, and (3) growth inhibition. In 1998, the U.S. EPA requested that the Interagency Coordinating Committee on the Validation of Alternative Methods (ICCVAM) evaluate FETAX.

Although FETAX appears to be capable of measuring key relevant developmental toxicity end points, including lethality, malformation, and growth, and estimating the dose–response relationships for these end points, based on the available data, the ICCVAM panel concluded that FETAX results were excessively variable, both within and between labs. The ICCVAM panel also concluded that FETAX was not sufficiently validated or optimized for inclusion in regulatory applications and it recommended further standardization of the assay to improve variability and expansion of the number of end points to identify developmental toxicants. Additional disadvantages of FETAX include that since frog embryos are only partially transparent, development and internal organ structures cannot be fully visualized. Frogs have also been shown to exhibit toxic responses different from those present in mammals (Fort et al., 1988, 1991, 1998).

An additional recent relevant development is that Public Law 103-43 directed the National Institute of Environmental Health Sciences (NIEHS, NIH) to develop and validate alternative methods that (1) reduce or eliminate use of animals in acute or chronic toxicity testing, (2) establish criteria for the validation and regulatory acceptance of alternative testing methods, and (3) recommend a process through which scientifically validated alternative methods can be accepted for regulatory use.

3.3.2   Advantages of Using Zebrafish to Assess Developmental Toxicity

The processes involved in embryogenesis have been well studied in zebrafish and functions of numerous genes have been shown to be highly conserved. Although transparent zebrafish embryos offer unique advantages for assessing compound effects on various developmental events, a comprehensive set of reference protocols for assessing developmental toxicity has not yet been developed.

As strong support for use of zebrafish as a model for assessing toxicity, we recently examined effects of 115 characterized compounds (Table 3.1) that were commercially available or provided by the National Cancer Institute (NCI). In this study, compounds were delivered by incubating animals in solution and we then assessed LC_{50}, developmental teratogenicity, and organ-specific toxicity. Teratogenicity results were remarkably similar to LD_{50} (compounds delivered in a single, controlled dose) results in mammals, underscoring the value of using zebrafish to predict toxicity in humans. A comparison of assay performance characteristics and cost for toxicity testing for model systems is shown in Table 3.2.

3.4   NEW GUIDELINES FOR CHEMICAL TESTING USING ZEBRAFISH

Supporting wider use of zebrafish for toxicity testing, the Organization for Economic Cooperation and Development (OECD), an international organization helping governments tackle the economic, social, and governance challenges of the globalized economy, is currently validating standards to assess chemical toxicity using the fish embryo toxicity (FET) test that has been developed from studies performed primarily using zebrafish (Groth et al., 1993, 1994; Schulte and Nagel, 1994; Lange et al., 1995; Roseth et al., 1996; Cheng et al., 2000; DIN, 2001; Nguyen and Janssen, 2001; Wiegand et al., 2001; Bachmann, 2002; Nagel, 2002; Ferrari et al., 2003; Hallare et al., 2004; Versonnen and Janssen, 2004; Versonnen et al., 2004; Braunbeck et al., 2005; Kammann et al., 2006). Since these guidelines are likely to become the gold standard for performing teratogenicity studies for chemicals and drugs, an overview of general methods is included. Although the draft test guidelines are under revision, the general principles are likely to remain largely unchanged.

3.4.1   General Principle of the Fish Embryo Test

The general principle of the test is based on chemical exposure of newly fertilized zebrafish embryos for up to 48 h, which reflects acute toxicity. Zebrafish embryos are individually exposed to a range of test concentrations in 24-well microtiter plates.

Table 3.1 Comparison of Toxicity in Zebrafish and Mammalian Models

Compound	Zebrafish		Mammalian models	
	LC_{50}/log LC_{50} (mg/L)	Target organs	LD_{50}/log LD_{50} (mg/kg)	Target organs
Didemnin B	6.2/0.79	Developmental	4.0/0.6 (mouse, i.v.), 4.0/0.6 (rat, oral)	Kidney
Merbarone	4.7/0.67	Teratogenic, liver, kidney, brain	55/1.74 (mouse, i.v.)	Liver, kidney
Fujisawa peptide	30/1.48	Teratogenic, heart, kidney	5.1/0.7 (rat, i.v.)	Lung, heart
Ecteinascidin	0.42	Teratogenic, kidney	Unknown	Nonspecific
4-Ipomeanol	94/1.97	Nonspecific	Unknown	Liver, kidney, lungs
Dexamethasone	324/2.51	Liver, gastrointestinal, kidney	410/2.61 (mouse, i.p.), 1800/3.26 (mouse, oral)	Liver, heart, kidney, gastrointestinal
Doxorubicin	30.3/1.51	Teratogenic, liver, cardiovascular, kidney	21/1.35 (mouse, i.v.), 12.60/1.10 (rat, i.v.), 9.4/0.97 (mouse, i.v.)	Liver, cardiovascular
Tacrine	11.13/1.04	Teratogenic, liver	20/1.3 (mouse, i.p.)	Liver
Vinblastine sulfate	90.9/1.96	Liver, gastrointestinal, developmental	305/2.48 (rat, oral), 220/2.34 (rat, i.p.), 15/1.18 (mouse, i.v.), 560/2.75 (mouse, i.p.)	Mutagenic, bone marrow depression, liver, developmental
5-Fluorouracil (5-FU)	3.3/0.52	Liver, kidney	18.9/1.2 (rabbit, oral), 230/2.36 (rat, oral), 115/2.06 (mouse, oral), 81/1.91 (mouse, i.v.)	Reproductive, liver, kidney
Staurosporine	0.012/− 1.92	Liver, kidney	16/1.2 (mouse, i.v.)	Unknown
Epirubicin	16.3/1.2	Cardiac, hemorrhage	16/1.2 (mouse, i.v.), 14.2/1.15 (rat, i.v.)	Carcinogenic, mutagenic, cardiac

Cyclophosphamide	650/2.8	Developmental, liver	315/2.5 (mouse, oral), 160/2.2 (rat, oral), 148/2.17 (rat, i.v.), 40/1.6 (rat, i.p.), 140/2.15 (mouse, i.v.), 315/2.5 (mouse, i.p.)	Teratogenic, liver
Methotrexate	454/2.6	Gastrointestinal, liver, kidney, teratogenic	180/2.3 (rat, oral), 15/1.18 (rat, i.p.), 94/1.97 (mouse, i.p.)	Gastrointestinal, liver, teratogenic
Actinomycin D	50/1.7	Liver, gastrointestinal, developmental	7.2/0.86 (rat, oral)	Mutagenic, respiratory, liver, developmental
Bleomycin	272.9/2.44	Heart, liver, gastrointestinal, smaller brain	168/2.23 (rat, i.p.)	Pulmonary, teratogenic
Melphalan	152.5/2.18	Gastrointestinal, liver, pancreas	13/1.11 (rat, oral)	Carcinogenic, mutagenic, bone marrow suppression, gastrointestinal, liver
Carboplatin	185.6/2.27	Liver, kidney	118/2.07 (mouse, i.p.), 343/2.54 (rat, oral), 61/1.79 (rat, i.v.), 89.4/1.95 (mouse, i.v.)	Bone marrow depression, kidney, liver, hearing loss, neurotoxicity
Cisplatin	3/0.48	Liver, kidney, teratogenic	20/1.3 (rat, oral), 8/0.90 (rat, i.v.), 6.4/0.81 (rat, i.p.), 327/2.51 (mouse, oral), 11/1.04 (mouse, i.v.), 6.6/0.82 mouse, i.p.)	Bone marrow depression, kidney, liver, hearing loss, neurotoxicity
Hydroxyurea	12160/4.08	Teratogenic, gastrointestinal	2000/3.3 (dog, oral)	Teratogenic, gastrointestinal, mutagenic, bone marrow depression
Thymidine	36300/4.56	Teratogenic, liver	2512/3.4 (mouse, i.p.)	Mutagenic, possible effects on fetus
Etoposide	176.7/2.25	Developmental, gastrointestinal	215/2.33 (mouse, oral)	Carcinogenic, mutagenic, developmental, gastrointestinal

(continued)

Table 3.1 (*Continued*)

Compound	Zebrafish LC$_{50}$/log LC$_{50}$ (mg/L)	Target organs	Mammalian models LD$_{50}$/log LD$_{50}$ (mg/kg)	Target organs
Chlorambucil	9.1/0.96	Gastrointestinal, liver, pancreas	17.7/1.25 (mouse, oral)	Carcinogenic, mutagenic, bone marrow suppression, gastrointestinal, liver
5-Azacytidine	6100/3.79	Teratogenic, liver, gastrointestinal	500/2.7 (mouse, oral), 229/2.36 (mouse, i.v.), 68/1.83 (mouse, i.p.)	Carcinogenic, teratogenic, liver, gastrointestinal
Busulfan	61.5/1.79	Liver, gastrointestinal	120/2.08 (mouse, oral)	Carcinogenic, bone marrow depression
Aminonicotinamide	479.5/2.68	Teratogenic, liver, gastrointestinal	35/1.54 (mouse, i.p.)	Teratogenic, mutagenic, liver
Aphidicolin	10.14/1.00	Liver, kidney	Unknown	Carcinogenic
Furazolidone	45/1.65	Teratogenic, liver, brain	300/2.48 (mouse, i.p.), 1782/3.25 (mouse, oral), 2336/3.37 (rat, oral)	Carcinogenic, mutagenic, teratogenic, liver, CNS
2-Chlorodeoxyadenosine	114.4/2.06	Gastrointestinal, kidney, teratogenic	Unknown	Immunotoxic, gastrointestinal, kidney, developmental
Topotecan	11.45/1.06	Teratogenic, liver	Unknown	Teratogenic, mutagenic
Mechlorethamine	96.5/1.98	Liver, teratogenic	10/1 (rat, oral), 100/2.0 (human, skin)	Carcinogenic, bone marrow suppression, liver, teratogenic
Ifosfamide	1174/3.06	Liver, kidney, teratogenic	1005/3.0 (mouse, oral), 143/2.16 (rat, oral), 190/2.28 (rat, i.v.), 140/2.15 (rat, i.p.), 338/2.53 (mouse, i.v.), 397/2.60 (mouse, i.p.)	Carcinogenic, mutagenic, liver, kidney

Compound				
Cycloheximide	8.4/0.93	Teratogenic, liver, brain, gastrointestinal	2/0.3 (rat, oral), 2/0.3 (rat, i.v.), 3.7/0.57 (rat, i.p.), 133/2.12 (mouse, oral), 150/2.18 (mouse, i.v.), 100/2.0 (mouse, i.p.)	Mutagenic, respiratory, CNS, kidney, liver
Tamoxifen	17/1.23	Liver, eye, CNS	218/2.33 (mouse, oral), 1190/3.08 (rat, oral), 62.50/1.8 (rat, i.v.), 575/2.76 (rat, i.p.), 62.5/1.8 (mouse, i.v.), 3100/3.49 (mouse, i.p.)	Carcinogenic, ocular, induce hypertrophy, reproductive, mutagenic
Clomiphene	4.5/0.65	Liver, red blood cells, gastrointestinal, brain	5750/3.76 (rat, oral), 1400/3.15 (mouse, oral)	Reproductive system
Flavopiridol	2.2/0.34	Teratogenic, liver, gastrointestinal	3/0.48 (rat, oral maximum tolerance dose)	Teratogenic, liver, gastrointestinal, spleen, bone marrow
Thalidomide	103.2/2.01	Teratogenic, liver, gastrointestinal	113/2.05 (rat, oral), 2000/3.30 (mouse, oral)	Teratogenic, liver, gastrointestinal
SU5416	1.0/0	Teratogenic, liver, cardiovascular	Unknown	Unknown
Col-3	1.0/0	Teratogenic, liver, gastrointestinal	75/1.87 (rat, oral)	Gastrointestinal, kidney, liver, bone marrow
Ethanol	11,180/4.0	Teratogenic, neuronal, craniofacial	7,000/3.8 (rat, oral), 1440/3.16 (rat, i.v.), 360/2.56 (rat, i.p.)	Liver, teratogenic, neuronal, craniofacial
Nicotine	161.7/2.21	Liver, kidney, brain, cardiovascular, teratogenic	188/2.27 (rat, oral), 1.0/0.00 (rat, i.v.), 30/1.48 (rat, i.p.), 24/1.38 (mouse, oral)	Liver, kidney, neurotoxicity, respiratory, cardiovascular
Caffeine	108/2.03	Behavioral, neuronal	127/2.1 (mouse, i.p.), 247/2.39 (rat, oral), 137/2.14 (mouse, oral)	Behavioral, neuronal
Pentylenetetrazole	>276/2.44	Liver	140/2.15 (rat, oral), 88/1.94 (mouse, oral)	CNS, cardiovascular

(continued)

Table 3.1 (*Continued*)

Compound	Zebrafish		Mammalian models	
	LC$_{50}$/log LC$_{50}$ (mg/L)	Target organs	LD$_{50}$/log LD$_{50}$ (mg/kg)	Target organs
Barbituric acid	166.4/2.22	Heart, eye, liver	900/2.95 (rat, oral)	CNS
Buspirone	126.6/2.1	Liver, gastrointestinal, CNS	196/2.29 (rat, oral), 655/2.82 (mouse, oral)	CNS, heart, gastrointestinal
Clozapine	11.45/1.06	Heart, gastrointestinal, liver	251/2.4 (rat, oral), 58/1.76 (rat, i.v.), 199/2.3 (mouse, oral), 58/1.76 (mouse, i.v.)	Heart, CNS
Thioridazine	10/1	Liver, heart, brain, yolk	1060/3.03 (rat, oral)	CNS
Imipramine	24/1.38	Liver, gastrointestinal, heart, swim bladder	305/2.48 (rat, oral), 15.9/1.2 (rat, i.v.), 79/1.9 (rat, i.p.), 188/2.27 (mouse, oral), 21/1.32 (mouse, i.v.), 51.6/1.71 (mouse, i.p.)	Heart, CNS
Primidone	>436/2.64	Gastrointestinal	1500/3.18 (rat, oral), 280/2.45 (mouse, oral)	Brain, blood
Haloperidol	18.8/1.27	Heart, jaw, brain	128/2.11 (rat, oral)	Blood, CNS
Carbamazepine	118/2.07	Heart, jaw, developmental	1957/3.29 (rat, oral)	Reproductive system, blood, nervous system
Desipramine HCl	116/2.06	Heart, jaw	871/2.94 (rat, oral), 29/1.46 (rat, i.v.), 48/1.68 (rat, i.p.), 448/2.65 (mouse, oral), 22/1.34 (mouse, i.v.), 85/1.93 (mouse, i.p.)	CNS, heart
trans-2-Phenylcyclopropylamine (tranylcypromine)	340/2.53	Heart, jaw, brain	Unknown	Heart, CNS
Dopamine	228/2.36	Heart, jaw, brain	Unknown	Heart, CNS

Chlorpromazine	107/2.02	Heart, jaw, brain	225/2.35 (rat, oral)	Reproductive, dermal
Fluoxetine	13.8/1.14	Liver, heart, gastrointestinal	452/2.66 (rat, oral), 248/2.39 (mouse, oral)	Liver, kidney
Thiothixene	38.4/1.58	Liver, pancreas	720/2.86(rat, oral), 400/2.60 (mouse, oral)	Heart
Naproxen	13.2/1.12	Liver, gastrointestinal	435/2.63 (mouse, i.v.), 534/2.73 (rat, oral), 1234/3.09 (mouse, oral)	Gastrointestinal
Ibuprofen	5.56/0.74	Liver, kidney	495/2.69 (mouse, i.p.), 636/2.80 (rat, oral), 626/2.80 (rat, i.p.), 740/2.87 (mouse, oral)	Kidney
Aspirin	100.9/2.0	Gastrointestinal, teratogenic, kidney, muscle	167/2.22 (mouse, i.p.), 1500/3.18 (rat, oral), 1100/3.04 (mouse, oral)	Gastrointestinal, kidney, ureter, cardiovascular, muscle
Acetaminophen	252/2.4	Liver	500/2.69 (mouse, i.p.), 1944/3.29 (rat, oral)	Liver, kidney, gastrointestinal
Indomethacin	3.58/0.55	Liver, gastrointestinal, kidney	12/1.08 (rat, oral), 50/1.70 (mouse, oral)	Liver, gastrointestinal, kidney
Colchicine	120/2.08	Liver, gastrointestinal	6.1/0.78 (rat, i.p.), 1.6/0.20 (mouse, i.v.), 5.9/0.77 (mouse, oral)	Mutagenic, teratogenic, gastrointestinal, bone marrow depression
Diclofenac	1.6/0.20	Developmental, liver, gastrointestinal	Not available	Liver, gastrointestinal
Cyclosporin A	69/1.83	Teratogenic, liver, cardiovascular, kidney	170/2.23 (mouse, i.v.), 1489/3.17 (rat, oral), 147/2.17 (rat, i.v.), 24/1.38 (rat, i.p.), 2803/3.45 (mouse, oral)	Kidney, ureter, bladder

(continued)

Table 3.1 (*Continued*)

Compound	Zebrafish LC$_{50}$/log LC$_{50}$ (mg/L)	Zebrafish Target organs	Mammalian models LD$_{50}$/log LD$_{50}$ (mg/kg)	Mammalian models Target organs
Diphenhydramine	89.3/1.95	Liver, kidney, teratogenic	500/2.69 (rat, oral), 35/1.54 (rat, i.v.), 114/2.06 (mouse, oral), 20/1.30 (mouse, i.v.), 56/1.75 (mouse, i.p.)	Carcinogenic, teratogenic
Terfenadine	7.1/0.85	Heart, liver	5/0.70 (rat, p.o.), 2000/3.3 (mouse, oral)	Heart
Famotidine	>674/2.83	Gastrointestinal, heart	>3000/3.48 (rat, p.o.)	No obvious toxicity
Cimetidine	>504/2.70	Heart, liver, gastrointestinal	5000/3.70 (rat, p.o.)	Genome toxic, liver, developmental toxic
Hydroxyzine	67.2/1.83	Gastrointestinal, heart	950/2.98 (rat, oral)	Oral clefts, orofacial defects, micromelia
Pyrilamine	51.3/1.71	Heart, liver	36/1.56 (rat, oral), 102/2.0 (mouse, i.p.)	Liver
Promethazine HCl	16.1/1.21	Heart, liver, gastrointestinal	15/1.17 (rat, i.v.), 50/1.70 (mouse, i.v.), 255/2.41 (mouse, oral), 170/2.23 (rat, oral), 160/2.20 (mouse, i.p.)	Liver
Ranitidine		Heart	60/1.78 (mouse, i.v.), 6610/3.82 (rat, oral)	Heart, liver
Nizatidine		Heart		Heart, liver
Enalapril	295.8/2.47	Heart, liver, craniofacial	2000/3.3 (rat, p.o.)	Developmental toxic
Captopril	499.5/2.7	Heart, liver, craniofacial	4266/3.36 (rat, p.o.), 1000/3.0 (mouse, i.v.)	Developmental toxic

Isoproterenol	>495.4/2.69	Heart, liver necrosis, gastrointestinal	800/2.90 (rat, p.o.)	Cardiovascular
Epinephrine	266.4/2.43	Heart, gastrointestinal	24/1.38 (rat, p.o.), 4/0.6 (mouse, i.p.)	Cardiovascular
Salbutamol	>478/2.68	Heart	660/2.82 (rat, oral)	Smooth muscle
Terbutaline	>548/2.74	Liver, heart, gastrointestinal, pancreas	>5000/3.70 (rat, p.o.)	
Lidocaine	70.2/1.85	Heart	317/2.50 (rat, p.o.), 25/1.4 (rat, i.v.), 133/2.12 (rat, i.p.), 292/2.47 (mouse, oral), 19.5/1.29 (mouse, i.v.), 105/2.02 (mouse, i.p.)	Cardiovascular, CNS
Verapamil	147.3/2.17	Heart, liver, gastrointestinal	108/2.03 (rat, p.o.)	Cardiovascular
Nifedipine	346/2.54	Liver, heart	1022/3.01 (rat, oral), 6/0.78 (rat, i.v.), 230/2.36 (rat, i.p.), 202/2.31 (mouse, oral), 4200/3.62 (mouse, i.v.), 185/2.27 (mouse, i.p.)	Cardiovascular
Digoxin	11.7/1.07	Heart, liver, gastrointestinal	28.27/1.45 (rat, oral), 25/1.4 (rat, i.v.), 4/0.6 (rat, i.p.), 415/2.62 (mouse, oral), 7.4/0.89 (mouse, i.p.), 40.6 (mouse, i.p.)	Heart, gastrointestinal
Quinidine	364/2.56	Brain, yolk, heart	263/2.42 (rat, oral), 23/1.36 (rat, i.v.), 140/2.15 (rat, i.p.), 173/2.24 (mouse, i.p.)	Heart
Procainamide	544/2.74		95/1.989 (rat, i.v.), 312/2.49 (mouse, oral), 103/2.01 (mouse, i.v.)	
Clonidine	9.3/0.97	Heart	157/2.20 (rat, oral), 108/2.03 (mouse, oral)	Cardiovascular, CNS, eye

(continued)

37

Table 3.1 (*Continued*)

Compound	Zebrafish		Mammalian models	
	LC$_{50}$/log LC$_{50}$ (mg/L)	Target organs	LD$_{50}$/log LD$_{50}$ (mg/kg)	Target organs
Atenolol	532/2.73	Heart	2000/3.3 (mouse, oral), 3000/3.48 (rat, oral), 98.7/1.99 (mouse, i.v.), 59/1.77 (rat, i.v.)	Developmental, cardiac
Metoprolol tartrate salt	411/2.61	Heart	5500/3.74 (rat, oral), 219/2.34 (rat, i.p.), 90/1.95 (rat, i.v.), 1500/3.18 (mouse, oral)	
Labetalol HCl	65.7/1.82	Heart	2114/3.33 (rat, oral), 107/2.03 (rat, i.p.), 53/1.72 (rat, i.v.), 1450/3.16 (mouse, oral), 114/2.06 (mouse, i.p.), 47/1.67 (mouse, i.v.)	
Timolol	866/2.94	Heart	1190/3.07 (mouse, oral), 900/2.95 (rat, oral)	
Propranolol	59.2/1.77	Heart, liver	565/2.75 (mouse, p.o.)	No obvious toxicity
Glipzide	17.8/1.25	GI, liver	1200/3.08 (rat, i.p.)	Endocrine
Troglitazone	0.80/−0.1	Liver, gastrointestinal	Unknown	Liver, gastrointestinal
Glyburide	>123.5/2.09	Liver, gastrointestinal	20/1.3 (rat, oral)	Pancreas, gastrointestinal
Acetazolamide	>444/2.65	Gastrointestinal, liver	2750/3.43 (rat, i.p.), 4500/3.65 (mouse, oral)	Kidneys, eyes
Amiloride	>532/2.73	Liver, heart	85/1.92 (oral, rat), 56/1.75 (mouse, oral)	Hyperkalemia, kidney, heart
Probenecid	14.3/1.16	Liver	1600/3.2 (rat, oral), 1666/3.22 (mouse, oral), 1000/3 (mouse, i.p.), 458/2.66 (moue, i.v.)	Liver, kidney

Furosemide	663/2.82	Liver	2700/3.43 (rat, oral), 800/2.9 9 (rat, i.p., i.v.), 2200/3.34 (mouse, oral), 308/2.49 (mouse, i.p.), 2000/3.3 (dog, oral), 800/2.9 (rabbit, oral), 400/2.6 (rabbit, i.v.)	Teratogenesis, liver, kidney
Chlorothiazide	592/2.77	No obvious toxicity	10000/4 (rat, oral), 1386/3.14 (rat, i.p.), 200/2.3 (rat, i.v.), 8000/3.9 (mouse, oral), 1400/3.15 (mouse, i.p.), 940/2.97 (mouse, i.v.), 1000/2 (dog, i.v.)	Skin and respiratory irritant
Copper sulfate	2/0.3	Liver, kidney, pancreas, teratogenic	50/1.7 (mouse, oral), 300/2.48 (rat, oral)	Kidney, liver, teratogenic
Geldanamycin	3.13/0.49	Liver	1.0/0 (mouse, i.p.)	Liver
AZT (3'-azido-3'-deoxythymidine)	1068/3.02	Teratogenic, liver	750/2.88 (rat, i.v.), 3000/3.48 (mouse, i.v.)	Carcinogenic, bone marrow depression, reproductive and teratogenic, immunotoxic, liver, kidney
17β-Estradiol	5.4/0.73	Teratogenic	1.2/0.08 (rat, i.m.)	Carcinogenic, bone marrow depression, teratogenic
Mifepristone	6.45/0.81	Heart, liver	5000/3.7 (rat, oral)	Thrombotic effect, heart, induce abortion
Retinoic acid (trans)	0.03/−1.52	Teratogenic, liver, gastrointestinal	78/1.89 (rat, i.v.)	Carcinogenic, teratogenic, liver
Coumarin	102.2/2	Cardiovascular, liver	293/2.47 (rat, oral), 196/2.29 (mouse, oral), 220/2.34 (mouse, i.p.), 242/2.38 (mouse, s.c.)	Liver

(continued)

40

Table 3.1 (*Continued*)

Compound	Zebrafish LC$_{50}$/log LC$_{50}$ (mg/L)	Zebrafish Target organs	Mammalian models LD$_{50}$/log LD$_{50}$ (mg/kg)	Mammalian models Target organs
Succinylcholine	238.2/2.38	Muscle, heart, liver	125/2.10 (mouse, oral), 2.1/0.32 (mouse, i.p.), 7.5/0.88 (mouse, s.c.), 0.28/−0.55 9 (mouse, i.v.), 0.8/−0.1 (rabbit, i.v.)	Muscle, heart, liver
Dichloroacetic acid (DCA)	72/1.85	Teratogenic, liver, kidney	1500/3.17 (rat, oral), 5640/3.75 (mouse, oral), 500/2.70 (rat, i.p.)	Liver, kidney, cardiovascular
Polychlorinated biphenyls (PCBs)	10/1	Liver, gastrointestinal	8000/3.9 (rat, i.v.), 1900/3.28 (mouse, oral)	Liver
Phenylurea dithiocarbamate	1.5	Teratogenic, liver	Unknown	Liver
Trithiophene	2.7/0.43	Liver, muscle	110/2.04 (rat, i.p.)	Liver
TCDD	0.032/−1.49	Cardiovascular, liver, kidney, teratogenic, growth inhibition, neuronal	0.020/−1.69 (rat, oral), 0.114/−0.94 (mouse, oral)	Cardiovascular, liver, kidney, teratogenic, neuronal
2,4-Dinitrotoluene (DNT)	23/1.4	Teratogenic, liver	268/2.4 (rat, oral), 790/2.9 (mouse, oral)	Teratogenic, carcinogenic, liver, reproductive
Alloxan	128/2.1	Pancreas, liver, heart	Unknown	Pancreas

Table 3.2 Comparison of Vertebrate Model Systems

Parameter	Rat	Rabbit	Zebrafish
Ex utero gestation	No	No	Yes
Embryogenesis	21 days	29 days	3 days
Embryos per mating	8–16	3–8	50–100
Transparent	No	No	Yes
Microplate analysis	No	No	Yes
Drug delivery	Inject	Inject	Inject/solution
Cost per assay	>$1000	>$1500	~$0.10

Lethal effects are compared to controls to identify LC_{50}, no observed effect concentration (NOEC), and lowest observed effect concentration (LOEC) values. The test uses a minimum of five concentrations and controls and 10 embryos per concentration in two to three independent replicates. To assess lethality, after treatment for 24 and 48 h, four end points are assessed: (1) coagulation of fertilized eggs, (2) somite formation, (3) detachment of the tail bud from the yolk sac, and (4) heartbeat.

3.4.2 Embryo Differentiation

Fertilized eggs undergo the first cleavage at ~15 min post fertilization and consecutive synchronous cleavages form 4-, 8-, 16-, and 32-cell blastomeres. At these stages, fertilized eggs can be clearly identified by development of a blastula. Unfertilized eggs that do not undergo cleavage or eggs exhibiting obvious irregularities during cleavage, such as vesicle formation and asymmetry, vesicle formation, or defective chorions, are discarded.

3.4.3 Test Solutions

Test concentrations are prepared by diluting stock solution and mixing or agitating the test substance in dilution water usually by mechanical means including stirring or ultrasonification.

3.4.4 Exposure Time and Duration of Test

The test is initiated directly after egg fertilization and embryos are immersed in test solutions before cleavage of the blastodisc begins. Approximately 40 eggs per concentration are transferred to dilution water no later than 60 min post fertilization. Under a stereomicroscope (30× magnification), fertilized eggs are separated from unfertilized eggs and then transferred to 24-well plates containing test solutions. Then, at ~120 min post fertilization, fertilized eggs are transferred to test chambers.

3.4.5 Test Concentrations

Normally, five compound concentrations are required to meet statistical requirements. The highest concentration tested is expected to result in 100% mortality and the lowest concentration tested is expected to cause no observable effects. A range-finding test is performed to determine the appropriate concentration range.

3.4.6 Controls

As an internal assay control, in each 24-well plate, 4 wells are filled with dilution water. If a solubilizing agent is used, an additional group, the same size as the internal control group (20 embryos), is exposed to the solubilizing agent in a separate 24-well plate.

3.4.7 Test End Points

A detailed description of the normal development of zebrafish embryos is available (Kimmel et al., 1995). Overall survival of embryos in controls, including solvent control, should be at least 90%. Presence of coagulation of embryos, irregularities in somite formation, nondetachment of the tail, and heartbeat are then recorded. Observation of end points after 24 and 48 h is ordinarily sufficient. Zebrafish are considered dead if one of the four end points is positive. After 24 and 48 h, the number of coagulated eggs is determined. Coagulated eggs are milky white and appear dark under the microscope. Formation of somites is examined after 24 and 48 h. After 24 h, about 20 somites normally have formed and embryos exhibit vigorous side-to-side contractions that may result in rotating movements within the chorion. Formation of somites is therefore a sufficient criterion for assessing lethality. After 24 and 48 h, detachment of the tail from the yolk following posterior elongation of the zebrafish body is recorded. After 48 h, presence of active heartbeat is recorded.

3.4.8 Results

For the zebrafish FET test, the cumulative percent mortality for each exposure period (24 and 48 h) is plotted against concentration. If the data obtained are inadequate for the use of standard methods of calculating the LC_{50}, the highest concentration causing no mortality (NOEC) and the lowest concentration producing 100% mortality should be used to approximate the LC_{50}.

3.5 CONCLUSIONS

The developmental toxicity field is increasingly seeking not just yes/no answers to questions regarding a chemical's potential to cause developmental toxicity, but it is now also demanding (1) more quantitative dose–response information,

(2) information on functional deficits, (3) expanded time frame for assessing developmental changes, and (4) more accurate kinetics and exposure assessment methods. Use of zebrafish for developmental toxicity studies will address several of these concerns.

REFERENCES

BACHMANN J. (2002). Development and validation of a teratogenicity screening test with embryos of the zebrafish (*Danio rerio*). PhD thesis, Technical University of Dresden, Germany.

BRAUNBECK T, BOETTCHER M, HOLLERT H, KOSMEHL T, LAMMER E, LEIST E, RUDOLF M, and SEITZ N (2005). Towards an alternative for the acute fish LC_{50} test in chemical assessment: the fish embryo toxicity test goes multi-species—an update. Altex 22(2): 87–102.

CHENG SH, WAI AWK, SO CH, and WU RSS (2000). Cellular and molecular basis of cadmium-induced deformities in zebrafish embryos. Environ Toxicol Chem 19: 3024–3031.

DIN (2001). German standard methods for the examination of water, waste water and sludge—Subanimal testing (group T)—Part 6: Toxicity to fish. Determination of the non-acute-poisonous effect of waste water to fish eggs by dilution limits (T 6). DIN 38415-6.

EISSES KT (1989). Teratogenicity and toxicity of ethylene glycol monomethyl ether (2-methoxyethanol) in *Drosophila melanogaster*: involvement of alcohol dehydrogenase activity. Teratog Carcinog Mutagen 9(5): 315–325.

FERRARI B, PAXEUS N, LO GIUDICE R, POLLIO A, and GARRIC J (2003). Ecotoxicological impact of pharmaceuticals found in treated wastewaters: study of carbamazepine, clofibric acid, and diclofenac. Ecotoxicol Environ Saf 55(3): 359–370.

FORT DJ, DAWSON DA, and BANTLE JA (1988). Development of a metabolic activation system for the frog embryo teratogenesis assay: *Xenopus* (FETAX). Teratog Carcinog Mutagen 8(5): 251–263.

FORT DJ, RAYBURN JR, DEYOUNG DJ, and BANTLE JA (1991). Assessing the efficacy of an Aroclor 1254-induced exogenous metabolic activation system for FETAX. Drug Chem Toxicol 14(1–2): 143–160.

FORT DJ, STOVER EL, PROPST TL, FAULKNER BC, VOLLMUTH TA, and MURRAY FJ (1998). Evaluation of the developmental toxicity of caffeine and caffeine metabolites using the frog embryo teratogenesis assay— *Xenopus* (FETAX). Food Chem Toxicol 36(7): 591–600.

GROTH G, KRONAUER K, and FREUNDT KJ (1994). Effects of *N,N*-dimethylformamide and its degradation products in zebrafish embryos. Toxicol In Vitro 8(3): 401–406.

GROTH G, SCHREEB K, HERDT V, and FREUNDT KJ (1993). Toxicity studies in fertilized zebrafish eggs treated with *N*-methylamine, *N,N*-dimethylamine, 2-aminoethanol, isopropylamine, aniline, *N*-methylaniline, *N,N*-dimethylaniline, quinone, chloroacetaldehyde, or cyclohexanol. Bull Environ Contam Toxicol 50(6): 878–882.

HALLARE AV, KOHLER HR, and TRIEBSKORN R (2004). Developmental toxicity and stress protein responses in zebrafish embryos after exposure to diclofenac and its solvent, DMSO. Chemosphere 56(7): 659–666.

HITCHCOCK DR, BLACK MC, and WILLIAMS PL (1997). Investigations into using the nematode *Caenorhabditis elegans* for municipal and industrial wastewater toxicity testing. Arch Environ Contam Toxicol 33(3): 252–260.

KAMMANN U, VOBACH M, and WOSNIOK W (2006). Toxic effects of brominated indoles and phenols on zebrafish embryos. Arch Environ Contam Toxicol 51(1): 97–102.

KIMMEL CB, BALLARD WW, KIMMEL SR, ULLMANN B, and SCHILLING TF (1995). Stages of embryonic development of the zebrafish. Dev Dyn 203(3): 253–310.

LANGE M, GEBAUER W, MARKL J, and NAGEL R (1995). Comparison of testing acute toxicity on embryo of zebrafish (*Brachydanio rerio*), and RGT-2 cytotoxicity as possible alternatives to the acute fish test. Chemosphere 30(11): 2087–2102.

NAGEL R (2002). DarT: the embryo test with the zebrafish *Danio rerio*—a general model in ecotoxicology and toxicology. Altex 19 (Suppl 1): 38–48.

NGUYEN LT and JANSSEN CR (2001). Comparative sensitivity of embryo–larval toxicity assays with African catfish (*Clarias gariepinus*) and zebra fish (*Danio rerio*). Environ Toxicol Chem 16(6): 566–571.

ROSETH S, EDVARDSSON T, BOTTEN TM, FUGLESTADT J, FONNUM F, and STENERSEN J (1996). Comparison of acute toxicity of process chemicals used in the oil refinery, tested with the diatom *Chaetoceros gracilis,* the flagellate *Isochrysis galbana,* and the zebrafish, *Brachydanio rerio*. Environ Toxicol Chem 15: 1211–1217.

SCHULTE C and NAGEL R (1994). Testing acute toxicity in embryo of zebrafish, *Brachydanio rerio* as alternative to the acute fish test: preliminary results. ATLA 22: 12–19.

VERSONNEN BJ and JANSSEN CR (2004). Xenoestrogenic effects of ethinylestradiol in zebrafish (*Danio rerio*). Environ Toxicol 19(3): 198–206.

VERSONNEN BJ, ROOSE P, MONTEYNE EM, and JANSSEN CR (2004). Estrogenic and toxic effects of methoxychlor on zebrafish (*Danio rerio*). Environ Toxicol Chem 23(9): 2194–2201.

WIEGAND C, KRAUSE E, STEINBERG C, and PFLUGMACHER S (2001). Toxicokinetics of atrazine in embryos of the zebrafish (*Danio rerio*). Ecotoxicol Environ Saf 49(3): 199–205.

Chapter 4

Assessment of Drug-Induced Cardiotoxicity in Zebrafish

Louis D'Amico[1], Wen Lin Seng[1], Yi Yang[2], and Willi Suter[2]

[1]*Phylonix, Cambridge, MA, USA*
[2]*Novartis Pharmaceuticals, East Hanover, NJ, USA*

4.1 INTRODUCTION

Cardiotoxicity is a major problem for hundreds of pharmaceutical agents, industrial chemicals, and naturally occurring products. In the pharmaceutical sector, several compounds have been shown to lengthen cardiac repolarization, leading to QT interval prolongation and torsades de pointes, which has the potential to cause sudden death. Previous research has shown that compounds capable of inducing repolarization abnormalities cause bradycardia in zebrafish (Chapter 5). In this chapter, we describe methods for assessing compound-induced cardiotoxicity in zebrafish and report results for 10 reference compounds.

4.2 ZEBRAFISH HEART

The zebrafish (*Danio rerio*) is a useful animal model system for studying cardiovascular development, genetics, and cardiotoxicity. Zebrafish transparency permits visual assessment of circulation and morphology. Zebrafish use gills for respiration and have a single-loop circulatory system. The heart consists of two chambers: an atrium that receives blood and a ventricle that pumps blood to the body. Both the mammalian and zebrafish heart share development of specialized chambers, outflow tracts to an intricate vasculature, valves to ensure directionality, specialized endothelial cells (endocardium) to drive a high-pressure system, and an electrical system to regulate rhythm. There is inflow of blood from a major vein to an atrium; the blood moves to a muscular ventricle for delivery to the aorta; valves are present to direct

Zebrafish: Methods for Assessing Drug Safety and Toxicity, Edited by Patricia McGrath.
© 2012 John Wiley & Sons, Inc. Published 2012 by John Wiley & Sons, Inc.

Table 4.1 Summary of Zebrafish Cardiotoxicity

Drug name	Concentration (µg/mL)	Gross morphology	Circulation	Pericardial edema	Thrombosis
DMSO (control)	1%	0%	0%	0%	0%
Clozapine	31	10% atrium and ventricle swollen	90% slow	40%	0%
Erythromycin	734	100% atrium and ventricle swollen	60% slow, 30% absent	100%	20% yolk, 10% pericardium
Quinidine	200	80% atrium and ventricle swollen	100% absent	30%	100% yolk, 40% ventricle
Terfenadine	20	50% atrial swelling caused compression of ventricle	100% absent	60%	80% yolk, 10% tail
Astemizole	4.6	40% atrium and ventricle swollen	80% absent, 20% slow	100%	100% yolk
Amiodarone	8.75	20% atrium and ventricle swollen	100% absent	50%	0%
Lidocaine	117	20% atrium and ventricle swollen	20% slow	10%	0%
Verapamil	98.2	0%	0%	30%	0%
Haloperidol	5.65	80% atrium and ventricle swollen with asynchrony between chambers	80% slow, 20% absent	60%	50% yolk
Clomipramine	56	70% swollen ventricle	80% absent, 20% slow	70%	10% yolk

Table 4.2 Comparison of Cardiotoxicity in Zebrafish and Humans

Drug name	Known adverse cardiotoxic side effects in humans	Cardiotoxicity observed in zebrafish	Abnormal AV ratio in zebrafish
Clozapine	Yes	Yes	Yes
Erythromycin	Yes	Yes	Yes
Quinidine	Yes	Yes	Yes
Terfenadine	Yes	Yes	Yes
Astemizole	Yes	Yes	Yes
Amiodarone	Yes	Yes	Yes
Lidocaine	Yes	Yes	Yes
Verapamil	Yes	Yes	Yes
Haloperidol	Yes	Yes	Yes
Clomipramine	Yes	Yes	No

blood flow, and the heartbeat is associated with pacemaker activity. The underlying development, patterning, genes, functions, and disease characteristics are similar to humans, making zebrafish a valuable animal model for studying cardiotoxicity.

4.3 SUMMARY OF CARDIOTOXICITY STUDY DESIGN AND RESULTS

The goals of this study were to assess the effects on zebrafish of short-term exposure to 10 cardiotoxic drugs. LC_{50} was calculated and then heart rate, rhythmicity, circulation, and morphology were assessed. For heart rate and rhythmicity, 2 day post fertilization (dpf) zebrafish were exposed to varying concentrations of compounds for 4 h. For morphology assessment, zebrafish were exposed continuously to compounds for 24 h. Compounds that had significant effects on each of the cardiac-specific end points were identified (Table 4.1). Results from this study were in good agreement with results observed in previously published data in zebrafish, other vertebrate models, and humans (Table 4.2). Together the results reported here demonstrate that (1) zebrafish exhibit a clear dose response to known cardiotoxic compounds, and (2) effects of known cardiotoxic drugs in zebrafish are strikingly similar to effects observed in primates and other preclinical animal models.

4.4 MATERIALS AND METHODS

4.4.1 Experimental Animals

Embryos were generated by natural pairwise mating, as described in *The Zebrafish Book* (Westerfield, 1993). Four to five pairs of adult zebrafish were set up for each mating, and, on average, 100–150 embryos per pair were generated. Embryos were maintained at 28°C in fish water (200 mg Instant Ocean Salt (Aquarium Systems, Mentor, OH) per liter of deionized water; pH 6.6–7.0 maintained with 2.5 mg/L Jungle pH Stabilizer (Jungle Laboratories Corporation, Cibolo, TX); conductivity 670–760 µS). Embryos were cleaned (dead embryos were removed) and sorted by developmental stage (Kimmel et al., 1995) at 6 and 24hpf. Because the embryo receives nourishment from an attached yolk sac, no feeding is required for 7dpf.

4.4.2 Compounds

Compounds included clozapine, erythromycin, quinidine, terfenadine, astemizole, amiodarone, lidocaine, verapamil, haloperidol, and clomipramine. Stock solutions for each compound were made in DMSO and stored at -20°C before use. Fish water containing 0.1% DMSO was used as vehicle control, terfenadine was used as a positive control for assessing heart rate, and Celebrex was used as a positive control for assessing heart function and morphology.

4.4.3 Determination of Concentration Range

Thirty, 2dpf hatched zebrafish embryos were distributed into six-well microplates in 3 mL fish water containing varying compound concentrations and exposed for 24 h. Untreated and 0.1% DMSO-treated zebrafish were used as controls. Five concentrations, one log apart, were tested initially: 0.1, 1.0, 10, 100, and 500 µM. If no lethality was observed in the initial trial, additional concentrations were tested. Best-fit concentration–response curve was generated using JMP statistical software; $Y = M_1 + \{(M_2 - M_1)/[1 + (X/M_3)^\wedge M_4]\}$, where $M_1 = $ maximum Y value (100% in this case), $M_2 = $ minimum Y value (0% in this case), $M_3 = $ the concentration corresponding to the value midway between M_1 and M_2 (i.e., LC_{50}), $M_4 = $ slope of the curve at M_3 (best fit, generating R^2 closest to 1), $X = $ concentration of compound, and $Y = $ percent lethality. LC_0 and $LC1_{10}$ were estimated using JMP statistical software (SAS Institute, Cary, NC). The appropriate concentrations were then used to perform for following assays.

4.4.4 Heart Rate

Ten hatched, 2dpf zebrafish embryos were distributed into six-well microplates in 3 mL fish water containing the compound at desired concentrations. Embryos were exposed for 4 h prior to heart rate assessment. After drug treatment, zebrafish were visually examined under a dissecting microscope (Zeiss Stemi 2000-C, Zeiss, Thornwood, NY) equipped with a SPOT Insight 10ZNC digital camera (Diagnostic Instruments, Inc., Sterling Heights, MI) at room temperature (22–24°C). To facilitate observation, zebrafish were placed in 2% methylcellulose. Atrial and ventricular rates (beats per min; bpm) were counted for 30 s with a stopwatch and a counter.

4.4.5 Rhythmicity

The ratio of atrial to ventricular beats per min was used to determine atrioventricular (AV) ratio; AV ratio >1 indicated irregular AV rhythmicity.

4.4.6 Circulation

Zebrafish heart development begins at early gastrulation. By 22hpf, the heart tube is beating, and around 24hpf, circulation begins. Movement of blood cells through the heart and major vessels, and the rate and pattern of blood flow were examined. Abnormal circulation included defects in circulation rate (no, slow, or fast circulation) and defects in circulatory pattern (circulation loop only observed in the head or trunk).

4.4.7 Cardiac Morphology

Drug treatment was performed as described in the concentration range determination assay. Cardiac morphology was examined under a Zeiss dissecting microscope equipped with a SPOT Insight 10ZNC digital camera. After compound treatment, representative images (at $8\times$ magnification) were acquired. The zebrafish heart has one atrium and one ventricle. In normal embryos, the atrium and ventricle are side by side; the blood flows into the atrium from the sinus venosus and out of the ventricle through the bulbus arteriosus. The heart is composed of concentric tubes of cells, an outer myocardial layer, and an inner endocardial layer. Morphological abnormalities, including the size of the heart and number and position of the heart chambers, were examined.

4.4.8 Pericardial Edema

Pericardial edema refers to the presence of an abnormal volume of fluid in the space surrounding heart chambers, which can be easily visualized in transparent zebrafish. Pericardial edema may be caused by venous or lymphatic obstruction, increased vascular permeability, heart failure, or kidney disease. To assess if drugs caused pericardial edema, we visually examined the presence of enlarged heart chambers.

4.4.9 Hemorrhage/Thrombosis

Hemorrhage may be caused by rupture of blood vessels in a specific tissue or organ. Thrombosis is defined as formation of a blood clot. In zebrafish, hemorrhage was visualized as a pool of blood in a tissue or organ. Thrombosis was observed as a stagnant blood flow or a blood clot in the cardinal vein.

4.4.10 Statistical Evaluation

To determine if a drug had a significant effect on heart rate, contractility, or rhythmicity, ANOVA followed by Dunnett's test (based on raw data) was used to identify concentrations that significantly altered heart rate; $P < 0.05$ was considered significant.

4.5 RESULTS

4.5.1 Heart Rate Assessment

Hatched 2dpf zebrafish were incubated with drugs for 4 h at 28°C. After incubation, zebrafish ($N = 10$) were visualized on a Zeiss dissecting scope, and ventricular contractions for 30 s were counted manually. The number of contractions was multiplied by 2 to calculate heart rate, reported in bpm.

To confirm that vehicle (0.1% DMSO) did not cause adverse effects, untreated normal and 0.1% DMSO-treated zebrafish were used as controls. Terfenadine was used both as a positive control drug and as a blinded test compound. Untreated zebrafish heart rate is highly consistent with <5% variation. 0.1% DMSO did not cause significant change in heart rate. Terfenadine caused significant ($P < 0.0001$) reduction in heart rate as expected. Drug effects on zebrafish heart rate are summarized in Fig. 4.1a and c.

4.5.2 Rhythmicity

The same zebrafish ($n = 10$) that were used for heart rate assessment were used to determine atrioventricular rhythmicity. Results showed that (1) untreated and 0.1% DMSO-treated controls exhibited synchronized atrioventricular rhythmicity (AV ratio = 1), and (2) positive control drug, terfenadine caused expected change in AV ratio (AV ratio = 2), indicating blockage. Summary of drug effects on zebrafish heart rhythmicity is shown in Fig. 4.1b and d.

4.5.3 Summary of Observed Zebrafish Cardiotoxicity on Function and Morphology

The most common drug-induced morphological defect was swelling of the heart chambers (Fig. 4.2). If the ventricle stopped beating, blood frequently pooled and clotted in the chamber. Circulation was frequently reduced or absent entirely. Note that since sufficient oxygenation is acquired by passive diffusion, this effect does not immediately kill zebrafish. Rates of pericardial edema were noted as well as the presence and location of thrombi. Most thrombi occur at the yolk just posterior to the entrance to the atrium.

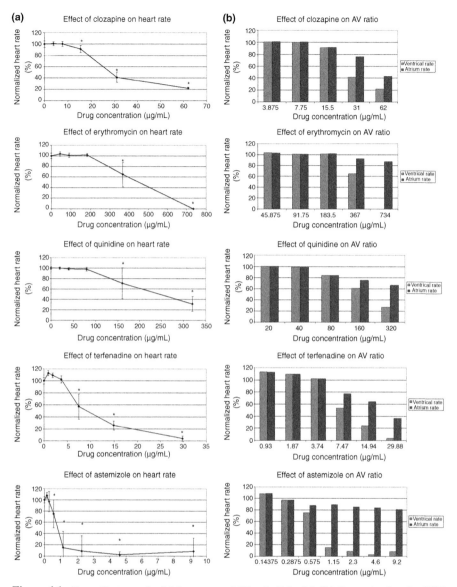

Figure 4.1 Drug effects on zebrafish heart rate and AV ratio. Zebrafish (2dpf) were incubated at 28°C with selected drugs at various concentrations for 4 h (a and c). The bar graph for each drug shows associated drug-induced changes in atrial contraction rates. The ventricle is more sensitive to the presence of cardiotoxic compounds and usually showed a more significant decrease in activity than the atrium. (b and d). Normal AV contraction ratios of 1:1 increased to 10:1 or higher with increasing drug concentration. A notable exception was clomipramine, which maintained a 1:1 AV ratio with increasing drug concentration despite a decrease in heart rate. ANOVA with a post-hoc Dunnett's test (based on raw data) was used to identify concentrations that significantly altered heart rate, $^*P < 0.05$ was considered significant, and error bars are equal to ± 1 SD.

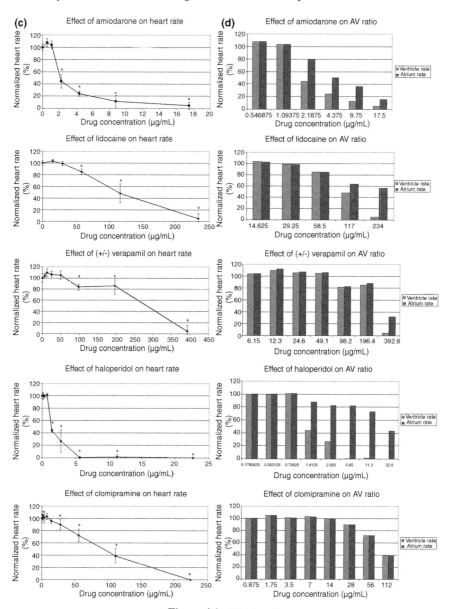

Figure 4.1 (*Continued*)

The percentage of zebrafish affected by drugs for all morphological cardiac end points and circulation is summarized in Table 4.1. As indicated above, swelling of the heart chamber, which led to abnormal circulation, was the most common effect. Thrombosis was the least common effect.

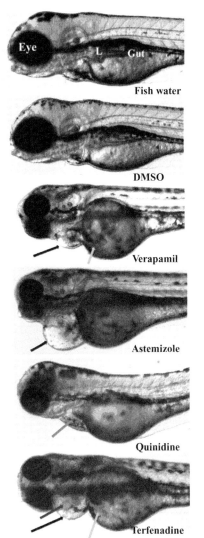

Figure 4.2 Cardiovascular morphological defects in drug-treated zebrafish. Two dpf zebrafish were treated with compounds for 24 h. Fish water and 1% DMSO-treated zebrafish exhibited normal morphology. L: liver. Black arrows: pericardial edema in verapamil-, astemizole-, and terfenadine-treated animals. Red arrow: ventricle swelling due to blood pooling in quinidine-treated animals. Blue arrow: ventricle compression due to atrial swelling in terfenadine-treated animals. Green arrow: stagnant blood flow in the cardinal vein in verapamil- and terfenadine-treated animals. (See the color version of this figure in Color Plates section.)

4.5.4 Comparison of Human and Zebrafish Cardiotoxicity

Drugs known to cause adverse cardiotoxic side effects in humans also caused cardiotoxicity in zebrafish (Table 4.2). Note that clomipramine did not cause an abnormal AV ratio, even at the highest test concentration.

4.6 CONCLUSIONS

1. Drugs known to cause cardiotoxic adverse side effects in humans show similar effects in zebrafish.
2. Zebrafish assays are rapid, reproducible, and quantitative.
3. Small amount of drug is required.
4. *In vivo* zebrafish cardiotoxicity assessment provides useful information complementing data from conventional hERG assays.

REFERENCES

KIMMEL CB, BALLARD WW, KIMMEL SR, ULLMANN B, and SCHILLING TF (1995). Stages of embryonic development of the zebrafish. Dev Dyn 203(3): 253–310.

WESTERFIELD M (1993). The Zebrafish Book: A Guide for the Laboratory Use of Zebrafish (*Danio rerio*). University of Oregon Press, Eugene, OR.

Chapter 5

Cardiotoxicity Studies in Zebrafish

David J. Milan[1] and Calum A. MacRae[2]

[1]*Cardiovascular Research Center and Cardiology Division, Massachusetts General Hospital, Boston, MA, USA*
[2]*Cardiovascular Division, Brigham and Women's Hospital, Boston, MA, USA*

5.1 INTRODUCTION

The major focus of regulatory attention in cardiotoxicity has been the risk of sudden death due to QT interval prolongation and torsades de pointes (TdP). This form of toxicity has led to the withdrawal of several agents from the U.S. market and has been estimated to cost the pharmaceutical industry close to 1 billion dollars per year. Life-threatening arrhythmia in the setting of QT prolongation occurs as a result of inherited mutations in ion channel genes or, in the case of drug-induced repolarization toxicity, as a consequence of drugs interacting with the function of these same channels (Camm et al., 2000; Keating and Sanguinetti, 2001). Despite their medical importance, drug-induced repolarization disorders and related arrhythmias remain difficult to predict.

Additional forms of cardiotoxicity including contractile dysfunction, other arrhythmias, valvulopathy, or pulmonary hypertension exist, but have been associated with agents targeting specific pathways that are more readily predicted. The mechanisms of many of these toxicities remain poorly understood, and range from on-target receptor-mediated effects to idiosyncratic responses. As the pathways targeted by pharmaceutical agents multiply, unpredicted toxicities will also become more common. Conventional preclinical testing in rodents and large animals, including nonhuman primates, can frequently detect common toxic effects. However, the total number of animals screened prior to clinical testing is often only a few hundred, and as a result, rarer toxic events, even if these are catastrophic, often will only be observed after extensive clinical use. Similarly, traditional preclinical toxicology studies usually focus on classic model organism strains where there is little genetic

Zebrafish: Methods for Assessing Drug Safety and Toxicity, Edited by Patricia McGrath.
© 2012 John Wiley & Sons, Inc. Published 2012 by John Wiley & Sons, Inc.

heterogeneity. Thus, toxicities that depend on drug interactions with specific genetic variations may never be observed prior to human trials, and systematic studies of gene–drug interactions are not currently performed. These limitations of conventional preclinical testing have led to efforts to develop *in vivo* models capable of high-fidelity and high-throughput studies of vertebrate drug response for predictive toxicology and toxicogenetics. Many groups have focused on the zebrafish, and in this chapter we will discuss the utility of the organism for such studies using repolarization toxicity as an example.

5.2 REPOLARIZATION TOXICITY

Repolarization is extremely complex, depending on the integrated effects of individual channels, receptors, cytoskeletal elements, and the membrane. There is also extensive regional electrical heterogeneity within the heart that has been shown to be an important contributor to arrhythmogenesis (Antzelevitch and Fish, 2001). Further, drugs that may be safe in isolation can perturb repolarization when given with other medications (Camm et al., 2000), through pharmacokinetic or pharmacodynamic drug–drug interactions. Finally, genetic variation has been shown to contribute to individual susceptibility to drug-induced arrhythmias, and may be much more common than has been appreciated (Roden, 2001). A tractable model system that enabled the systematic identification of genes responsible for such variation would offer many advantages.

Virtually all of the drugs causing repolarization toxicity have been shown to inhibit the rapid component of the repolarizing potassium current (I_{Kr}) *in vitro*. This current is conducted by a channel composed of multiple subunits, including the proteins KCNH2 and KCNE2. Traditional *in vitro* assays of I_{Kr} are limited by the heterologous nature of existing systems, the absence of many channel components and interacting proteins that are important in differentiated cardiomyocytes, low throughput, and the inability to detect drug–drug interactions that depend on other organs (Eckardt et al., 1998; Yang et al., 2001). Animal models, while more physiologic, have an even lower throughput, restricting their ability to screen systematically at scale for drug–drug interactions which may involve a large proportion of the current formulary.

The zebrafish heart exhibits both a complex repertoire of ion channels and functioning metabolism within only 24 h of fertilization (Baker et al., 1997). The embryo's optical properties facilitate the evaluation of heart rate and rhythm. In addition, profoundly abnormal cardiac function is tolerated by larval fish that may survive for 4–5 days without a circulation (Warren and Fishman, 1998).

5.3 INITIAL SCREENING: BRADYCARDIA

In initial work to evaluate 100 small molecules for their effects on embryonic physiology, we observed that compounds known to cause QT prolongation and

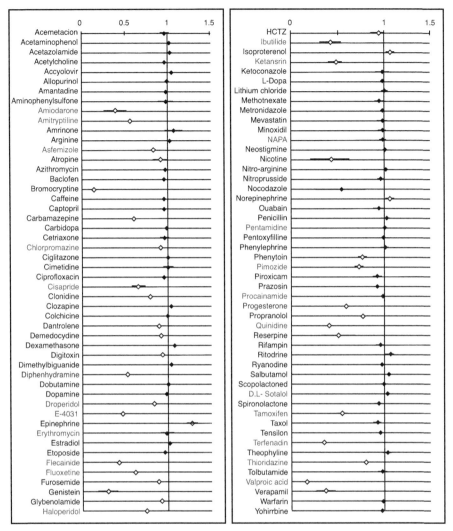

Figure 5.1 Screening 100 compounds for heart rate effects in zebrafish. Heart rates are normalized to vehicle controls. Mean heart rate and standard deviation are shown. Compounds listed in red prolong the QT interval in humans and have been associated with TdP. Blue color indicates association with QT prolongation, but not TdP in humans. Open diamonds signify a statistically significant difference from control ($p < 0.05$). (Figure adapted from Milan et al., 2003.) (See the color version of this figure in Color Plates section.)

TdP in humans consistently resulted in bradycardia and/or atrioventricular block in the zebrafish. Figure 5.1 shows the effects of 100 biologically active small molecules on zebrafish heart rate. Thirty-six compounds caused bradycardia, of which 18 were known in humans to result in QT prolongation and TdP. Five of the 23 compounds (erythromycin, N-acetylprocainamide (NAPA), pentamidine, procainamide, and sotalol) that cause QT prolongation or TdP did not cause bradycardia in the assay. Where physicochemical characteristics predicted limited absorption (a logarithm of the octanol/water partition coefficient less than 1), microinjection was undertaken to bypass this problem and in each case bradycardia was observed. Vehicle alone showed no significant effect on heart rate. Drug effects are potentiated by mutation of the zebrafish KCNH2 (Fig. 5.2b and c), suggesting

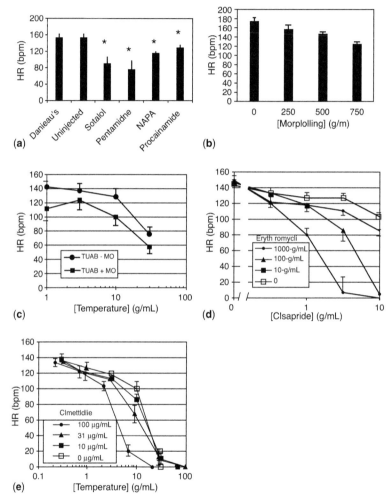

Figure 5.2 Mean heart rates and standard deviations are shown throughout. (a) The effects of drug microinjection on HR. An asterisk denotes statistically significant difference from control ($p < 0.05$). (b) The effect of ZF KCNH2 antisense morpholino on HR. (c) The additive effect of KCNH2 morpholino and terfenadine. (d) Interaction between erythromycin and cisapride. Atrial heart rate in bpm is plotted against increasing doses of cisapride for each concentration of erythromycin shown in the box. (e) Interaction between cimetidine and terfenadine. Atrial heart rate in bpm is plotted against increasing doses of terfenadine for each concentration of cimetidine shown in the box. (Figure adapted from Milan et al., 2003.)

that drug effects are mediated via I_{Kr} inhibition. Of note, submaximal doses of two I_{Kr} blocking drugs have been shown to have additive properties.

Classic drug–drug interactions have also been reproduced in this model. For example, it was possible to recapitulate the potentiating effect of erythromycin on cisapride, with a 10-fold decrease in the ED_{50} for heart rate for cisapride observed with increasing doses of erythromycin. Several other drug–drug interactions were reproduced including that reported between cimetidine and terfenadine (Fig. 5.2e). Taken together, these data suggest that the embryonic zebrafish model of QT drug toxicity may prove useful for studying individual drug toxicities, drug–drug interactions, and the pharmacogenetics of this drug response.

Bradycardia has been previously reported as a result of I_{Kr} blockade in several experimental and clinical situations (Eckardt et al., 1998; De Clerck et al., 2002). This effect has been attributed to action potential prolongation and slowing of phase 4 depolarization in pacemaker tissue (Verheijck et al., 1995). Of the 23 known I_{Kr} blockers tested in our initial screen, 22 caused bradycardia when injection of hydrophilic compounds was performed. Erythromycin is the one false negative result in this assay. We deduced that erythromycin was absorbed by observing its interaction with cisapride (Fig. 5.2d). Erythromycin prolongs the human QT interval and can cause TdP at least in injectable form (Antzelevitch et al., 1996). It is conceivable that the rare events associated with erythromycin may represent the effects of the carrier, though QT prolongation resulting from CYP3A4-mediated effects of erythromycin on other drugs is well documented. The recapitulation of such drug–drug interactions demonstrates a major advantage of the zebrafish model (Roden, 2000). In humans, these interactions have been shown to be pharmacokinetic: the inhibition of hepatic metabolism by one drug resulting in increased levels of the other.

This simple assay is useful as a screen, presenting an opportunity to test not only large numbers of molecules, but also their quantitative interactions with a throughput superior to current methods. Some of the effects seen in the fish (as in other models) may reflect interactions with targets other than KCNH2. Clearly, not all drugs that result in bradycardia do so through I_{Kr} blockade; for example, propranolol and clonidine both reduce heart rate in this assay.

5.4 HIGH-RESOLUTION ASSAYS OF REPOLARIZATION

Given the lack of specificity of the simple heart rate assay, more direct assessment of cardiac electrical events, including action potential duration, at the larval stage is critical if the zebrafish is to become a robust tool for the exploration of drug-induced cardiotoxicity. We therefore developed optical voltage mapping for high-resolution electrophysiological analysis of the zebrafish heart.

Using this technology it is possible to record cardiac action potentials in the zebrafish mutant *breakdance* that carries a mutation in *KCNH2* (Chen et al., 1996; Langheinrich et al., 2003), the major subunit responsible for I_{Kr} (Bezzina et al., 2003;

Pfeufer et al., 2005; Newton-Cheh et al., 2007). Homozygous *breakdance* embryos display striking increases in APD compared to wild-type siblings (Fig. 5.3a). There was also evidence of action potential "triangulation," a hallmark of arrhythmic risk characterized by replacement of the plateau phase of the action potential with slow, linear repolarization (Fig. 5.3a). We observed spontaneous early afterdepolarizations (EADs), the triggers of TdP (Fig. 5.3b), in the $bkd^{-/-}$ zebrafish hearts (Nerbonne and Kass, 2005). Optically measured action potential durations correlate well with reported intracellular recordings from wild-type zebrafish ventricular myocytes (Arnaout et al., 2007).

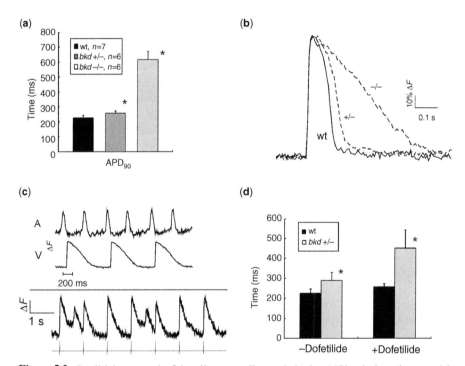

Figure 5.3 Parallels between zebrafish and human cardiac repolarization. (a) Ventricular action potential durations (APD) in wild-type (wt) and *breakdance* heterozygotes ($+/-$) and homozygotes ($-/-$) at 6 days post fertilization. * denotes $p < 0.05$. (b) Typical ventricular action potentials are displayed for wild-type (wt), *breakdance* heterozygote ($+/-$), and homozygote ($-/-$) embryos. The heterozygote action potential is subtly prolonged, while the homozygote recording shows marked action potential prolongation. Vertical calibration bar denotes 20% $\Delta F/F_0$ and horizontal bar denotes 100 ms. (c) *Upper panel*: Simultaneous atrial and ventricular voltage recordings from a *breakdance* ($-/-$) heart showing the mechanism of 2:1 atrioventricular block: action potentials are so prolonged in the ventricle that alternate atrial impulses encroach on the refractory plateau of the previous ventricular repolarization. *Lower panel*: EADs (arrows) observed in *breakdance* ($-/-$) embryos during ventricular pacing; the pacing train is shown below the action potential recording. EADs appear as spontaneous depolarizations occurring before repolarization is complete and prior to the subsequent paced beat. (d) Heterozygote *breakdance* embryos display increased sensitivity to I_{Kr} blockade (10 nM dofetilide). (Figure adapted from Milan et al., 2009.)

We were able to confirm heightened sensitivity of heterozygote *breakdance* mutants to I_{Kr} blockade that would be predicted from current understanding of repolarization reserve (Roden, 2004). At baseline, *breakdance* heterozygotes display minimal action potential prolongation, but exposure to the potent and specific I_{Kr} blocker dofetilide at 10 nM (similar to circulating plasma levels of the drug in clinical use) prolonged their action potentials (Fig. 5.3c). In order to establish that other mechanisms of repolarization toxicity are accessible in the zebrafish model, we also exposed zebrafish to sea anemone toxin (ATX-II), a toxin that interferes with sodium channel inactivation (Catterall and Beress, 1978). This resulted in a dose-dependent increase in the action potential duration, extending the parallels with human QT interval to include the LQT3 syndrome (Fig. 5.3d).

Subsequent work using optical mapping has also confirmed that novel repolarization genes such as NOS1AP can be efficiently studied in the zebrafish. These data suggest that ongoing work by us, and others, to define novel predictive algorithms for cardiotoxicity is likely to be useful in preclinical testing. Recently, we have obtained evidence that pharmacogenetic discovery in the zebrafish may predict human repolarization genes. In a shelf screen using staged assays (heart rate, atrioventricular block, and optical mapping) of insertional mutants, we implicated 15 novel genes in cardiac repolarization, one of which has recently been identified as a modifier of the human QT interval (Milan et al., 2009).

5.5 FUTURE DIRECTIONS

By building assays that exploit the throughput and the fidelity of the zebrafish, it should be feasible to define a truly predictive preclinical toxicology that encompasses not only complex effects on native cardiac biology, but also absorption, distribution metabolism, and excretion. Clearly, it will be vital to establish robust correlations between each phenotype in larval zebrafish and the cognate clinical trait, but it is evident from work in repolarization to date that the fish may be at least as representative as traditional models. The throughput feasible in multiwell plates opens up *in vivo* screening to a scale of investigation previously only accessible *in vitro*. Preliminary observations also suggest that gender effects observed in human drug-induced cardiotoxicity may be accessible (Pham and Rosen, 2002). Possible differences in the mechanisms or scope of action of different "I_{Kr} blockers" and drug interaction mechanisms can be addressed with this whole animal model. The development of sensitized or resistant zebrafish reporter strains will improve the specificity of this system. Finally, human pharmacogenetic studies of cardiac repolarization have been limited to the evaluation of candidate genes, as more powerful segregation-based family studies of drug responses are not feasible (Sesti et al., 2000; Roden and George, 2002; Splawski et al., 2002; Yang et al., 2002). The ease of genetic manipulation in the zebrafish should allow the unbiased identification of inherited modifiers of drug responses in phenotype-driven screens. By combining assays of increasing resolution in a tiered approach it is possible to optimize throughput and biological representation, and thus allow predictive toxicology to

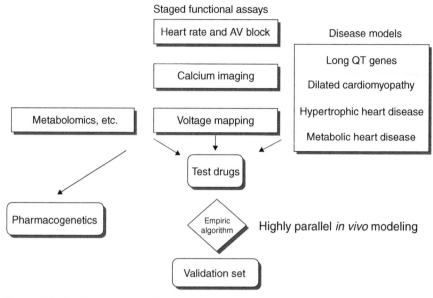

Figure 5.4   Combination assays offer optimized sensitivity and specificity for predictive toxicology. A major advantage of high-throughput assays in the zebrafish is the feasibility of combining multiple assays in series or in parallel to generate a final predictive algorithm with optimized sensitivity and specificity for robust predictive toxicology. The tractability of the fish is such that this might include disease modeling to incorporate critical predisposing proarrhythmic traits, the testing of drug combinations to explore drug–drug interactions, and stepwise assays of integrated physiology to generate quantitative metrics of risk. The final optimized assay algorithm can then be prospectively tested for its predictive utility.

move earlier into the drug discovery pipeline (Fig. 5.4). Ultimately, it is possible that drug discovery and toxicology will be performed in parallel *in vivo* allowing optimization of therapeutic and toxic effects.

REFERENCES

ANTZELEVITCH C and FISH J (2001). Electrical heterogeneity within the ventricular wall. Basic Res Cardiol 96(6): 517–527.

ANTZELEVITCH C, SUN ZQ, ZHANG ZQ, and YAN GX. (1996). Cellular and ionic mechanisms underlying erythromycin-induced long QT intervals and torsade de pointes. J Am Coll Cardiol 28(7): 1836–1848.

ARNAOUT R, FERRER T, HUISKEN J, SPITZER K, STAINIER DY, TRISTANI-FIROUZI M, and CHI NC (2007). Zebrafish model for human long QT syndrome. Proc Natl Acad Sci USA 104(27): 11316–11321.

BAKER K, WARREN KS, YELLEN G, and FISHMAN MC (1997). Defective "pacemaker" current (Ih) in a zebrafish mutant with a slow heart rate. Proc Natl Acad Sci USA 94(9): 4554–4559.

BEZZINA CR, VERKERK AO, BUSJAHN A, JERON A, ERDMANN J, KOOPMANN TT, BHUIYAN ZA, WILDERS R, MANNENS MM, TAN HL, LUFT FC, SCHUNKERT H, and WILDE AA (2003). A common polymorphism in KCNH2 (HERG) hastens cardiac repolarization. Cardiovasc Res 59(1): 27–36.

CAMM AJ, JANSE MJ, RODEN DM, ROSEN MR, CINCA J, and COBBE SM. (2000). Congenital and acquired long QT syndrome. Eur Heart J 21(15): 1232–1237.

CATTERALL WA and BERESS L. (1978). Sea anemone toxin and scorpion toxin share a common receptor site associated with the action potential sodium ionophore. J Biol Chem 253(20): 7393–7396.

CHEN JN, HAFFTER P, ODENTHAL J, VOGELSANG E, BRAND M, VAN EEDEN FJ, FURUTANI-SEIKI M, GRANATO M, HAMMERSCHMIDT M, HEISENBERG CP, JIANG YJ, KANE DA, KELSH RN, MULLINS MC, and NUSSLEIN-VOLHARD C (1996). Mutations affecting the cardiovascular system and other internal organs in zebrafish. Development 123: 293–302.

DE CLERCK F, VAN DE WATER A, D'AUBIOUL J, LU HR, VAN ROSSEM K, HERMANS A, and VAN AMMEL K (2002). *In vivo* measurement of QT prolongation, dispersion and arrhythmogenesis: application to the preclinical cardiovascular safety pharmacology of a new chemical entity. Fundam Clin Pharmacol 16(2): 125–140.

ECKARDT L, HAVERKAMP W, BORGGREFE M, and BREITHARDT G (1998). Experimental models of torsade de pointes. Cardiovasc Res 39(1): 178–193.

KEATING MT and SANGUINETTI MC (2001). Molecular and cellular mechanisms of cardiac arrhythmias. Cell 104(4): 569–580.

LANGHEINRICH U, VACUN G, and WAGNER T (2003). Zebrafish embryos express an orthologue of HERG and are sensitive toward a range of QT-prolonging drugs inducing severe arrhythmia. Toxicol Appl Pharmacol 193(3): 370–382.

MILAN DJ, PETERSON TA, RUSKIN JN, PETERSON RT, and MACRAE CA (2003 Mar 18). Drugs that induce repolarization abnormalities cause bradycardia in zebrafish. Circulation 107(10): 1355–1358.

MILAN DJ, KIM AM, WINTERFIELD JR, JONES IL, PFEUFER A, SANNA S, ARKING DE, AMSTERDAM AH, SABEH KM, MABLY JD, ROSENBAUM DS, PETERSON RT, CHAKRAVARTI A, KÄÄB S, RODEN DM, and MACRAE CA (2009 Aug 18). Drug-sensitized zebrafish screen identifies multiple genes, including GINS3, as regulators of myocardial repolarization. Circulation 120(7): 553–559.

NERBONNE JM and KASS RS (2005). Molecular physiology of cardiac repolarization. Physiol Rev 85(4): 1205–1253.

NEWTON-CHEH C, GUO CY, LARSON MG, MUSONE SL, SURTI A, CAMARGO AL, DRAKE JA, BENJAMIN EJ, LEVY D, D'AGOSTINO R.B. SR., HIRSCHHORN JN, and O'DONNELL CJ (2007). Common genetic variation in KCNH2 is associated with QT interval duration: the Framingham Heart Study. Circulation 116(10): 1128–1136.

PFEUFER A, JALILZADEH S, PERZ S, MUELLER JC, HINTERSEER M, ILLIG T, AKYOL M, HUTH C, SCHOPFER-WENDELS A, KUCH B, STEINBECK G, HOLLE R, NABAUER M, WICHMANN HE, MEITINGER T, and KAAB S (2005). Common variants in myocardial ion channel genes modify the QT interval in the general population: results from the KORA study. Circ Res 96(6): 693–701.

PHAM TV and ROSEN MR (2002). Sex, hormones, and repolarization. Cardiovasc Res 53(3): 740–751.

RODEN DM (2000). Acquired long QT syndromes and the risk of proarrhythmia. J Cardiovasc Electrophysiol 11(8): 938–940.

RODEN DM (2001). Pharmacogenetics and drug-induced arrhythmias. Cardiovasc Res 50(2): 224–231.

RODEN DM (2004). Drug-induced prolongation of the QT interval. N Engl J Med 350(10): 1013–1022.

RODEN DM and GEORGE AL JR., (2002). The genetic basis of variability in drug responses. Nat Rev Drug Discov 1(1): 37–44.

SESTI F, ABBOTT GW, WEI J, MURRAY KT, SAKSENA S, SCHWARTZ PJ, PRIORI SG, RODEN DM, GEORGE A.L. JR., and GOLDSTEIN SA (2000). A common polymorphism associated with antibiotic-induced cardiac arrhythmia. Proc Natl Acad Sci USA 97(19): 10613–10618.

SPLAWSKI I, TIMOTHY KW, TATEYAMA M, CLANCY CE, MALHOTRA A, BEGGS AH, CAPPUCCIO FP, SAGNELLA GA, KASS RS, and KEATING MT (2002). Variant of SCN5A sodium channel implicated in risk of cardiac arrhythmia. Science 297(5585): 1333–1336.

VERHEIJCK EE, VAN GINNEKEN AC, BOURIER J, and BOUMAN LN (1995). Effects of delayed rectifier current blockade by E-4031 on impulse generation in single sinoatrial nodal myocytes of the rabbit. Circ Res 76(4): 607–615.

WARREN KS and FISHMAN MC. (1998). "Physiological genomics": mutant screens in zebrafish. Am J Physiol 275(1 Pt 2): H1–H7.

YANG P, KANKI H, DROLET B, YANG T, WEI J, VISWANATHAN PC, HOHNLOSER SH, SHIMIZU W, SCHWARTZ PJ, STANTON M, MURRAY KT, NORRIS K, GEORGE A.L. JR., and RODEN DM (2002). Allelic variants in long-QT disease genes in patients with drug-associated torsades de pointes. Circulation 105(16): 1943–1948.

YANG T, SNYDERS D, and RODEN DM (2001). Drug block of I_{Kr}: model systems and relevance to human arrhythmias. J Cardiovasc Pharmacol 38(5): 737–744.

Chapter 6

In Vivo Recording of the Adult Zebrafish Electrocardiogram

David J. Milan[1] and Calum A. MacRae[2]

[1]*Cardiovascular Research Center and Cardiology Division, Massachusetts General Hospital, Boston, MA, USA*
[2]*Cardiovascular Division, Brigham and Women's Hospital, Boston, MA, USA*

6.1 INTRODUCTION

QT prolongation and the attendant risk of the malignant arrhythmia, torsades de pointes, have led to the post marketing withdrawal of multiple medications from the U.S. market. Since drug-induced cardiotoxicity is difficult to predict, it has become a major focus for regulatory scrutiny, and is an area of intense pharmaceutical and academic research (Roden, 2004; Camm, 2005). The cardiac toxicities of new drugs are one of the most important areas of investigation in new drug development. Indeed, more drugs were withdrawn from the US market due to concerns regarding QT prolongation than for any other single cause (Roden, 2004; Camm, 2005).

Cardiac repolarization is the result of the integrated effects of channels, receptors, and cytoskeletal proteins. Further complexity arises from substantial local differences in repolarization within the heart. Drug-induced cardiotoxicity is often difficult to predict as many of the responsible agents perturb repolarization only in combination with other drugs. These interactions may be pharmacokinetic or pharmacodynamic in nature. Inherited variation is known to affect the diathesis toward drug-induced arrhythmias and contributes to limited predictive utility of simple *in vitro* systems (Yang et al., 2002).

Other issues that make drug-induced cardiotoxicity so difficult to predict prior to human use are the low frequency of arrhythmic events, the increasing complexity of drug–drug interactions, the low specificity of cellular hERG assays, and the low throughput of truly representative animal models. Multiple new assays have been introduced to the field in the last few years, highlighting not only to the importance of cardiotoxicity, but also the limited predictive utility current techniques (Bass et al., 2005).

Zebrafish: Methods for Assessing Drug Safety and Toxicity, Edited by Patricia McGrath.
© 2012 John Wiley & Sons, Inc. Published 2012 by John Wiley & Sons, Inc.

Animal models of repolarization toxicity have proven difficult to develop, because of interspecies differences in repolarization and the incomplete understanding of the mechanisms of repolarization. We previously demonstrated that zebrafish embryos develop bradycardia in response to the drugs known to prolong the QT interval in humans, and in subsequent work, detailed below, we have gone on to demonstrate that these drugs also prolong the QT in adult zebrafish (Milan et al., 2006).

Over the past decade, the zebrafish has become established as a major model organism for biomedical research. First pioneered in developmental biology laboratories (Shin and Fishman, 2002), this role has been consolidated by the acquisition of genetic and genomic resources including the sequencing of the zebrafish genome. The zebrafish is ideal for the study of the cardiovascular system as the heart is readily observed and the entire circulation is within 72 h of fertilization (Sehnert et al., 2002). Though many cardiovascular mutants have been identified, standard rodent or large animal phenotyping tools are not readily available in zebrafish due to the small size of the heart and its aqueous environment. Stable electrocardiograms have been documented from large fish species including tuna, carp, and trout, and *ex vivo* cardiac electrophysiology has been widely studied (Ueno et al., 1986; Korsmeyer et al., 1997). We developed a method for the reproducible recording of the adult zebrafish ECG (electrocardiogram), and employed this technique to study the effects of QT prolonging drugs on cardiac repolarization (Milan et al. 2006).

6.2 OPTIMIZATION OF ZEBRAFISH ELECTROCARDIOGRAM RECORDING

Our early efforts to record the zebrafish electrocardiogram were hindered by hypoxia and gill motion. Nonoxygenated zebrafish die in a rapid sequence of bradycardia, heart block, and asystole within 30 min (Fig. 6.1, squares). To

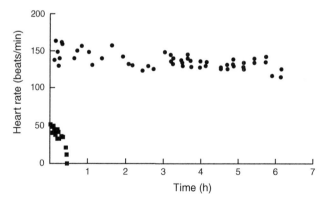

Figure 6.1 Oral perfusion of the zebrafish viability results for up to 6 h as assessed by stable RR and QT values as shown here for a typical zebrafish. Nonperfused fish succumb to hypoxia and expire within 30 min. (Figure adapted from Milan et al., 2006.)

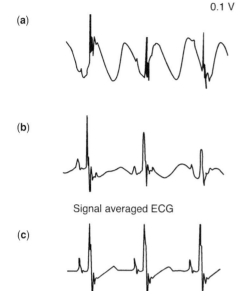

0.1 V

100 ms

(a)

(b)

Signal averaged ECG

(c)

Signal averaged ECG + CTX

Figure 6.2 (a) Initial attempts at electrocardiographic zebrafish recordings were contaminated by skeletal muscle artifacts, especially gill motion that can be seen as a sinusoidal baseline signal. (b) Signal averaging allowed some reduction of this artifact as seen in this 50-beat average. (c) Further reduction in electrocardiographic artifacts was achieved with the combination of signal averaging and paralysis with μ-conotoxin GIIIB (10-beat average). (Figure adapted from Milan et al., 2006.)

prevent this, we had to construct a perfusion system compatible with electrocardiographic recording outside of the normal aqueous environment for several hours. Recordings from fish whose gills were superfused with oxygenated and buffered medium did not exhibit significant RR or QT interval variation (Fig. 6.1, circles).

This system allowed recordings for several hours, but did not eliminate electromyographic noise, primarily from gill motion (Fig. 6.2a). Signal averaged electrocardiograms were still inadequate for the reproducible measurement of cardiac repolarization parameters (Fig. 6.2b). However, we were able to reduce noise sufficiently for high fidelity recording with the intraperitoneal injection of conotoxin, and consequent skeletal muscle paralysis (Fig. 6.2c).

Optimal signal filtering used a high-pass cutoff value of 0.5 Hz in order to eliminate excessive baseline wander, but such filtering would be predicted to blunt the terminal T wave. To assess this, we performed duplicate measurements in five fish, varying the high-pass cutoff value. Reducing the cutoff frequency from 0.5 to 0.1 Hz resulted in a 1.5% increase in the measured QT interval ($p = 0.2$). Based on this minimal effect, and the improved recordings with the higher value, we chose a cutoff of 0.5 Hz for the remainder of our work. T-wave morphologies vary between individual fish presumably as a result of subtle changes in the electrode placement or the orientation of the heart.

Table 6.1 Baseline Electrocardiographic Intervals of the Adult Male Zebrafish in the Presence (+) or Absence (−) of μ-Conotoxin GIIIB (CTX)

	PR	QRS	QT	RR
Males (−) CTX ($n = 5$)	52 ± 19	26 ± 3	229 ± 11	348 ± 43
Males (+) CTX ($n = 58 - 110$)	66 ± 14	34 ± 11	242 ± 54	398 ± 77

All values are presented in ms.

6.3 BASIC INTERVALS

Once we had optimized our ex aqua system to allow reproducible electrocardiographic recordings, we were able to obtain technically satisfactory QT measurements in over 80% of zebrafish tested (Fig. 6.2c). Synchronized video microscopy and electrocardiography was employed to demonstrate the precise correlation of electrical and mechanical events (PQRS sequence with atrial and ventricular contraction). The mean heart rate of the adult male zebrafish is 151 ± 30 bpm (beats per minute), and the baseline intervals are described in Table 6.1. Conotoxin did not affect heart rate, QRS, or QT intervals, but did affect the PR interval prolonging it reproducibly by about 27%.

We obtained normal zebrafish QT and RR data from a total of 110 zebrafish. Each point in Fig. 6.3 is a three-beat average of QT and RR for a given fish. The data were

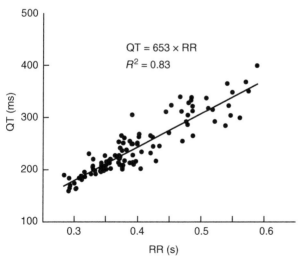

QT = 653 × RR
$R^2 = 0.83$

Figure 6.3 The dependence of the QT interval on the basic cycle length is shown in this graph derived from recordings of 110 zebrafish. Each point represents the average of three consecutive RR and QT measurements for a given fish. The relationship is nearly linear with $R^2 = 0.83$ for the equation QT = 634 × RR. (Figure adapted from Milan et al., 2006.)

best fit by the equation $QT = QTc \times RR^{1.05}$ with a QTc of 634 ($R^2 = 0.84$) (Fig. 6.3). As the simpler linear equation $QT = 634 \times RR$ provided a nearly equivalent fit ($R^2 = 0.83$), we used this latter estimate for all subsequent analyses of zebrafish QT duration.

6.4 DRUG EFFECTS

Zebrafish were perfused with seven clinically relevant drugs, four of which prolong the QT interval in humans (astemizole, haloperidol, pimozide, and terfenadine) and three control drugs that do not prolong the QT interval (clonidine, penicillin, and propranolol). Exposure to the known QT prolonging agents increased the QTc in each case as follows: astemizole $18 \pm 9\%$ ($p = 0.009$), haloperidol $16 \pm 11\%$ ($p = 0.019$), pimozide $17 \pm 9\%$ ($p = 0.005$), and terfenadine $11 \pm 6\%$ ($p = 0.015$). Drugs not known to prolong cardiac repolarization did not significantly prolong the QTc interval: clonidine $1 \pm 6\%$ ($p = 0.5$), penicillin $-1 \pm 2\%$ ($p = 0.46$), and propranolol $-6 \pm 6\%$ ($p = 0.054$). All of the results are summarized in Fig. 6.4. The RR intervals increased for each drug as follows: astemizole $16 \pm 6\%$ ($p = 0.006$), haloperidol $38 \pm 14\%$ ($p = 0.004$), pimozide $9 \pm 13\%$ ($p = 0.19$), terfenadine $18 \pm 15\%$ ($p = 0.07$), clonidine $9 \pm 11\%$ ($p = 0.11$), penicillin $4 \pm 6\%$ ($p = 0.12$), and propranolol $19 \pm 18\%$ ($p = 0.04$). Drug-induced QT prolongation was dose dependent as shown for astemizole (Fig. 6.5) Typical QT prolongation responses are shown in Fig. 6.6 for the cardiotoxic drugs tested.

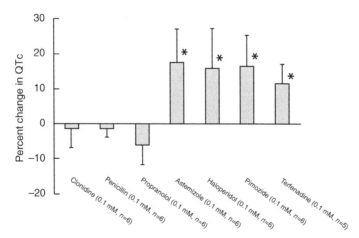

Figure 6.4 QT prolonging drugs astemizole, haloperidol, pimozide, and terfenadine all caused an increase in the corrected QT interval. Control drugs clonidine, penicillin, and propranolol did not cause a significant change in the QTc. Statistically significant differences ($p < 0.05$) are indicated by an asterisk. (Figure adapted from Milan et al., 2006.)

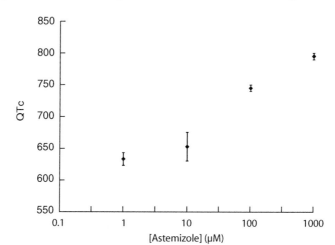

Figure 6.5 Administration of increasing doses of astemizole resulted in a dose-dependent lengthening of the corrected QT interval. The data demonstrate a characteristic sigmoidal dose–response curve. (Figure adapted from Milan et al., 2006.)

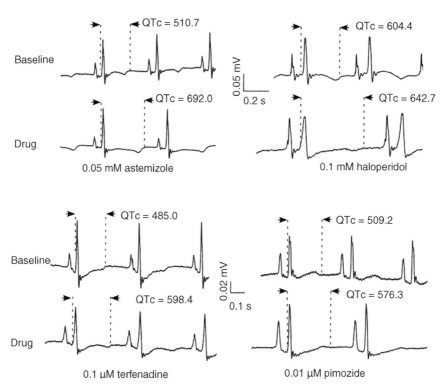

Figure 6.6 Sample tracings from representative zebrafish. For each fish, a baseline recording is shown above the recording obtained after drug treatment. Dotted lines indicate the QT interval and the QTc measurement is indicated for each tracing. (Figure adapted from Milan et al., 2006.)

ECG recording electrode placement

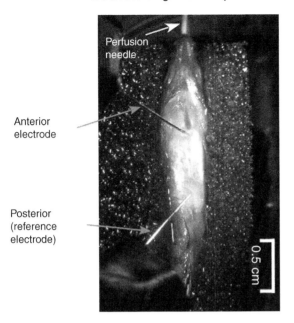

Anterior
electrode

Posterior
(reference
electrode)

Figure 6.7 Photograph of the experimental setup. The ventral surface of the adult zebrafish is visible. There is a perfusion needle resting in the mouth of the fish (white arrow) and the recording electrodes are seen in their standard locations (red arrows). (See the color version of this figure in Color Plates section.) (Figure adapted from Milan et al., 2006.)

6.5 CONCLUSIONS

The ex aqua techniques we have described enable the rapid assessment of the adult zebrafish electrocardiogram and overcome the challenges of hypoxia and gill motion. The resultant zebrafish electrocardiograms are highly reproducible, with fundamental intervals that are remarkably similar to those observed in adult humans. The stable system we have designed allows the study of the effects of pharmacological agents including those known human cardiotoxins. In a series of test drugs, we have found an excellent correlation between QT effects in human and in the zebrafish. Even less well characterized effects on repolarization such as the shortening of the QT interval observed with propranolol were confirmed (Sarma et al., 1988; Endresen and Amlie, 1990).

There is tremendous variation in the basal characteristics of cardiac electrophysiology across the vertebrate phyla with even simple heart rate ranging from 20 to 800 bpm. Interspecies differences in repolarization have been attributed to the variable contributions of multiple ionic currents to the integrated cardiac action potential. Data from a wide range of genetic models have established the complexity of ion channel biology, with role not only in passive charge movements but also in many distinctive signaling pathways (Olgin and Verheule, 2002). These findings suggest that electrocardiographic characteristics may not correlate well with the species' ultimate predictive utility for human physiology and pharmacology.

Limitations of the current adult model of drug-induced QT prolongation include the lack of aqueous solubility of all medications, and the relatively slow throughput compared with *in vitro* technologies. Nevertheless, when used in combination with a higher throughput technique, such our previously published embryonic zebrafish assay, this adult zebrafish model will allow direct assessment of the QT interval. The methods described here add to the available assays for detecting drug-induced cardiotoxicity in a model with remarkably similar heart rate and repolarization time as humans. Recording the electrocardiogram of the adult zebrafish will permit the cost-effective screening of QT prolonging drugs.

REFERENCES

BASS AS, TOMASELLI G, BULLINGHAM R 3RD, and KINTER LB (2005). Drugs effects on ventricular repolarization: a critical evaluation of the strengths and weaknesses of current methodologies and regulatory practices. J Pharmacol Toxicol Methods 52(1): 12–21.

CAMM AJ (2005). Clinical trial design to evaluate the effects of drugs on cardiac repolarization: current state of the art. Heart Rhythm 2(2 Suppl): S23–S29.

ENDRESEN K and AMLIE JP (1990). Effects of propranolol on ventricular repolarization in man. Eur J Clin Pharmacol 39(2): 123–125.

KORSMEYER KE, LAI NC, SHADWICK RE, and GRAHAM JB (1997). Heart rate and stroke volume contribution to cardiac output in swimming yellowfin tuna: response to exercise and temperature. J Exp Biol 200(Pt 14): 1975–1986.

MILAN DJ, PETERSON TA, RUSKIN JN, PETERSON RT, and MACRAE CA (2003 Mar 18). Drugs that induce repolarization abnormalities cause bradycardia in zebrafish. Circulation 107(10): 1355–1358.

OLGIN JE and VERHEULE S (2002). Transgenic and knockout mouse models of atrial arrhythmias. Cardiovasc Res 54(2): 280–286.

RODEN DM (2004). Drug-induced prolongation of the QT interval. N Engl J Med 350(10): 1013–1022.

SARMA JS, VENKATARAMAN K, SAMANT DR, and GADGIL UG (1988). Effect of propranolol on the QT intervals of normal individuals during exercise: a new method for studying interventions. Br Heart J 60(5): 434–439.

SEHNERT AJ, HUQ A, WEINSTEIN BM, WALKER C, FISHMAN M, and STAINIER DY (2002). Cardiac troponin T is essential in sarcomere assembly and cardiac contractility. Nat Genet 31(1): 106–110.

SHIN JT and FISHMAN MC (2002). From zebrafish to human: modular medical models. Annu Rev Genomics Hum Genet 3: 311–340.

UENO S, YOSHIKAWA H, ISHIDA Y, and MITSUDA H (1986). Electrocardiograms recorded from the body surface of the carp, *Cyprinus carpio*. Comp Biochem Physiol A Comp Physiol 85(1): 129–133.

YANG P, KANKI H, DROLET B, YANG T, WEI J, VISWANATHAN PC, HOHNLOSER SH, SHIMIZU W, SCHWARTZ PJ, STANTON M, MURRAY KT, NORRIS K, GEORGE A.L. JR., and RODEN DM (2002). Allelic variants in long-QT disease genes in patients with drug-associated torsades de pointes. Circulation 105(16): 1943–1948.

Chapter 7

Hematopoietic and Vascular System Toxicity

Alison M. Taylor and Leonard I. Zon

Stem Cell Program and Division of Hematology/Oncology, Children's Hospital Boston and Dana Farber Cancer Institute, Harvard Medical School

7.1 INTRODUCTION

Research in recent decades has demonstrated a high level of conservation of hematopoietic and vascular development among all vertebrates, including zebrafish (de Jong and Zon, 2005; Baldessari and Mione, 2008). Not only are regulatory networks conserved, but work from our lab also demonstrates that the effect of chemical compounds on zebrafish blood, stem cells, and vasculature can be extrapolated to mouse and other mammals (North et al., 2007). Furthermore, small size, high fecundity, and optically clear development make the zebrafish ideal for high-throughput screening and chemical testing. These data support the use of zebrafish for assessing drug safety and toxicity for the hematopoietic and vascular systems. This chapter provides a review of hematopoiesis and vasculogenesis in the zebrafish, as well as methods for assays available to test compound-induced toxicity on blood and endothelial development.

7.2 HEMATOPOIESIS AND VASCULAR DEVELOPMENT IN THE ZEBRAFISH

7.2.1 Hematopoietic and Endothelial Progenitors

Primitive blood and vascular cells originate from the mesoderm of the embryo and form clusters called blood islands (Liao et al., 1997). Bone morphogenic protein (BMP), among other signals, is responsible for generating blood islands from the mesoderm tissue (Lengerke et al., 2008). Blood islands form in both the cephalic and

Zebrafish: Methods for Assessing Drug Safety and Toxicity, Edited by Patricia McGrath.
© 2012 John Wiley & Sons, Inc. Published 2012 by John Wiley & Sons, Inc.

ventral mesoderm, leading to primitive blood cell production in the rostral blood islands (RBIs) and intermediate cell mass (ICM), respectively. All of the blood island cells are initially positive for expression of scl, gata2, and lmo2, transcription factors crucial for driving endothelial and hematopoietic cell fate (Dooley et al., 2005). By five somites, the cells also express hematopoietic-specific markers (gata1, pu.1) or endothelial-specific markers (flk1, fli1) (Davidson and Zon, 2004).

7.2.2 Primitive Hematopoiesis

Hematopoiesis occurs in two waves in vertebrates. The first primitive wave occurs in the ICM and the RBI of the zebrafish embryo, and erythrocytes and myeloid cells are formed. Primitive hematopoiesis continues until definitive hematopoiesis begins at 36hpf, when hematopoietic stem cells (HSCs) are created and subsequently all blood lineages can be derived.

Primitive erythropoiesis occurs in the ICM. First, blood and endothelial precursors form bilateral stripes. At five somites, the endothelial precursors express flk1, whereas the blood precursors express gata1 (Davidson and Zon, 2004). By 10 somites, the blood precursors begin to express erythroblast markers such as embryonic globins (Galloway et al., 2005). As development progresses, cells from these stripes migrate toward the midline and fuse to form the ICM at 18hpf (Detrich et al., 1995). Gata1 positive cells enter circulation by 24hpf (Galloway et al., 2005) and mature into erythrocytes after several days (Detrich et al., 1995). Concurrently, endothelial cells also differentiate to form the vasculature (described below).

Blood islands that form in the RBI are the source of primitive myeloid cells. RBI hematopoietic progenitors, unlike those in the ICM, express pu.1, a myeloid marker (Lieschke et al., 2002). Pu.1 expression decreases by 18 somites, but other myeloid markers come on at this time point, including cebp1, fms, l-plastin, and lysozyme-c (Davidson and Zon, 2004). Myeloperoxidase (mpo) expression can soon be detected in granulocytes as well, and this expression lasts throughout development (Bennett et al., 2001).

7.2.3 Definitive Hematopoiesis

The first wave of definitive hematopoiesis is transient and occurs in the posterior blood island (PBI) between 24 and 48hpf. At this point, erythroid myeloid progenitors (EMPs) arise. These cells are not hematopoietic stem cells, as they do not have self-renewal potential and cannot differentiate into lymphoid cells (Bertrand et al., 2008). The second wave of definitive hematopoiesis involves generation of HSCs in the aorta–gonad–mesonephros (AGM) region of the zebrafish embryo. Here, cells expressing runx1, c-myb, and CD41, hematopoietic stem cell markers, can be detected by 36hpf in the dorsal aorta (Lin et al., 2005; North et al., 2007). Currently, definitive HSCs are thought to derive from the ventral wall of the aorta, termed "hemogenic endothelium." Runx1 and c-myb positive cells are capable of differentiating into all blood lineages. During development, the site of definitive hematopoiesis moves from the AGM at 36hpf to the caudal hematopoietic tissue (CHT) (Murayama et al., 2006). By 3–4 days post fertilization (dpf), all hematopoiesis occurs in the thymus and pronephros (kidney). Blood cells of all types develop in the kidney in between the renal tubules, referred to as "kidney marrow." One exception is T cells, which originate from the kidney but are educated in the thymus (Traver et al., 2003a).

7.2.4 Differentiated Blood Cells

Hematopoietic cell types can be differentiated based on morphological features when cells are stained with May-Grunwald and Giemsa stains (Fig. 7.1). In addition, some blood cell types can be identified by other staining methods described in this chapter. Each lineage can also be characterized by specific gene expression profiles.

7.2.5 Lymphocytes

Zebrafish have both B cells and T cells that express many of the same genes as mammalian lymphocytes (Langenau et al., 2004) and are morphologically very similar (Fig. 7.1). Study of rag1 expression patterns demonstrates that B-cell lymphopoiesis begins in the pancreas by 4dpf, but later occurs in the kidney (Danilova and Steiner, 2002). T-cell progenitors, on the other hand, are first seen in the thymi at 3–4dpf as determined by expression of ikaros, lck, rag1, rag 2, and other genes

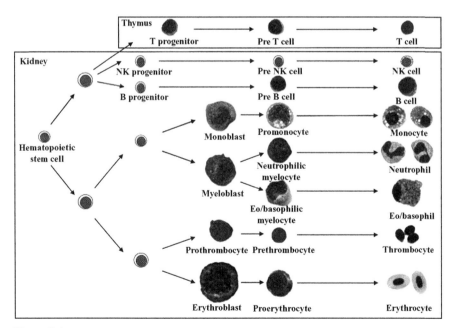

Figure 7.1 Hematopoiesis in the adult zebrafish. All blood cells are derived from a hematopoietic stem cell. Images are cells stained with May-Grunwald and Giemsa stains (Traver et al., 2003a).

(Langenau et al., 2004; Trede et al., 2004). T cells develop in the thymi throughout adulthood, and mature T cells can also be found in the kidney.

7.2.6 Thrombocytes

The zebrafish equivalents to platelets are called thrombocytes, and unlike platelets these cells maintain their nuclei (Fig. 7.1) (Jagadeeswaran et al., 1999). Thrombocytes and their precursors, prothrombocytes and prethrombocytes, are found to express high levels of CD41. CD41 expression studies have determined that thrombocytes do not develop in primitive or early definitive hematopoiesis, but begin to appear at 2–3dpf (Lin et al., 2005).

7.2.7 Granulocytes

Zebrafish myeloid cell types differ slightly from mammalian myeloid cell types. In zebrafish, three myeloid cell types are present: macrophages, neutrophils, and a third cell type that has characteristics of both basophils and eosinophils (Bennett et al., 2001) (Fig. 7.1). Macrophages are derived from monocytes, whereas the other cell types are derived from a myeloblast precursor (Traver et al., 2003a). Neutrophils have granules that are mpo positive, as well as segmented nuclei. The basophil/eosinophil population can be characterized by a nonsegmented nucleus (Bennett et al., 2001).

7.2.8 Erythrocytes

During zebrafish erythroid differentiation, a proerythroblast can be distinguished by its large nucleus and is positive for gata1 and globin expression (Brownlie et al., 2003). In contrast to mammalian red blood cells, when zebrafish erythroblasts differentiate, the nucleus condenses and is retained (Fig. 7.1). These cells maintain expression of gata1 and globins.

7.2.9 Vascular Development

The vascular system develops in parallel to hematopoiesis and HSC formation. Cells from the blood islands form both HSCs and precursors termed angioblasts (Dooley et al., 2005). Angioblasts differentiate into endothelial cells that line the blood vessels. This differentiation is induced by the vascular endothelial growth factors (VEGFs) (Lawson and Weinstein, 2002). VEGF can signal through multiple receptors, including flk1, to induce differentiation and formation of the dorsal aorta (DA) and posterior cardinal vein (PCV). After formation, these two vessels must be specified. The DA and all artery precursors express ephrinB2, notch5, flk1, deltaC, and tbx20. On the other hand, the PCV and all venous precursors express EphB4 receptor tyrosine kinase and flt4. Notch5 and flt4 are both induced by VEGF signaling, though how

VEGF activates these two pathways differentially is unknown (Lawson and Weinstein, 2002; Covassin et al., 2006). Once the dorsal aorta is specified, angiogenic sprouting of the intersomitic vessels (ISVs) can occur. After ISVs are created, new vessels sprout off the PCV. Other vessels develop throughout the body as well, providing blood flow to all organ systems (Baldessari and Mione, 2008).

7.3 MORPHOLOGICAL AND FUNCTIONAL ASSAYS TO ASSESS TOXICITY

There are many assays available to test the zebrafish hematopoietic and vascular systems. Here we describe methods to look at blood cell morphology and function, as well as vascular function. If a compound induces a toxic effect on the blood or vascular systems, the function and morphology will likely be affected. When treating zebrafish embryos with drugs for these experiments, treatment should begin at one to five somites, when hematopoietic and endothelial progenitors are first specified. Drug-treated animal samples should also be compared to untreated and vehicle controls.

7.3.1 Embryonic

During early zebrafish development, there are several methods of assessing development of the hematopoietic and vascular systems. By 24hpf, the first erythroblasts enter circulation (de Jong and Zon, 2005) and can be visualized in a standard dissection scope. The number of blood cells in circulation increases as definitive hematopoiesis begins. To assay the morphology of these cells, peripheral blood can be removed from the embryo for visualization. Cells can then be stained with May-Grunwald and Giemsa stains to allow cell identification by morphology.

Isolating Peripheral Blood from the Zebrafish Embryo (Chan et al., 1997)

1. Submerge the zebrafish embryo in $0.9\times$ phosphate-buffered saline (PBS).
2. Using a sharp scalpel or razor, cut across the tail.
3. Allow the blood cells to collect in PBS.
4. To collect, use a glass pipette.

Staining with May-Grunwald and Giemsa

1. Stain with May-Grunwald for 10 min.
2. Rinse with water.
3. Stain with Giemsa diluted 1:5 for 15 min.
4. Rinse with water. Look at slides to confirm that the staining is visible. If not, repeat step 3.
5. Allow the slides to dry before analysis, which may include differential cell counts to determine percent of total for each blood cell type.

7.3.2 Adult Blood Cell Morphology

There are several ways to assay blood cells in the adult zebrafish kidney. One possibility is to section the fish, stain with hematoxylin and eosin, and look at general kidney morphology. In a normal kidney, the hematopoietic cells will be distributed among the renal tubules (Fig. 7.2a). If the fish has developed a myeloproliferative disorder, the hematopoietic cells will have taken over the kidney and fewer tubules will be apparent (Le et al., 2007). Alternatively, if the fish has been irradiated, the kidney will have few hematopoietic cells when compared to healthy fish (Traver et al., 2004). Analysis of kidney sections can provide information about gross hematopoietic development, but does not provide specific information about blood cell number, blood cell types present in the hematopoietic system, or individual cell morphology. To look at individual cell morphology and percentages

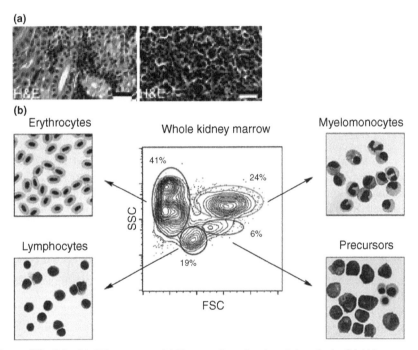

Figure 7.2 Zebrafish kidney marrow. (a) Hematoxylin and eosin staining of zebrafish kidney marrow sections (Le et al., 2007). A healthy kidney (left) comprised of hematopoietic cells scattered in between the renal tubules. A fish with myeloproliferative disorder (right) has blood cells overtaking the kidney marrow. (b) Flow cytometry analysis of zebrafish whole kidney marrow (Traver et al., 2003b). Four populations can be separated after flow cytometry sorting by forward scatter and side scatter—erythrocytes, lymphocytes, myelomonocytes, and precursors. Erythrocytes are present in a low FSC population. Lymphocytes are found in a low SSC and intermediate FSC population. Myelomonocytes fall into a high FSC and SSC population. Finally, precursors of these cell types have high FSC and intermediate SSC properties. (See the color version of this figure in Color Plates section.)

of blood cells, peripheral blood or whole kidney marrow can be isolated and visualized by cytospins.

Isolating Peripheral Blood from the Adult Zebrafish (Chan et al., 1997)

1. Anesthetize adult zebrafish in 4% 4 g/L tricaine.
2. Puncture the heart with a glass microhematocrit tube to collect blood cells.
3. Wash the cells in 0.9× PBS before characterization.

Isolating Kidney Marrow from the Adult Zebrafish

1. Euthanize adult zebrafish in 10–20% 4 g/L tricaine.
2. Dissect fish and remove kidney (LeBlanc et al., 2007).
3. Place kidney in 300–500 μL of 0.9× PBS and 5% fetal bovine serum solution.
4. Triterate using 1000 μL pipette tip until kidney marrow has been disaggregated.
5. Pass through a 40 μm filter.

Cytospins

1. Dilute cells 1:10 in 0.4% filtered trypan blue. Trypan blue is incorporated by dead cells, so they appear blue under the microscope.
2. Count live cells using a hemacytometer or other method.
3. Dilute (or concentrate) cells so that there are 100,000–200,000 cells in 600 μL.
4. In a cytospin cytocentrifuge, spin down 200 μL for 3 min at 300 rpm onto glass slides.
5. Wait overnight for slides to dry.
6. Stain with May-Grunwald and Giemsa as described above.

Although cytospin analysis provides morphological data, flow cytometry analysis is an alternative approach to identifying the percentage of kidney marrow cells of each blood cell type. Using flow cytometry, cells can be analyzed by light scatter characteristics, specifically forward scatter (FSC) and side scatter (SSC). FSC is directly proportional to cell size, whereas SSC is directly proportional to cellular granularity (Traver et al., 2003b). When whole kidney marrow is analyzed using these characteristics, four populations can be identified (Fig. 7.2b). A population with low FSC contains mature erythrocytes. A population with both high FSC and SSC contains myeloid cells and monocytes/macrophages. Lymphoid cells fall into a population with low side scatter and intermediate forward scatter characteristics. Myeloid, lymphoid, and erythroid precursors fall into a population in between the myeloid and lymphoid populations. More recent analysis of CD41 expressing cells demonstrates that HSCs, as well as thrombocytes, are present in the lymphoid gate (Lin et al., 2005). An advantage of the flow cytometry approach is that cells can also be sorted into the four populations using flow-activated cell sorting (FACS). Overall, flow cytometry of zebrafish whole kidney marrow correlates with data seen from cytospin analysis, while allowing study of much higher blood cell numbers (Traver et al., 2003b).

Flow Cytometry Analysis

1. Isolate kidney marrow as described above.
2. Add 1 μL propidium iodide (PI), which is incorporated by dead cells, to filtered marrow cells.
3. Perform analysis on a flow cytometer. Analyze PI levels and light scatter characteristics.
4. Remove PI positive cells (dead cells) from analysis.
5. Create gates based on forward and side scatter characteristics.

7.3.3 Functional HSC and Blood Progenitor Assays

Looking at gross morphology or blood cell number in the adult zebrafish does not prove that HSCs are unaffected by chemical treatment. Unlike many mammalian systems, it is not possible to sort out HSCs from the zebrafish kidney marrow using stem cell-specific antibodies. However, the presence of HSCs can still be tested by functional transplantation assays. As shown in mammalian models (Uchida et al., 1994), zebrafish HSCs are capable of reconstituting the entire hematopoietic system. In particular, they can rescue the blood system in a fish treated with a lethal dose of irradiation (Traver et al., 2004) and contribute to reconstitution of blood cells in a sublethally irradiated fish (White et al., 2008). If a chemical-treated adult fish has maintained functional HSCs, then it will also be able to rescue the hematopoietic system. This assay may primarily be useful for testing a select number of compounds.

Transplant Assay

1. Chemically treat adult glo-fish (available from S-D Tropical and Segrest Farms) (or an alternate transgenic fish with fluorophore-labeled blood cells) for the desired length of time. Alternatively, cells can be treated *ex vivo* at step 3 for up to 4 h.
2. Irradiate wild-type untreated recipient fish at 30 Gy evenly split between one and two days pre-transplant.
3. Remove kidney marrow cells as described above from chemically treated fish.
4. Count cells using a hemacytometer or an alternate approach.
5. Via intracardial or retro-orbital injection, inject 100,000 kidney marrow cells into each irradiated recipient fish (2 days post irradiation).
6. Four weeks post transplant, isolate kidney marrow from recipient fish.
7. Analyze recipient fish for fluorophore (i.e., GFP) positive cells in the marrow. In particular, analyze the light scatter characteristics of GFP positive cells. If HSCs in the chemical treated fish or cells were functional, then GFP positive cells will be present in the population (White et al., 2008).

7.3.4 Confocal Microangiography

Brant Weinstein and colleagues developed microangiography for assaying flow in the vasculature of zebrafish embryos (Weinstein et al., 1995). By visualizing blood flow in the embryo, any vasculature defect can be identified and localized. This technology has also been developed using quantum dots (Rieger et al., 2005), but here we will discuss microangiography using fluorescent beads. This approach was used to develop the "Interactive Atlas of Zebrafish Vascular Anatomy" (Isogai et al., 2001), and a more detailed methodology can be found at this site.

Confocal Microangiography

1. Fluorescein-tagged carboxylated latex beads, 0.01 μm in diameter, are diluted 1:1 in a 2% BSA solution, sonicated, and spun for 2 min at full speed.
2. One- to seven-day-old zebrafish embryos/larvae are anesthetized with tricaine.
3. Beads are injected into the sinus venosus (in a 1–3dpf embryo) or directly into the cardiac ventricle (4–7dpf embryo).
4. Beads will spread throughout the entire fish within minutes, but will begin to cluster. Subsequently, confocal imaging must be performed immediately. Embryos are mounted in methylcellulose or agarose. Images of optical sections are collected with 2–5 μm between sections.

7.3.5 RNA and Protein Level

Another approach to toxicity testing in the hematopoietic and endothelial systems is by determining RNA and protein levels of key markers in these cells. In this section, we will describe methods used to analyze RNA and protein levels in the zebrafish. For each blood and endothelial cell type, we will describe specific genes and tools that are used to identify the cell of interest. For toxicity testing, these assays can be used to ensure that blood and vascular development is not delayed or blocked after treatment with compound.

7.3.5.1 Assays RNA level is commonly assessed in zebrafish embryos by whole mount *in situ* hybridization (ISH). ISH involves first developing an antisense dig-labeled riboprobe against your RNA of interest. Templates for this process can often be clones of your gene of interest, readily available for all discussed RNAs at zfin.org. The probe generated is then hybridized in the whole mount embryo, and staining localizes RNA expression to specific cell types. A similar approach has also been used on adult zebrafish tissue sections (Smith et al., 2008). The protocol below, describing probe preparation and whole mount *in situ* hybridization of the embryo, is described in more detail in Thisse and Thisse (2008).

Making an Antisense Probe (Paffett-Lugassy and Zon, 2005)

1. Digest template plasmid to linearize DNA. Confirm complete digest by running on an agarose gel. Alternatively, generate a PCR product of your template.

2. Purify the DNA.
3. Perform an RNA polymerase reaction.
4. Treat with DNase I.
5. Purify RNA probe and maintain at $-80°C$.

In Situ Hybridization: Embryos (Thisse and Thisse, 2008)

1. Embryos are fixed at specific stages in development in 4% PFA at 4°C overnight. If the embryos have developed pigment (are older than 24hpf), then bleaching is required. If not, skip to step 5.
2. Wash embryos 1× in PBS with 0.1% Tween (PBT).
3. Incubate at room temperature in a 3% H_2O_2/0.8% KOH solution until the pigment has disappeared. This will take from 1 to 20 min, depending on embryo stage.
4. Wash embryos 1× in PBT for 5 min.
5. Refix in 4% PFA for at least 20 min.
6. Wash 1–2× for 5 min each in PBT.
7. Dehydrate embryos in 100% methanol for 3× at 5 min each at room temperature.
8. Incubate embryos for at least 2 h at $-20°C$. Alternatively, they can be kept at this temperature for several months.
9. Rehydrate embryos into PBT by 5 min incubations in the following solutions: 75% MeOH in PBS, 50%, 25%, and 100% PBS.
10. Wash embryos 4× for 5 min each in PBT.
11. Permeabilize with proteinase K (10 µg/mL) at room temperature. Length of this incubation is dependent on embryo stage, and detailed information can be found in Thisse and Thisse (2008).
12. After this reaction is stopped, refix embryos in PFA for at least 20 min.
13. Wash embryos 4× for 5 min each in PBT.
14. Incubate for 2–5 h at 70°C in prehybridization buffer (50% formamide, 5× SSC, 0.5 mg/mL torula yeast RNA, 50 mg/mL heparin, 0.1% Tween in sterilized water).
15. Dilute probe 0.8–1.5 ng/µL in prehybridization buffer.
16. Incubate embryos in probe overnight at 70°C.
17. Remove probe and gradually change to 2× SSC by the following 15 min washes at 70°C: 75% Hybe in 2× SSC, 50%, 25%, and 100% 2× SSC.
18. Wash 2× for 30 min in 0.2× SSC at 70°C.
19. Gradually change to PBT by washing for 10 min in the following solutions: 75% 0.2× SSC in PBT, 50%, 25%, and 100% PBT.
20. Incubate for 3 h at room temperature in blocking buffer solution (100 mg BSA and 1 mL 100% lamb/sheep serum in 50 mL PBT).
21. Add anti-dig antibody to blocking solution and incubate embryos overnight at 4°C.
22. Wash embryos in PBT 6× for 15 min each at room temperature shaking.

23. Prestain for 5 min in staining buffer (100 mM Tris (pH 9.5), 50 mM $MgCl_2$, 100 mM NaCl, 0.5% Tween 20 in water).
24. Make staining mix (225 µL of 50 mg/mL nitro blue tetrazolium chloride (NBT) and 175 µL of 50 mg/mL 5-bromo-4-chloro-3-indolyl phosphate, toluidine salt (BCIP) in 50 mL staining buffer).
25. Stain until color develops in relevant blood and vascular tissues.
26. Wash off with PBT.
27. Maintain in 4% PFA.

Protein levels are difficult to assay as antibodies against zebrafish proteins are not well developed. If there is an antibody that recognizes a zebrafish protein of interest, immunohistochemical analysis can be performed on whole mount embryos or adult kidney sections (Vasilyev et al., 2009). An alternative approach to looking at protein level involves transgenic models. Many transgenics have been created where a fluorophore has been attached to a protein of interest or, more commonly, is expressed under the control of a promoter for a gene of interest. These fluorophores include, but are not limited to, GFP, dsRED, and mCherry. Fluorophore levels can be assayed using fluorescence or confocal microscopy. Alternatively, levels can also be measured using antibodies or RNA probes that recognize the fluorophore. Beyond these methods, there are alternative experimental approaches specific to one or more blood cell types. These are described below.

7.3.5.2 *Hematopoietic Stem Cells*
Definitive HSCs express c-myb, runx1, and low levels of CD41, so HSC number and location can be determined by runx1 and c-myb ISH. Probes for runx1 and c-myb RNA demonstrate that HSCs first appear at 36hpf in the AGM, and exist throughout zebrafish development and life. Since c-myb also stains primitive blood cells in the ICM, adjacent to the AGM, staging embryos at 36–40hpf is important for this assay.

To assay protein level, some transgenic lines have been developed with promoters of these genes. A CD41-gfp transgenic (Lin et al., 2005) can be used to visualize HSCs in the AGM and CHT of the developing embryo (Kissa et al., 2008). Myb-GFP (North et al., 2007) and runx1-EGFP (Lam et al., 2009) transgenics have also been recently created.

7.3.5.3 *Erythrocytes*
Embryonic erythrocytes, from both primitive and definitive hematopoiesis, express gata1 and embryonic globins $\alpha e3$ and $\beta e1$. These RNAs can be visualized from five somites (de Jong and Zon, 2005). In the adult zebrafish, erythrocytes express gata1 and adult globins $\alpha a1$ and $\beta a1$. Gata1 transgenic lines, where GFP (Long et al., 1997) or dsRED (Vogeli et al., 2006) is under control of the gata1 promoter, can be used to visualize erythrocytes. In addition to these tools, o-dianisidine staining can be used for looking at hemoglobin levels. Presence of hemoglobin enhances o-dianisidine oxidation by hydrogen peroxide (Paffett-Lugassy and Zon, 2005).

o-Dianisidine Staining (Detrich et al., 1995)

1. Collect embryos aged 40–48hpf. Dechorionate if necessary.
2. Incubate embryos for 30–60 min in the dark in *o*-dianisidine and peroxide mix (0.6 mg/mL *o*-dianisidine, 0.01 M sodium acetate (pH 4.5), 0.65% hydrogen peroxide, 40% ethanol).
3. Remove *o*-dianisidine and rinse 2× with PBT.
4. Fix in 4% PFA.
5. Positive orange staining represents erythrocyte cells.

7.3.5.4 Myeloid Cells Numerous myeloid markers have been identified in zebrafish (Bennett et al., 2001). L-Plastin is present in early myelomonocytic progenitors and seen in embryos up to 5dpf. Pu.1 is expressed during primitive myelopoiesis, appearing from 16hpf to about 28hpf. Lysozyme C expression is localized to the myeloid population, and lysozyme-C dsRED and EGFP transgenic fish have been created (Hall et al., 2007). Myeloperoxidase labels neutrophils, and mpo staining is observed starting in the ICM at 18hpf before spreading across the embryo. Mpo expression persists in adult neutrophils, as seen in adult zebrafish kidneys (Bennett et al., 2001).

Myeloid cell-specific methods are also available. For example, Sudan Black, a lipid stain that stains granulocyte granules, can be used to stain myeloid cells. In addition, granulocytes in the fish express myeloid peroxidases, and peroxidase levels can specifically be measured.

Sudan Black Staining of Myeloid Cells (Le Guyader et al., 2008)

1. Fix embryos in 4% PFA for 2 h.
2. Rinse in PBS.
3. Incubate in Sudan Black stain for 20 min.
4. Wash repeatedly in 70% EtOH.
5. Gradually change to rehydrate to PBT.

Peroxidase Activity Assay (Le Guyader et al., 2008)

1. Synthesize fluorescein isothiocyanate (FITC) and Cyanine 3 (Cy3) conjugated tyramides as described (Xenbase).
2. Fix embryos in 4% PFA for 2 h.
3. Wash in PBS.
4. Incubate in the dark for 10–30 min in the following:
   a. 1/100 tyramide in PBS.
   b. 0.1 mol/L imidazole.
   c. 0.001% hydrogen peroxide.
5. Wash 3× for 10 min each in PBT.
6. Stop reaction by incubating in a 2% hydrogen peroxide (in PBT) solution for 30 min.

7.3.5.5 Lymphoid Cells T cells can be assayed by looking at rag1, rag2, and ikaros expression beginning at 4dpf in the thymus. Rag2:EGFP (Langenau et al., 2004) and RAG2:dsRED (Langenau et al., 2005) transgenic lines label T-cell populations with fluorescent markers. B-cell lymphocytes are not as well characterized, but can be identified with immunoglobin gene expression levels in the pancreas and kidney (Danilova and Steiner, 2002).

7.3.5.6 Thrombocytes Thrombocytes and their precursors express CD41 at high levels. (It is important to note that zebrafish HSCs also express CD41, but at lower levels.) CD41 positive cells can be detected in circulation by 2–3dpf, and are present in the adult kidney. CD41 levels can be assayed by ISH or with a CD41-eGFP transgenic line (Lin et al., 2005).

7.3.5.7 Vasculature Flk1 and fli1 are expressed in endothelial precursors from five somites. They appear as two stripes until merging and forming the ICM by 18hpf. Flk1 and fli1 are also expressed in differentiated vascular cells. By 24hpf, ISVs have formed, and express flk1 and fli1. fli:eGFP (Lawson and Weinstein, 2002) and flk:eGFP (Jin et al., 2005) transgenic lines have been created. Specification of the DA and PCV can also be analyzed by ISH at 24–28hpf. Numerous genes are expressed in the dorsal aorta, including ephrinB2, notch5, and deltaC. In contrast, Flt4 and ephB4 are expressed in the PCV.

7.4 SUMMARY

Hematopoietic and vascular development is highly conserved among vertebrate organisms, including the zebrafish. This chapter has summarized many methods that are available to assay the zebrafish hematopoietic and vascular systems. When analyzing toxic effects of compounds, these assays provide numerous approaches to confirm functional blood and vascular development in embryonic and adult zebrafish.

ACKNOWLEDGMENT

We would like to thank Michelle Lin for critical review of this manuscript.

REFERENCES

Baldessari D and Mione M (2008). How to create the vascular tree? (Latest) help from the zebrafish. Pharmacol Ther 118(2): 206–230.

Bennett CM, Kanki JP, Rhodes J, Liu TX, Paw BH, Kieran MW, Langenau DM, Delahaye-Brown A, Zon LI, Fleming MD, and Look AT (2001). Myelopoiesis in the zebrafish, Danio rerio. Blood 98(3): 643–651.

BERTRAND JY, KIM AD, TENG S, and TRAVER D (2008). CD41$^+$ cmyb$^+$ precursors colonize the zebrafish pronephros by a novel migration route to initiate adult hematopoiesis. Development 135(10): 1853–1862.

BROWNLIE A, HERSEY C, OATES AC, PAW BH, FALICK AM, WITKOWSKA HE, FLINT J, HIGGS D, JESSEN J, BAHARY N, ZHU H, LIN S, and ZON L (2003). Characterization of embryonic globin genes of the zebrafish. Dev Biol 255(1): 48–61.

CHAN FY, ROBINSON J, BROWNLIE A, SHIVDASANI RA, DONOVAN A, BRUGNARA C, KIM J, LAU BC, WITKOWSKA HE, and ZON LI (1997). Characterization of adult alpha- and beta-globin genes in the zebrafish. Blood 89(2): 688–700.

COVASSIN LD, VILLEFRANC JA, KACERGIS MC, WEINSTEIN BM, and LAWSON ND (2006). Distinct genetic interactions between multiple Vegf receptors are required for development of different blood vessel types in zebrafish. Proc Natl Acad Sci USA 103(17): 6554–6559.

DANILOVA N and STEINER LA (2002). B cells develop in the zebrafish pancreas. Proc Natl Acad Sci USA 99(21): 13711–13716.

DAVIDSON AJ and ZON LI (2004). The 'definitive' (and 'primitive') guide to zebrafish hematopoiesis. Oncogene 23(43): 7233–7246.

DE JONG JL and ZON LI (2005). Use of the zebrafish system to study primitive and definitive hematopoiesis. Annu Rev Genet 39: 481–501.

DETRICH HW 3RD, KIERAN MW, CHAN FY, BARONE LM, YEE K, RUNDSTADLER JA, PRATT S, RANSOM D, and ZON LI (1995). Intraembryonic hematopoietic cell migration during vertebrate development. Proc Natl Acad Sci USA 92(23): 10713–10717.

DOOLEY KA, DAVIDSON AJ, and ZON LI (2005). Zebrafish scl functions independently in hematopoietic and endothelial development. Dev Biol 277(2): 522–536.

GALLOWAY JL, WINGERT RA, THISSE C, THISSE B, and ZON LI (2005). Loss of gata1 but not gata2 converts erythropoiesis to myelopoiesis in zebrafish embryos. Dev Cell 8(1): 109–116.

HALL C, FLORES MV, STORM T, CROSIER K, and CROSIER P (2007). The zebrafish lysozyme C promoter drives myeloid-specific expression in transgenic fish. BMC Dev Biol 7: 42.

ISOGAI S, HORIGUCHI M, and WEINSTEIN BM (2001). The vascular anatomy of the developing zebrafish: an atlas of embryonic and early larval development. Dev Biol 230(2): 278–301.

JAGADEESWARAN P, SHEEHAN JP, CRAIG FE, and TROYER D (1999). Identification and characterization of zebrafish thrombocytes. Br J Haematol 107(4): 731–738.

JIN SW, BEIS D, MITCHELL T, CHEN JN, and STAINIER DY (2005). Cellular and molecular analyses of vascular tube and lumen formation in zebrafish. Development 132(23): 5199–5209.

KISSA K, MURAYAMA E, ZAPATA A, CORTES A, PERRET E, MACHU C, and HERBOMEL P (2008). Live imaging of emerging hematopoietic stem cells and early thymus colonization. Blood 111(3): 1147–1156.

LAM EY, CHAU JY, KALEV-ZYLINSKA ML, FOUNTAINE TM, MEAD RS, HALL CJ, CROSIER PS, CROSIER KE, and FLORES MV (2009). Zebrafish runx1 promoter-EGFP transgenics mark discrete sites of definitive blood progenitors. Blood 113(6): 1241–1249.

LANGENAU DM, FENG H, BERGHMANS S, KANKI JP, KUTOK JL, and LOOK AT (2005). Cre/lox-regulated transgenic zebrafish model with conditional myc-induced T cell acute lymphoblastic leukemia. Proc Natl Acad Sci USA 102(17): 6068–6073.

LANGENAU DM, FERRANDO AA, TRAVER D, KUTOK JL, HEZEL JP, KANKI JP, ZON LI, LOOK AT, and TREDE NS (2004). *In vivo* tracking of T cell development, ablation, and engraftment in transgenic zebrafish. Proc Natl Acad Sci USA 101(19): 7369–7374.

LAWSON ND and WEINSTEIN BM (2002). Arteries and veins: making a difference with zebrafish. Nat Rev Genet 3(9): 674–682.

LE X, LANGENAU DM, KEEFE MD, KUTOK JL, NEUBERG DS, and ZON LI (2007). Heat shock-inducible Cre/Lox approaches to induce diverse types of tumors and hyperplasia in transgenic zebrafish. Proc Natl Acad Sci USA 104(22): 9410–9415.

LEBLANC J, BOWMAN TV, and ZON L (2007). Transplantation of whole kidney marrow in adult zebrafish. J Vis Exp (2): 159.

LE GUYADER D, REDD MJ, COLUCCI-GUYON E, MURAYAMA E, KISSA K, BRIOLAT V, MORDELET E, ZAPATA A, SHINOMIYA H, and HERBOMEL P (2008). Origins and unconventional behavior of neutrophils in developing zebrafish. Blood 111(1): 132–141.

LENGERKE C, SCHMITT S, BOWMAN TV, JANG IH, MAOUCHE-CHRETIEN L, McKINNEY-FREEMAN S, DAVIDSON AJ, HAMMERSCHMIDT M, RENTZSCH F, GREEN JB, ZON LI, and DALEY GQ (2008). BMP and Wnt specify hematopoietic fate by activation of the Cdx-Hox pathway. Cell Stem Cell 2(1): 72–82.

LIAO W, BISGROVE BW, SAWYER H, HUG B, BELL B, PETERS K, GRUNWALD DJ, and STAINIER DY (1997). The zebrafish gene cloche acts upstream of a flk-1 homologue to regulate endothelial cell differentiation. Development 124(2): 381–389.

LIESCHKE GJ, OATES AC, PAW BH, THOMPSON MA, HALL NE, WARD AC, HO RK, ZON LI, and LAYTON JE (2002). Zebrafish SPI-1 (PU.1) marks a site of myeloid development independent of primitive erythropoiesis: implications for axial patterning. Dev Biol 246(2): 274–295.

LIN HF, TRAVER D, ZHU H, DOOLEY K, PAW BH, ZON LI, and HANDIN RI (2005). Analysis of thrombocyte development in CD41-GFP transgenic zebrafish. Blood 106(12): 3803–3810.

LONG Q, MENG A, WANG H, JESSEN JR, FARRELL MJ, and LIN S (1997). GATA-1 expression pattern can be recapitulated in living transgenic zebrafish using GFP reporter gene. Development 124(20): 4105–4111.

MURAYAMA E, KISSA K, ZAPATA A, MORDELET E, BRIOLAT V, LIN HF, HANDIN RI, and HERBOMEL P (2006). Tracing hematopoietic precursor migration to successive hematopoietic organs during zebrafish development. Immunity 25(6): 963–975.

NORTH TE, GOESSLING W, WALKLEY CR, LENGERKE C, KOPANI KR, LORD AM, WEBER GJ, BOWMAN TV, JANG IH, GROSSER T, FITZGERALD GA, DALEY GQ, ORKIN SH, and ZON LI (2007). Prostaglandin E2 regulates vertebrate haematopoietic stem cell homeostasis. Nature 447(7147): 1007–1011.

PAFFETT-LUGASSY NN and ZON LI (2005). Analysis of hematopoietic development in the zebrafish. Methods Mol Med 105: 171–198.

RIEGER S, KULKARNI RP, DARCY D, FRASER SE, and KOSTER RW (2005). Quantum dots are powerful multipurpose vital labeling agents in zebrafish embryos. Dev Dyn 234(3): 670–681.

SMITH A, ZHANG J, GUAY D, QUINT E, JOHNSON A, and AKIMENKO MA (2008). Gene expression analysis on sections of zebrafish regenerating fins reveals limitations in the whole-mount *in situ* hybridization method. Dev Dyn 237(2): 417–425.

THISSE C and THISSE B (2008). High-resolution *in situ* hybridization to whole-mount zebrafish embryos. Nat Protoc 3(1): 59–69.

TRAVER D, HERBOMEL P, PATTON EE, MURPHEY RD, YODER JA, LITMAN GW, CATIC A, AMEMIYA CT, ZON LI, and TREDE NS (2003a). The zebrafish as a model organism to study development of the immune system. Adv Immunol 81: 253–330.

TRAVER D, PAW BH, POSS KD, PENBERTHY WT, LIN S, and ZON LI (2003b). Transplantation and *in vivo* imaging of multilineage engraftment in zebrafish bloodless mutants. Nat Immunol 4(12): 1238–1246.

TRAVER D, WINZELER A, STERN HM, MAYHALL EA, LANGENAU DM, KUTOK JL, LOOK AT, and ZON LI (2004). Effects of lethal irradiation in zebrafish and rescue by hematopoietic cell transplantation. Blood 104(5): 1298–1305.

TREDE NS, LANGENAU DM, TRAVER D, LOOK AT, and ZON LI (2004). The use of zebrafish to understand immunity. Immunity 20(4): 367–379.

UCHIDA N, AGUILA HL, FLEMING WH, JERABEK L, and WEISSMAN IL (1994). Rapid and sustained hematopoietic recovery in lethally irradiated mice transplanted with purified Thy-1.1lo Lin-Sca-1 + hematopoietic stem cells. Blood 83(12): 3758–3779.

VASILYEV A, LIU Y, MUDUMANA S, MANGOS S, LAM PY, MAJUMDAR A, ZHAO J, POON KL, KONDRYCHYN I, KORZH V, and DRUMMOND IA (2009). Collective cell migration drives morphogenesis of the kidney nephron. PLoS Biol 7(1): e9.

VOGELI KM, JIN SW, MARTIN GR, and STAINIER DY (2006). A common progenitor for haematopoietic and endothelial lineages in the zebrafish gastrula. Nature 443(7109): 337–339.

WEINSTEIN BM, STEMPLE DL, DRIEVER W, and FISHMAN MC (1995). Gridlock, a localized heritable vascular patterning defect in the zebrafish. Nat Med 1(11): 1143–1147.

WHITE RM, SESSA A, BURKE C, BOWMAN T, LEBLANC J, CEOL C, BOURQUE C, DOVEY M, GOESSLING W, BURNS CE, and ZON LI (2008). Transparent adult zebrafish as a tool for *in vivo* transplantation analysis. Cell Stem Cell 2(2): 183–189.

Chapter 8

Hepatotoxicity Testing in Larval Zebrafish

Adrian Hill

Evotec (UK) Ltd, Abingdon, Oxfordshire, UK

8.1 INTRODUCTION: THE LARVAL ZEBRAFISH MODEL

When zebrafish research was in its infancy, the scientific community initially focused on genetics and development and soon discovered a high degree of conservation compared to humans (Postlethwait et al., 2000). During the next few decades, the number of academic researchers working in this field grew rapidly and the zebrafish was established as a key model system for the study of disease and toxicology with mechanisms closely echoing those of mammalian model species (Hill et al., 2005; Zon and Peterson, 2005; Lieschke and Currie, 2007; Barros et al., 2008; Hill, 2008a). Today, around 3,500 zebrafish peer-reviewed publications are released each year.

In addition to academic research, the zebrafish has also recently gained attention from the pharmaceutical sector as a tool for the evaluation of novel drug candidates, in particular as a way of assessing compounds for toxicity and safety liabilities early in the drug discovery process. As noted in the previous chapters, the transparent nature of the larvae and the ability to screen in a microtiter plate format with small amounts (single milligrams) of compounds, as well as other advantages, make the zebrafish an ideal model. Most importantly, physiological, morphological, and histological similarities to mammals and humans can be observed when zebrafish larvae are treated with organ-specific toxicants and hepatotoxic compounds are no exception.

8.2 LIVER DEVELOPMENT

Hepatocyte precursors have been identified in zebrafish during mid-somitogenesis; approximately 16 h post fertilization (hpf) (Korzh et al., 2001). One early marker,

Zebrafish: Methods for Assessing Drug Safety and Toxicity, Edited by Patricia McGrath.
© 2012 John Wiley & Sons, Inc. Published 2012 by John Wiley & Sons, Inc.

ceruloplasmin (a liver-specific multicopper oxidase gene), can be visualized at this time asymmetrically on the left-hand side of the endoderm up to 32hpf. Together with other pan-endodermal markers such as foxA1, foxA2, and foxA3 (Odenthal and Nusslein-Volhard, 1998) and prox1 (Glasgow and Tomarev, 1998; Ho et al., 1999) and hhex, two established liver markers in amniotes (Oliver et al., 1993; Keng et al., 1998; Sosa-Pineda et al., 2000), these cells, derived from the primitive gut, gradually form the liver. This has been investigated and characterized using a transgenic line of zebrafish that expressed GFP in the gut (Field et al., 2003). Hepatocyte aggregation between 24 and 28hpf first leads to a thickening in the intestinal rod, followed by an outward projection to the left side and a looping of the intestinal primordium. By 50hpf, liver tissue is easily recognized and a period of significant growth begins. In addition, the tissue that eventually connects the liver to the intestinal primordium becomes the polarized epithelium of the bile duct. Like mammals, the zebrafish liver produces bile, which is stored in gall bladder (Pack et al., 1996). Although there are other striking similarities to its mammalian counterpart, there are exceptions. In particular, there are no true portal lobules in the zebrafish liver.

During vascularization, endothelial cells labeled with Tie2-GFP have been shown to partially encapsulate the liver bud and subsequently start to invade it around 60hpf. The mechanisms by which this occurs have not been fully investigated although the aryl hydrocarbon receptor nuclear translocator (ARNT), a basic helix–loop–helix–PAS heterodimeric transcription factor commonly associated with detoxification, may be involved along with other gene pathways (Hill et al., 2009). This was noted as part of an initial characterization of the zebrafish *arnt2* null mutant in which, among various other abnormalities, dilated liver sinusoids were found to merge abnormally to form an extensive, labyrinth-like network of vascular channels. Previous mouse models have also implicated AHR and ARNT as well as other dimerization partners such as hif-1α in the normal development of the murine vasculature and liver (Walisser et al., 2004a, 2004b), but further investigation is required. By 72hpf, vascularization is essentially complete and the liver becomes perfused with blood shortly after (Pack et al., 1996; Isogai et al., 2001).

Embryogenesis is essentially complete by this time and the digestive system is fully functional. Even though the zebrafish embryo can survive purely on a reserve of yolk during the first 4–5 days of development, it is ready to begin feeding. Development of a physiologically functional liver is therefore very rapid in comparison to other vertebrate models. With the exception of minor differences, the general anatomy, organization, cellular composition, and function of a healthy adult zebrafish liver are virtually the same as in mammals (Hinton and Couch, 1998), and the early embryonic stages of hepatogenesis are similar to that of mice (Duncan, 2003; Field et al., 2003; Ober et al., 2003). Likewise, with regard to disease phenotypes, the histopathology of cholestasis, fatty liver (steatosis), and neoplasia is also comparable (Spitsbergen et al., 2000; Amatruda et al., 2002; Amali et al., 2006).

8.3   HEPATIC GENE KNOCKDOWN AND MUTATION

Genes identified in the zebrafish model are well conserved among other vertebrates and as a result a plethora of zebrafish mutants have been identified in order to help understand their homologs in mammals and humans (Haffter et al., 1996). *N*-Ethyl-*N*-nitrosurea (ENU) has been a common mutagen used in zebrafish studies and has led to hundreds of mutations in genes essential for numerous processes including pattern formation, morphogenesis, organogenesis, and differentiation. Morpholino oligonucleotides (MOs) (Kemper et al., 2003) and various mutant zebrafish studies have also been key in allowing scientists to rapidly advance our knowledge of the functional definition of genes required for normal development of the liver.

When injected into 1–4-cell stage zebrafish embryos, morpholinos have been shown to successfully knock down gene expression by binding to and blocking translation of specific mRNA (Nasevicius and Ekker, 2001). MOs are antisense nucleic acid analogs that have ribosides converted to morpholines (C_4H_9NO) and a phosphorodiamidate intersubunit linkage instead of phosphorodiester linkage (Summerton and Weller, 1997). By injecting complementary sequences of *hhex*, these MOs cause a significant reduction of liver size by 50hpf (Wallace et al., 2001), indicating that *hhex* may play an important role in liver formation. Similarly, loss of Hhex function in mutant mouse studies has shown that it impedes growth and differentiation of an initially established liver diverticulum (Keng et al., 2000; Martinez Barbera et al., 2000). In addition, the *pescadillo* (*pes*) gene, initially identified in an insertional mutagenesis screen in zebrafish, appears to play an important role during the growth phase (Allende et al., 1996). *pes* mutants exhibit a smaller liver at 120hpf, although there is no obvious difference to wild-type larvae at 72hpf. This suggests liver development becomes arrested during this time and supports the conclusion that expression of *pes* precedes phases of active proliferation in the liver and gut (Allende et al., 1996).

As well as being useful to identify genes involved in development and detoxification, mutants are often useful models for studying human diseases (van Heyningen, 1997; Zon, 1999; Barut and Zon, 2000; Dooley and Zon, 2000; Amatruda et al., 2002). For example, mutations on genes including *lumpazi*, *gammler*, and *tramp* lead to liver necrosis (Chen et al., 1996) and the *beefeater* mutation exhibits liver necrosis and impaired glycogen utilization, as seen in human glycogen storage diseases (Pack et al., 1996). Sadler et al. (2005) also identified three mutants that had phenotypes resembling different liver diseases. One mutant of a novel gene named *foie gras* developed large, lipid-filled hepatocytes and therefore resembled humans with fatty liver disease. Hepatomegaly and vesicle-filled hepatocytes were observed in the second mutant that lacked *vps18*, a class C vacuolar protein sorting gene. This also exhibited defects in the bile canaliculi and had marked biliary paucity, indicating that this gene functioned to traffic vesicles to the hepatocyte apical membrane and also could be involved in the development of the intrahepatic biliary tree. This phenotype is comparable to that reported for individuals with arthrogryposis–renal dysfunction–cholestasis (ARC), a syndrome attributed to mutation of another class C vps gene. In contrast, a third mutant involving the tumor suppressor gene *nf2* manifested extrahepatic choledochal cysts in the common bile duct, similar to those observed in humans. In addition, in a further study, genes that underlie Alagille syndrome, a pediatric disorder that results in a variety of abnormalities including a paucity of intrahepatic bile ducts, play an important role in zebrafish biliary development (Lorent et al., 2004). These examples therefore illustrate the utility of zebrafish as a suitable model vertebrate for studying liver development, disease, and potentially toxic insult.

8.4 HEPATOTOXICITY TESTING IN DRUG DISCOVERY

Regulatory legislation exists to ensure that any new drug administered to patients has passed a standardized series of toxicity and safety assessments, but those that fail and are identified late in development using current techniques, due to toxic liabilities, are very costly to the pharmaceutical industry and would have needlessly been exposed to numerous test animals during its evaluation. Currently, there is no *in vivo* or *in vitro* model available that can fully assess novel candidate drug compounds for safety and toxicity liabilities before being progressed into clinical development. In particular, for certain kinds of organ toxicity such as hepatotoxicity, regulatory animal testing has shown the poorest correlation with humans (Olson et al., 2000) and as a result is the most frequent reason cited for labeling drugs with a black box warning and for withdrawal of approved drugs from the market such as iproniazid and troglitazone (Fung et al., 2001).

Various different types of hepatic injuries including those caused by alcohol abuse, infection, and toxic insult all can cause a similar pattern of histological degeneration and ultimately lead to cirrhosis (Shin and Fishman, 2002). However, the underlying mechanisms that lead to liver failure are poorly understood.

Liver disease has been on the increase since the 1970s and was responsible for over 15,000 deaths in 2007 in the United Kingdom according to the British Liver Trust (http://www.britishlivertrust.org.uk) and UK National Statistics (www.statistics.gov.uk). In comparison, about 27,500 deaths related to liver disease were recorded for 2005 in the United States (Kung et al., 2008). Whereas choledochal cysts and ARC syndrome are relatively rare, nearly 1–2% of U.S. citizens have fatty liver disease and steatosis is found in about a quarter of the U.S. population (Neuschwander-Tetri and Caldwell, 2003) making it among the most common hepatic pathologies in the developed world (Clark et al., 2001; Neuschwander-Tetri and Caldwell, 2003). In addition, due to the association of liver disease with obesity, this number is predicted to increase further as the number of obese individuals in the United States approaches one-third of the population (CDC, 2007). Notwithstanding the importance of these figures, it is also necessary to understand that as people can survive with 70% liver damage, there is a substantial burden of morbidity from liver disease with the associated economic costs.

Drug-induced liver injury (DILI) is the most common cause of death from acute liver failure and accounts for approximately 13% of cases of acute liver failure in the

United States (Ostapowicz et al., 2002). It is also the most common adverse drug response that leads to the failure of otherwise promising drug candidates during preclinical or clinical development and also the withdrawal or restriction of prescription drugs use after initial approval (Navarro and Senior, 2006; Abboud and Kaplowitz, 2007). Idiosyncratic DILI induced by a single drug is a rare event occurring in less than 1 per 10,000–100,000 of subjects who take the medication and as such the risk factors and pathogenesis are poorly understood (Navarro and Senior, 2006; Abboud and Kaplowitz, 2007; Uetrecht, 2007). Likewise, the symptoms exhibited by DILI sufferers can be diverse and unpredictable: acute liver disease, overt acute or subacute liver failure, prolonged jaundice and disability, or alternatively predominantly asymptomatic. A necessity therefore exists to devise ways to help identify these compounds.

Conventional cytotoxicity assays have less than 25% sensitivity for the detection of hepatotoxins (O'Brien et al., 2003) and even the newest high-content cell assays still fail in detecting the more complex hepatotoxic mechanisms, such as cholestasis and those of an idiosyncratic nature (O'Brien et al., 2006). Therefore, assessment of organ toxicity frequently involves full histopathological assessment in rodents and commonly requires the use of higher order species such as macaque monkeys or even human subjects before their toxic liabilities can be fully ascertained. It can therefore be concluded that the traditional approaches for identifying hepatotoxicants are insufficient and new technologies and models need to be developed.

8.5 PHENOTYPIC-BASED LARVAL ZEBRAFISH HEPATOTOXICITY SCREENS

Recently, the zebrafish has been evaluated as a potential model vertebrate to be used to identify compounds with hepatotoxic liability early in the lead optimization stage of drug discovery, a time when large compound scale-up and whole animal studies have not commenced (Hill, 2008a, 2008b). This is possible because these larval studies can be performed with single milligrams of compound in microtiter plates, essentially allowing high-content *in vivo* information to be gathered in an *in vitro* format. In addition, as a whole organism, the zebrafish larva should also be able to capture toxicity associated with toxic metabolites that are unlikely to be synthesized *in vitro* (Hill, 2008a). In this way, pharmaceutical companies should be able to advance lead candidate drugs for further development with a low likelihood of toxicity after having made only a small investment in compound manufacture and typical ADME *in vitro* screens.

As described previously, one of the major advantages of this model is that the zebrafish larvae are virtually transparent. Therefore, the intention of recent studies has been to concentrate on creating a medium-throughput assay based on morphological endpoints of toxicity that are assessable using transmitted light without the need for dissection. When viewed dorsolaterally with a stereomicroscope, the liver of a 5–6dpf larvae is situated posterior to the pericardium and predominantly anterior to the gut, although they tend to overlap in this view and in transverse section liver tissue can

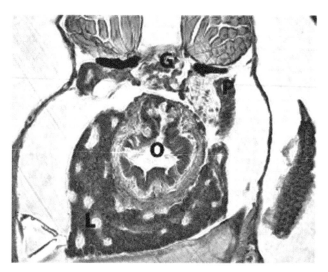

Figure 8.1 Transverse section of a wild-type Tu zebrafish larvae at 120hpf. Larva embedded in JB-4 resin, sectioned, and stained with Giemsa. O, esophagus; L, liver; G, glomerulus; P, pronephric duct. (See the color version of this figure in Color Plates section.)

appear to surround the esophagus (Fig. 8.1). The organ is fairly globular in structure, has a clearly recognizable periphery against the neighboring tissues, and is perfused with circulating blood cells. When exposed to a hepatotoxicant, changes to liver morphology can be evaluated using a method such as those described below by Hill (2008b).

8.5.1 Phenotypic Screening Methods

Tuebingen zebrafish larvae are arrayed in 24-well culture plates containing fresh $0.3\times$ Danieau's solution (the supportive medium used to maintain the zebrafish; $1\times$ stock: 58 mM NaCl, 0.7 mM KCl, 0.4 mM $MgSO_4$, 0.6 mM $Ca(NO_3)_2$, 5.0 mM HEPES, pH 7.1) and allowed to acclimate. Meanwhile stock solutions of test compounds are produced for a six-point concentration range (3–1000 μM). Sixteen larvae per concentration are statically treated with test compound alongside relevant controls for 2 days from 72hpf, at $28.5 \pm 0.5°C$ in a humidity-controlled environment, and checked for dead larvae after 24 h. At 120hpf (after a 48 h incubation), larvae are anaesthetized in MS222 (tricaine) and screened for three specific phenotypic endpoints of hepatotoxicity: liver abnormality (tissue degradation), changes in liver size (hepatomegaly), and yolk retention (an endpoint of liver function, as yolk is utilized through the liver and this is diminished if the liver is impaired).

8.5.2 Phenotypic Screening Validation

In order to assess the zebrafish hepatotoxicity model, Hill et al. (2008) selected a variety of well-known liver toxicants, toxic compounds that did not cause adverse effects directly to the liver, and some nontoxic controls. When assessed using the described methods, a distinct change in hepatic morphology was reported for certain compounds. In affected larvae, the organ appeared darker than controls with a brown and gray coloration, the texture of the tissue had become amorphous, and the outer edge of the organ was more diffuse, making it more difficult to distinguish liver from the surrounding tissues, especially in cases where secondary or gross toxicity was also induced (Fig. 8.2). Similar phenotypes have also been reported in other studies such as for brefeldin A (Zhang et al., 2003). Although these characteristics are typical for necrotic tissue when viewed at this magnification, histology would need to be

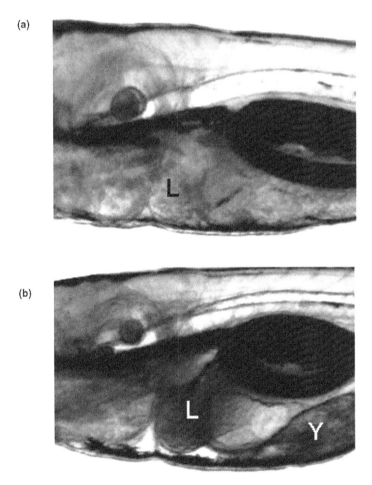

Figure 8.2   Phenotypic identification of hepatotoxicity at 120hpf. (a) Control larvae exhibiting a clear, healthy liver at 120hpf. (b) Larva treated with a hepatotoxin for 48 h exhibiting tissue degradation, hepatomegaly, and yolk retention. L, Liver; Y, yolk retention. (See the color version of this figure in Color Plates section.)

performed on all larvae to confirm this classification on a case-by-case basis. In this study, the overall correlation with mammalian *in vivo* data was good at 82%, with sensitivity (hepatotoxic-specific compounds correctly identified) at 88% and specificity (nonhepatotoxic compounds correctly identified) at 67% (Hill et al., 2008). Of these compounds, tamoxifen and danazol, both of which require metabolism in mammals to cause toxicity, and troglitazone, which was withdrawn from the market (Fung et al., 2001), were correctly identified in the zebrafish model, indicating that zebrafish larvae may have the equivalent detoxification pathways. Supporting evidence for this comes from numerous studies that have been cataloging zebrafish metabolizing enzymes including various cytochrome P450s such as CYP3A, CYP1A, CYP19, and CYP26 (Carney et al., 2004; Bresolin et al., 2005; Tseng et al., 2005; Rubenstein, 2006). A couple of nonhepatotoxic compounds that also caused adverse effects on the zebrafish liver were attributed to secondary toxicity induced when the LOEC (lowest observed effect concentration) was exceeded, such as for nephrotoxicity in the case of gentamycin, but this needed further investigation. In addition, certain toxicants including ketoconazole and sodium valproate were incorrectly identified as nontoxic, demonstrating again that this phenotypic model alone may not be infallible. However, when bioanalytical techniques based upon LC–MS were later incorporated into this assay, these compounds were shown to have only achieved very low body burdens of compound in the larvae and hence should have been unclassified pending additional tests with either higher aqueous concentrations or microinjection, to ensure there were sufficient levels to cause an adverse effect. The need for bioanalysis has also been discussed in the context of other zebrafish assays such as for developmental toxicity (Gustafson et al., 2008; Hill, 2008b) and cardiotoxicity (Hill, 2009) and at present no single physiochemical property has been identified that can suitably predict the uptake of different classes of compounds into zebrafish larvae (Doshna et al., 2009). In addition to demonstrating when uptake has been poor, bioanalysis has also been useful to identify those compounds that are readily absorbed and only cause toxicity at a body burden well in excess of the expected therapeutic window for that drug. When these factors were later considered for the hepatotoxicity assay, the sensitivity, specificity and overall predictivity were revised to 97, 77 and 91% respectively (Jones et al. 2009).

Some further examples of zebrafish phenotypic hepatotoxicity screens are shown in Table 8.1. Included are compounds that only cause overt adverse effects to the liver as part of a severe acutely toxic response, such as for astemizole and haloperidol that affect cardiac function as a primary endpoint of toxicity, and therefore hepatotoxicity must be interpreted accordingly.

8.6 SECONDARY AND MECHANISTIC LIVER ASSAYS

Although the phenotypic assay is a valuable, relatively high-throughput assay for the determination of hepatotoxic liability, as the endpoints are indiscriminant for specific modes of action, additional screens may be utilized to help elucidate the underlying mechanisms involved.

Table 8.1 Comparison of Compound-Induced Hepatotoxicity in Zebrafish and Mammals

Compound	Zebrafish toxicity	Mammalian toxicity
Hepatotoxic compounds		
α-Naphthyl isocyanate	Liver necrosis	Liver toxic to mammals
Amiodarone	Liver necrosis, hepatomegaly, yolk retention	Severely liver toxic to humans
Aspirin	Liver necrosis	Liver toxic to mammals
Danazol	Liver necrosis, yolk retention, intestinal toxicity	Severely liver toxic to humans after metabolism
Furazolidone	Liver necrosis, yolk retention	Moderately liver toxic to humans
Ibuprofen	Liver necrosis	Liver toxic to mammals
Naproxen	Liver necrosis	Liver toxic to mammals
Ridogrel	Liver necrosis, yolk retention	Liver toxic to mammals
Tamoxifen	Liver necrosis	Moderately liver toxic to humans after metabolism
Troglitazone	Liver necrosis	Liver toxic to humans (withdrawn from market)
Valproic acid	Liver necrosis	Liver toxic to mammals
Wyeth 14643	Liver necrosis	Liver toxic to mammals
Diasatine	Non-liver toxic	Liver toxic to humans but not mammals
Ketoconazole	Non-liver toxic (but limited uptake)	Liver toxic to mammals
Sodium valproate	Non-liver toxic (but limited uptake)	Severely liver toxic to humans after metabolism
Control compounds		
Astemizole	Non-liver toxic, acutely toxic (cardiotoxic)	Non-liver toxic, causes QT prolongation
Biotin	Non-liver toxic	Non-liver toxic
HP-β-CD	Non-liver toxic	Non-liver toxic
Haloperidol	Non-liver toxic, acutely toxic (cardiotoxic)	Non-liver toxic, causes QT prolongation
Metformin	Non-liver toxic	Non-liver toxic
Paracetamol	Non-liver toxic	Non-liver toxic (except at high concentration)
Saccharin	Non-liver toxic	Non-liver toxic
Sucrose	Non-liver toxic	Non-liver toxic
Gentamycin	Liver necrosis, hepatomegaly	Non-liver toxic; a nephrotoxin
Praziquantel	Liver necrosis	Non-liver toxic; an antihelminthic

8.6.1 Reporter Assays

Enzymes in the zebrafish larval liver and gut including biotin and carboxylase have been successfully measured and used as a further assessment of liver function. Using such an enzyme reporter assay, carbamate and merbarone, two mammalian liver toxicants, have been demonstrated to cause similar organ-specific effect in the zebrafish as in mice and humans (Zhang et al., 2003).

8.6.2 Liver Histopathology

Histology can be performed on large numbers of zebrafish larvae (Hill et al., 2001) and hence support for the more gross readouts obtained from other screens can be achieved relatively simply. However, the main drawback to these investigations is the time it takes to process and assess the samples. In liver mutant studies the change in liver morphology previously described, namely, the altered size, shape, color, and texture of the tissue, has been investigated, and at the cellular level degeneration of hepatocytes, compressed sinusoids, and pooling of erythrocytes have been verified (Pack et al., 1996). Likewise, larvae exposed to a variety of toxicants have also exhibited similar changes (unpublished data). In addition, use of TUNEL staining has also proved useful for the identification of apoptosis (Amali et al., 2006) and specific conditions such as hepatosteatosis have been visualized in zebrafish using stains such as Oil Red O that labels fat deposits. For example, when zebrafish are treated with gamma-hexachlorocyclohexane, fat droplets and fibrosis are induced that phenocopy the effects seen in mammals (Braunbeck et al., 1990), and similarly after treatment with thioacetamide, steatosis has also been identified (Amali et al., 2006).

When assessing compound toxicity in adult zebrafish, more traditional mammalian techniques can be adapted such as the administration of drugs by injection or oral intubation. Histopathology would then be a valid option to test for hepatotoxicity but in addition serum samples can also be taken to measure the levels of certain liver function enzymes such as alanine transaminase (Murtha et al., 2003).

8.6.3 Biomarkers

With the availability of vast amounts of molecular and genetic information, new techniques now facilitate analysis of the effects of toxic compounds on a large array of specific biomarker molecules. In recent years, biomarker approaches for assessing organ-specific toxicities have been investigated in rodent models (Chisholm and Dolphin, 1996) and human cells and to a lesser extent in zebrafish. Hepatotoxicity in particular can benefit from such investigations due to the idiosyncratic nature of many hepatotoxicants. As well as identifying new mechanisms of toxicity to advance of understanding of each condition, it is hoped that personalized therapies may also become possible. In addition, use of a biomarker approach instead of a full physiological assessment of toxic effects may enable refinement of current toxicity testing regimes such that animals are dosed with lower amounts of chemicals for a shorter duration, thereby reducing the potential for suffering. To date, zebrafish proteomic and transcriptomic studies investigating pharmaceuticals have revealed certain similarities to mammalian models and verified the presence of existing mammalian biomarkers such as glutathione S-transferase and aldehyde dehydrogenase (unpublished data), but more validation is still required.

8.7 CONCLUSIONS

Hepatotoxicity is one of the main causes of drug attrition in the pharmaceutical industry but of the assays currently being performed, no single assay or battery of screens is able to reliably predict all hepatotoxic compounds. Therefore, although *in vitro* assays have significantly improved, many candidate drugs only get identified as toxic in the most strongly regulated mammalian models, or once in use by humans. Zebrafish hepatotoxicity models have shown great promise and have become more widely accepted in recent years. Good correlation with mammals, the ability to identify toxic metabolites, and drug–drug interactions associated with CYP3A4, make this model a valuable addition to other traditional screens, but more validation is still required as various classes of drugs are yet to be evaluated and ways to further our understanding of the underlying mechanisms of liver toxicity in zebrafish are warranted.

REFERENCES

ABBOUD G and KAPLOWITZ N (2007). Drug-induced liver injury. Drug Saf 30(4): 277–294.

ALLENDE ML, AMSTERDAM A, BECKER T, KAWAKAMI K, GAIANO N, and HOPKINS N (1996). Insertional mutagenesis in zebrafish identifies two novel genes, pescadillo and dead eye, essential for embryonic development. Genes Dev 10(24): 3141–3155.

AMALI AA, REKHA RD, LIN CJ, WANG WL, GONG HY, HER GM, and WU JL (2006). Thioacetamide induced liver damage in zebrafish embryo as a disease model for steatohepatitis. J Biomed Sci 13(2): 225–232.

AMATRUDA JF, SHEPARD JL, STERN HM, and ZON LI (2002). Zebrafish as a cancer model system. Cancer Cell 1(3): 229–231.

BARROS TP, ALDERTON WK, REYNOLDS HM, ROACH AG, and BERGHMANS S (2008). Zebrafish: an emerging technology for *in vivo* pharmacological assessment to identify potential safety liabilities in early drug discovery. Br J Pharmacol 154(7): 1400–1413.

BARUT BA and ZON LI (2000). Realizing the potential of zebrafish as a model for human disease. Physiol Genomics 2(2): 49–51.

BRAUNBECK T, GORGE G, STORCH V, and NAGEL R (1990). Hepatic steatosis in zebra fish (*Brachydanio rerio*) induced by long-term exposure to gamma-hexachlorocyclohexane. Ecotoxicol Environ Saf 19(3): 355–374.

BRESOLIN T, DE FREITAS REBELO M, and CELSO DIAS BAINY A (2005). Expression of PXR, CYP3A and MDR1 genes in liver of zebrafish. Comp Biochem Physiol C Toxicol Pharmacol 140(3–4): 403–407.

CARNEY SA, PETERSON RE, and HEIDEMAN W (2004). 2,3,7,8-Tetrachlorodibenzo-*p*-dioxin activation of the aryl hydrocarbon receptor/aryl hydrocarbon receptor nuclear translocator pathway causes developmental toxicity through a CYP1A-independent mechanism in zebrafish. Mol Pharmacol 66(3): 512–521.

CDC (2007). Behavioral Risk Factor Surveillance System Survey Data. U.S. Department of Health and Human Services, Centers for Disease Control and Prevention, Atlanta, GA.

CHEN JN, HAFFTER P, ODENTHAL J, VOGELSANG E, BRAND M, VAN EEDEN FJ, FURUTANI-SEIKI M, GRANATO M, HAMMERSCHMIDT M, HEISENBERG CP, JIANG YJ, KANE DA, KELSH RN, MULLINS MC, and NUSSLEIN-

VOLHARD C (1996). Mutations affecting the cardiovascular system and other internal organs in zebrafish. Development 123: 293–302.

CHISHOLM JW and DOLPHIN PJ (1996). Abnormal lipoproteins in the ANIT-treated rat: a transient and reversible animal model of intrahepatic cholestasis. J Lipid Res 37(5): 1086–1098.

CLARK DW, LAYTON D, WILTON LV, PEARCE GL, and SHAKIR SA (2001). Profiles of hepatic and dysrhythmic cardiovascular events following use of fluoroquinolone antibacterials: experience from large cohorts from the Drug Safety Research Unit Prescription-Event Monitoring database. Drug Saf 24(15): 1143–1154.

DOOLEY K and ZON LI (2000). Zebrafish: a model system for the study of human disease. Curr Opin Genet Dev 10(3): 252–256.

DOSHNA C, BENBOW J, DEPASQUALE M, OKERBERG C, TURNQUIST S, STEDMAN D, CHAPIN R, SIVARAMAN L, WALDRON G, NAVETTA K, BRADY J, BANKER M, CASIMIRO-GARCIA A, HILL A, JONES M, BALL J., and Aleo M (2009). Multi-phase analysis of uptake and toxicity in zebrafish: relationship to compound physical–chemical properties. 48th Annual Meeting of the Society of Toxicology, Baltimore, MD, March 15–19, 2009.

DUNCAN SA (2003). Mechanisms controlling early development of the liver. Mech Dev 120(1): 19–33.

FIELD HA, OBER EA, ROESER T, and STAINIER DY (2003). Formation of the digestive system in zebrafish. I. Liver morphogenesis. Dev Biol 253(2): 279–290.

FUNG M, THORNTON A, MYBECK K, HSIAO-HUI W, HORNBUCKLE K, and MUNIZ E (2001). Evaluation of the characteristics of safety withdrawal of prescription drugs from worldwide pharmaceutical markets: 1960–1999. Drug Inf J 35: 293–317.

GLASGOW E and TOMAREV SI (1998). Restricted expression of the homeobox gene prox 1 in developing zebrafish. Mech Dev 76(1–2): 175–178.

GUSTAFSON A, WEISER T, CLEMANN N, HOSSAINI A, JANAITIS C, BLUEMEL J, DELONGEAS JL, and HILL AJ (2008). Validation of zebrafish as a model for screening teratogenicity. 47th Annual Meeting of the Society of Toxicology, Seattle, WA, March 16–20, 2008.

HAFFTER P, GRANATO M, BRAND M, MULLINS MC, HAMMERSCHMIDT M, KANE DA, ODENTHAL J, VAN EEDEN FJ, JIANG YJ, HEISENBERG CP, KELSH RN, FURUTANI-SEIKI M, VOGELSANG E, BEUCHLE D, SCHACH U, FABIAN C, and NUSSLEIN-VOLHARD C (1996). The identification of genes with unique and essential functions in the development of the zebrafish, *Danio rerio*. Development 123: 1–36.

HILL A, BALL J, JONES M, DODD A, MESENS N, and VANPARYS P (2008). Implementation of zebrafish toxicity testing between *in vitro* and *in vivo* models to advance candidate selection. 29th Annual Meeting of the American College of Toxicology, Tucson, AZ, November 9–12, 2008.

HILL AJ (2008a). Zebrafish in drug discovery: bridging the gap between *in vitro* and *in vivo* methodologies. Preclinical World 121–123.

HILL AJ (2008b). Zebrafish use in drug discovery. 47th Annual Meeting of the Society of Toxicology, Seattle, WA, March 16–20, 2008.

HILL AJ, DUMOTIER B, and TRAEBERT M (2009). Cardiotoxicity testing in zebrafish: relevance of bioanalysis. Pharmacol Toxicol Methods 62(2): e22–e23.

HILL AJ, HEIDEN TC, HEIDEMAN W, and PETERSON RE (2009). Potential roles of Arnt2 in zebrafish larval development. Zebrafish 6(1): 79–91.

HILL AJ, HOWARD CV, and COSSINS AR (2001). Efficient embedding technique for preparing small specimens for stereological volume estimation: zebrafish larvae. J Microsc 206: 179–181.

HILL AJ, TERAOKA H, HEIDEMAN W, and PETERSON RE (2005). Zebrafish as a model vertebrate for investigating chemical toxicity. Toxicol Sci 86(1): 6–19.

HINTON DE and COUCH JA (1998). Architectural pattern, tissue and cellular morphology in livers of fishes: relationship to experimentally-induced neoplastic responses. EXS 86: 141–164.

HO CY, HOUART C, WILSON SW, and STAINIER DY (1999). A role for the extraembryonic yolk syncytial layer in patterning the zebrafish embryo suggested by properties of the hex gene. Curr Biol 9(19): 1131–1134.

ISOGAI S, HORIGUCHI M, and WEINSTEIN BM (2001). The vascular anatomy of the developing zebrafish: an atlas of embryonic and early larval development. Dev Biol 230(2): 278–301.

JONES M, BALL JS, DODD A, and HILL AJ (2009) Comparison between zebrafish and Hep G2 assays for the predictive identification of hepatotoxins. Toxicology 262(1): 13–14.

KEMPER EM, van ZANDBERGEN AE, CLEYPOOL C, MOS HA, BOOGERD W, BEIJNEN JH, and van TELLINGEN O (2003). Increased penetration of paclitaxel into the brain by inhibition of P-glycoprotein. Clin Cancer Res 9(7): 2849–2855.

KENG VW, FUJIMORI KE, MYINT Z, TAMAMAKI N, NOJYO Y, and NOGUCHI T (1998). Expression of Hex mRNA in early murine post implantation embryo development. FEBS Lett 426(2): 183–186.

KENG VW, YAGI H, IKAWA M, NAGANO T, MYINT Z, YAMADA K, TANAKA T, SATO A, MURAMATSU I, OKABE M, SATO M, and NOGUCHI T (2000). Homeobox gene Hex is essential for onset of mouse embryonic liver development and differentiation of the monocyte lineage. Biochem Biophys Res Commun 276(3): 1155–1161.

KORZH S, EMELYANOV A, and KORZH V (2001). Developmental analysis of ceruloplasmin gene and liver formation in zebrafish. Mech Dev 103(1–2): 137–139.

KUNG H-C, HOYERT D, XU J, and MURPHY S (2008). Deaths: Final Data for 2005. National Vital Statistics Reports. Department of Health and Human Services, Centers for Disease Control and Prevention.

LIESCHKE GJ and CURRIE PD (2007). Animal models of human disease: zebrafish swim into view. Nat Rev Genet 8(5): 353–367.

LORENT K, YEO SY, ODA T, CHANDRASEKHARAPPA S, CHITNIS A, MATTHEWS RP, and PACK M (2004). Inhibition of Jagged-mediated Notch signaling disrupts zebrafish biliary development and generates multi-organ defects compatible with an Alagille syndrome phenocopy. Development 131(22): 5753–5766.

MARTINEZ BARBERA JP, CLEMENTS M, THOMAS P, RODRIGUEZ T, MELOY D, KIOUSSIS D, and BEDDINGTON RS (2000). The homeobox gene Hex is required in definitive endodermal tissues for normal forebrain, liver and thyroid formation. Development 127(11): 2433–2445.

MURTHA JM, QI W, and KELLER ET (2003). Hematologic and serum biochemical values for zebrafish (*Danio rerio*). Comp Med 53(1): 37–41.

NASEVICIUS A and EKKER SC (2001). The zebrafish as a novel system for functional genomics and therapeutic development applications. Curr Opin Mol Ther 3(3): 224–228.

NAVARRO VJ and SENIOR JR (2006). Drug-related hepatotoxicity. N Engl J Med 354(7): 731–739.

NEUSCHWANDER-TETRI BA and CALDWELL SH (2003). Nonalcoholic steatohepatitis: summary of an AASLD Single Topic Conference. Hepatology 37(5): 1202–1219.

OBER EA, FIELD HA, and STAINIER DY (2003). From endoderm formation to liver and pancreas development in zebrafish. Mech Dev 120(1): 5–18.

O'BRIEN PJ, IRWIN W, DIAZ D, HOWARD-COFIELD E, KREJSA CM, SLAUGHTER MR, GAO B, KALUDERCIC N, ANGELINE A, BERNARDI P, BRAIN P, and HOUGHAM C (2006). High concordance of drug-induced human hepatotoxicity with *in vitro* cytotoxicity measured in a novel cell-based model using high content screening. Arch Toxicol 80(9): 580–604.

O'BRIEN PJ, SLAUGHTER MR, BIAGINI C, DIAZ D, GAO B, IRWIN W, KREJSA C, HOUGHAM C, ABRAHAM V, and HASKINS JR (2003). Predicting drug-induced human hepatotoxicity with *in vitro* cytotoxicity assays. Proceedings Tox. 2003 London, UK.

ODENTHAL J and NUSSLEIN-VOLHARD C (1998). fork head domain genes in zebrafish. Dev Genes Evol 208(5): 245–258.

OLIVER G, SOSA-PINEDA B, GEISENDORF S, SPANA EP, DOE CQ, and GRUSS P (1993). Prox 1, a prospero-related homeobox gene expressed during mouse development. Mech Dev 44(1): 3–16.

OLSON H, BETTON G, ROBINSON D, THOMAS K, MONRO A, KOLAJA G, LILLY P, SANDERS J, SIPES G, BRACKEN W, DORATO M, VAN DEUN K, SMITH P, BERGER B, and HELLER A (2000). Concordance of the toxicity of pharmaceuticals in humans and in animals. Regul Toxicol Pharmacol 32(1): 56–67.

OSTAPOWICZ G, FONTANA RJ, SCHIODT FV, LARSON A, DAVERN TJ, HAN SH, MCCASHLAND TM, SHAKIL AO, HAY JE, HYNAN L, CRIPPIN JS, BLEI AT, SAMUEL G, REISCH J, and LEE WM (2002). Results of a prospective study of acute liver failure at 17 tertiary care centers in the United States. Ann Intern Med 137(12): 947–954.

PACK M, SOLNICA-KREZEL L, MALICKI J, NEUHAUSS SC, SCHIER AF, STEMPLE DL, DRIEVER W, and FISHMAN MC (1996). Mutations affecting development of zebrafish digestive organs. Development 123: 321–328.

POSTLETHWAIT JH, WOODS IG, NGO-HAZELETT P, YAN YL, KELLY PD, CHU F, HUANG H, HILL-FORCE A, and TALBOT WS (2000). Zebrafish comparative genomics and the origins of vertebrate chromosomes. Genome Res 10(12): 1890–1902.

RUBENSTEIN A (2006). Zebrafish assays for drug toxicity screening. Expert Opin Drug Metab Toxicol 2: 231–240.

SADLER KC, AMSTERDAM A, SOROKA C, BOYER J, and HOPKINS N (2005). A genetic screen in zebrafish identifies the mutants vps18, nf2 and foie gras as models of liver disease. Development 132(15): 3561–3572.

SHIN JT and FISHMAN MC (2002). From zebrafish to human: modular medical models. Annu Rev Genomics Hum Genet 3: 311–340.

SOSA-PINEDA B, WIGLE JT, and OLIVER G (2000). Hepatocyte migration during liver development requires Prox1. Nat Genet 25(3): 254–255.

SPITSBERGEN JM, TSAI HW, REDDY A, MILLER T, ARBOGAST D, HENDRICKS JD, and BAILEY GS (2000). Neoplasia in zebrafish (*Danio rerio*) treated with 7,12-dimethylbenz[*a*]anthracene by two exposure routes at different developmental stages. Toxicol Pathol 28(5): 705–715.

SUMMERTON J and WELLER D (1997). Morpholino antisense oligomers: design, preparation, and properties. Antisense Nucleic Acid Drug Dev 7(3): 187–195.

TSENG HP, HSEU TH, BUHLER DR, WANG WD, and HU CH (2005). Constitutive and xenobiotics-induced expression of a novel CYP3A gene from zebrafish larva. Toxicol Appl Pharmacol 205(3): 247–258.

UETRECHT J (2007). Idiosyncratic drug reactions: current understanding. Annu Rev Pharmacol Toxicol 47: 513–539.

van HEYNINGEN V (1997). Model organisms illuminate human genetics and disease. Mol Med 3(4): 231–237.

WALISSER JA, BUNGER MK, GLOVER E, and BRADFIELD CA (2004a). Gestational exposure of Ahr and Arnt hypomorphs to dioxin rescues vascular development. Proc Natl Acad Sci USA 101(47): 16677–16682.

WALISSER JA, BUNGER MK, GLOVER E, HARSTAD EB, and BRADFIELD CA. (2004b). AT Patent ductus venosus and dioxin resistance in mice harboring a hypomorphic Arnt allele. J Biol Chem 279(16): 16326–16331.

WALLACE KN, YUSUFF S, SONNTAG JM, CHIN AJ, and PACK M (2001). Zebrafish hhex regulates liver development and digestive organ chirality. Genesis 30(3): 141–143.

ZHANG C, FREMGEN T, and WILLETT C (2003). Zebrafish: an animal model for toxicological studies. Wiley pp. 1.7.1–1.7.18.

ZON LI (1999). A new model for human disease. Genome Res 9: 99–100.

ZON LI and PETERSON RT (2005). *In vivo* drug discovery in the zebrafish. Nat Rev Drug Discov 4(1): 35–44.

Chapter 9

Whole Zebrafish Cytochrome P450 Assay for Assessing Drug Metabolism and Safety

Chunqi Li, Liqing Luo, Jessica Awerman, and Patricia McGrath

Phylonix, Cambridge, MA, USA

9.1 INTRODUCTION

Cytochrome P450 (CYP) enzymes catalyze the majority of known drug metabolizing reactions and many clinically relevant drug–drug interactions are associated with CYP inhibition or induction. Because of their genetic and physiological similarity to humans, zebrafish show promise as a predictive animal model for assessing drug metabolism and safety. Several zebrafish CYP genes that either have high homology to human genes or cause catalyzing reactions similar to those in mammals, including CYP3A65, CYP1A1, A19, B19, 2K6, 3C1, 2J1, and CYP26D1, have been cloned and characterized. Zebrafish CYP3A65 has been confirmed as a CYP3A ortholog. In this research, using a commercially available human CYP3A4-specific substrate, we developed a microplate-based whole zebrafish CYP3A4 functional activity assay. Specificity of the assay was initially confirmed using azamulin, a mammalian CYP3A4 inhibitor, dexamethasone and rifampicin, CYP3A4 inducers, and α-naphthoflavone (ANF), a non-CYP3A4 inhibitor/inducer. We further validated the zebrafish model by assessing five additional mammalian CYP3A4 inhibitors, disopyramide, erythromycin, fluvoxamine, omeprazole, and cimetidine, and six mammalian CYP3A4 inducers, carbamazepine, hydrocortisone, prednisone, pregnenolone-16α-carbonitrile, lovastatin, and phenytoin. Overall successful prediction rate was 87% (13/15): 100% for inhibition (6/6) and 75% for induction (6/8). These results demonstrate that zebrafish exhibit comparable CYP metabolism profiles as mammals supporting use of the whole zebrafish microplate assay as a preliminary screen for assessing drug safety.

Zebrafish: Methods for Assessing Drug Safety and Toxicity, Edited by Patricia McGrath.
© 2012 John Wiley & Sons, Inc. Published 2012 by John Wiley & Sons, Inc.

9.2 BACKGROUND AND SIGNIFICANCE

9.2.1 CYPs Are Responsible for Drug Metabolism and Drug Toxicity

Many clinically relevant drug–drug interactions are associated with inhibition or induction of specific CYP enzymes (Gomez-Lechon et al., 2003; Daly, 2004; Luo et al., 2004; Vermeir et al., 2005) and modification of CYP activity can profoundly affect therapeutic efficacy leading to life-threatening toxicity. The U.S. Food and Drug Administration (FDA) now requires CYP assessment prior to drug approval and it has published industry guidance for performing studies to assess drug–drug interaction. In well-described examples, a first drug inhibits a CYP enzyme that metabolizes a second coadministered drug, slows clearance of the second drug, and causes toxic accumulation. CYP-dependent drug–drug interaction can also occur when a first drug induces CYP gene expression, leading to increased CYP enzyme activity, accelerated clearance, and reduced drug efficacy (Wienkers, 2001; Yueh et al., 2005). CYP3A4, CYP2D6, and CYP2C9, the most abundant hepatic P450 enzymes, metabolize more than 70% of frequently prescribed drugs and safety profiling usually focuses on these enzymes (Davidson, 2000; Gomez-Lechon et al., 2003). Examples of costly, high-profile drug withdrawals due to CYP-related drug–drug interactions include terfenadine and astemizole (antihistamines), mibefradil (antihypertensive), and cerivastatin (statin). Although additional problematic drugs, including cimetidine (antacid), are still commercially available, due to complications from CYP-dependent drug–drug interactions, many of these are rapidly losing market share.

9.2.2 Similarity of Drug Metabolism and CYPs in Humans and Zebrafish

As a general defense against toxic chemicals, zebrafish exhibit mammalian-equiv-alent mechanisms, including induction of xenobiotic enzymes and increase in reactive oxygen species (ROS) (Wiegand et al., 2000; Dong et al., 2001; Carney et al., 2004). Several zebrafish CYP genes, including CYP3A65, CYP1A1, A19, B19, 2K6, 3C1, 2J1, and CYP26D1, which have human homology, exhibit similar catalyzing reactions (Miranda et al., 1993; Collodi et al., 1994; Trant et al., 2001; Bresolin et al., 2005; Gu et al., 2005; Tseng et al., 2005; Wang-Buhler et al., 2005; Corley-Smith et al., 2006; Rubinstein, 2006; Wang et al., 2007). In our recent studies, we confirmed that CYP1A1, 4A4, A19, and B19 are expressed in zebrafish (Parng et al., 2002) and we determined that, similar to effects in mice and humans, drug toxicity is dose responsive (Semino et al., 2002). A human CYP3A ortholog, designated CYP3A65, is expressed in zebrafish liver and intestine. Similar to the response of CYP3A4 to these drugs in humans, CYP3A65 is upregulated in zebrafish by both rifampicin and dexamethasone. TCDD, the environmental toxin, also upregulates zebrafish CYP3A65 gene expression (Tseng et al., 2005). Similar to results in humans (Bertilsson et al., 1998), another study found that pregnelone-16α-carbonitrile, a synthetic steroid, upregulated zebrafish CYP3A gene expression, whereas nifedipine did not. In addition, the antibiotic clotrimazole, a human CYP3A inducer, increased zebrafish CYP3A expression (Bresolin et al., 2005). These findings suggest that zebrafish CYP3A65 exhibits the same functions as human CYP3A4.

9.2.3 CYP Experimental Models

During the past decade, a number of CYP assays for assessing drug metabolism have been developed (Luo et al., 2004). Most of these approaches rely on use of subcellular liver fractions or human hepatocytes. However, results using these *in vitro* assays have been shown to differ from results *in vivo* (Gomez-Lechon et al., 2003; Vermeir et al., 2005). CYP assays in rat and mouse models, including LPTA® *CYP3A4-luc* transgenic mice, have recently been investigated; however, reliance on complex technologies and low throughput make them largely unsuitable for drug screening (Zhang et al., 2003, 2004; Gonzalez and Yu, 2006).

Although LC/UV is generally considered to be the gold standard for assessing drug metabolism, this method is relatively slow, expensive, and low throughput. A second widely used method relies on synthetic fluorogenic CYP substrates. Although adapted for high-throughput screening, these substrates are not pharmaceutically relevant, and, given the complexity of CYP active sites, in order to fully explore their potential for CYP inhibition or induction, test compounds may require use of five separate fluorogenic substrates. Furthermore, results using fluorogenic substrates to assess a wide range of CYP3A4 inhibitors were less predictive than results using LC/UV (Gomez-Lechon et al., 2003; Vermeir et al., 2005; Yueh et al., 2005; Cali et al., 2006).

For recently developed luminogenic assays, CYP enzyme activity is coupled to the light generating reaction of firefly luciferase (Cali et al., 2006). These *in vitro* assays were initially designed to measure CYP activity for recombinant CYPs, liver microsomes, or hepatocytes. Compared to HPLC and radiochemical-based methods, luminogenic CYP assays are rapid and safe. Compared to fluorogenic methods, this approach offers high sensitivity and low interference between optical properties of test compound and substrate. These homogenous assays are robust tools for high-throughput screening for early drug discovery. Development of new rapid assay methods and animal models for assessing drug metabolism and predicting potential drug interactions are needed. A rapid and robust whole zebrafish CYP assay, amenable to automation in a microplate format (Serbedzija et al., 2003), will speed up drug metabolism and safety profiling and reduce the possibility of market withdrawal or termination in late stages of drug development.

9.3 MATERIALS AND METHODS

9.3.1 Zebrafish Handling

Phylonix AB zebrafish were generated by natural pairwise mating in our aquaculture facility, as described by Westerfield (1993). Four to five pairs were set up for each mating; on average, 50–100 zebrafish per mating were generated. Zebrafish were maintained in fish water (5 g of Instant Ocean Salt with 3 g of $CaSO_4$ in 25 L of distilled water) at 28°C for 24 h before sorting for viability. Because the zebrafish embryo receives nourishment from an attached yolk sac, no additional maintenance was required.

9.3.2 Reagents and Drugs

CYP3A4 chemiluminogenic luciferin-BE substrate and luciferin detection reagent were purchased from Promega (Madison, WI). CYP inhibitors and inducers and controls (azamulin, dexamethasone, rifampicin, α-naphthoflavone, disopyramide, erythromycin, fluvoxamine, omeprazole, and cimetidine, carbamazepine, hydrocortisone, prednisone, pregnenolone-16α-carbonitrile, lovastatin, and phenytoin, and DMSO) were obtained from Sigma (St. Louis, MO).

9.3.3 Drug Treatment

Stock solutions of chemicals/drugs were diluted directly in fish water. Two dpf zebrafish were treated with five drug concentrations (0.01, 0.1, 1, 10, and 100 µM) for 24 h. Ten zebrafish per concentration were exposed in a total volume of 1 mL (ratio of 100 µL/zebrafish) using 24-well plates. Zebrafish were maintained in test medium throughout the experiments. Controls with and without vehicle solvent were included in all experiments. To protect compounds from light-induced decomposition, experiments were carried out at a constant temperature (28°C) in the dark.

9.3.4 Zebrafish CYP3A4 Functional Activity Assay

After drug treatment, zebrafish were transferred to untreated, white, opaque, flat-bottom 96-well microplates, one zebrafish per well. Excess fish water was removed by pipetting and then 25 µL Millipore filtered water and 25 µL of 400 mM KH_2PO4–100 µM luciferin-BE substrate were added to each well. Microplates were incubated at 37°C for 30 min and 50 µL of reconstituted luciferin detection reagent was added. Final KPO_4 concentration was 200 mM and final luciferin-BE substrate concentration was 50 µM. Samples with no substrate were used as background control. After mixing on an orbital shaker for 10 s or gently tapping the plate, luminescence was recorded with a microplate reader. Net chemiluminescence was calculated by subtracting background luminescence signal from the no substrate negative control wells. Experiments were performed at least three times and final results were expressed as mean ± SD.

9.3.5 IC$_{50}$ and EC$_{50}$ Estimation

Best-fit concentration–response curves for CYP3A4 inhibition or induction were generated using the linear regression function of JMP statistical software (SAS, Cary,

NC) and drug concentrations for 50% inhibition (IC_{50}) and 50% induction (EC_{50}) were estimated based on best-fit concentration–response curves.

9.3.6 Statistics

Since multiple concentrations were assessed, ANOVA was used to determine if drug effect was significant ($P < 0.05$). Dunnett's test, a multiple pairwise comparison test, was then performed to identify concentrations that exhibited significant effects.

9.3.7 Vertebrate Animal Care and Safety

The Office of Laboratory Animal Welfare (OLAW), National Institutes of Health (NIH), approved our Animal Welfare Assurance effective through February 2012. We euthanize zebrafish of all ages by overexposure to tricaine methanesulfonate. This procedure is consistent with the American Veterinary Medical Association's (AVMA) Panel on Euthanasia.

9.4 RESULTS

Using a commercially available human CYP3A4-specific chemiluminescent substrate, we developed a microplate-based zebrafish whole animal functional assay to assess drug metabolism and drug safety. Specificity of the zebrafish CYPA65 (CYP3A4) assay was initially validated by assessing (a) mammalian CYP3A4 inhibitors, (b) mammalian CYP3A4 inducers, and (c) a no-effect compound. The assay was further validated using five additional mammalian CYP3A4 inhibitors and six mammalian CYP3A4 inducers. Overall prediction success rate was 87% (13/15): 100% for inhibition (6/6) and 75% for induction (6/8). We established that treating 2dpf zebrafish for 24 h was optimum. Our results demonstrate that zebrafish exhibit comparable CYP drug metabolism profiles as mammals, supporting use of the whole zebrafish microplate CYP3A4 functional activity assay for preliminary drug screening.

9.4.1 Whole Zebrafish CYP Microplate Assay Development

In an initial pilot investigation, we found that similar to CYP3A4 response in humans, CYP3A4 functional activity was upregulated in zebrafish treated with dexamethasone at 10 and 50 μM concentrations. Our results were consistent with a recent report that showed that a low dose of dexamethasone (10 μM) enhanced zebrafish CYP3A65 transcription, whereas a high dose (100 μM) did not (Tseng et al., 2005). We initially established specificity of the *in vivo* microplate-based zebrafish functional assay by quantifying levels of CYP3A65 (CYP3A4) activity in zebrafish treated with azamulin, a known mammalian CYP3A4 inhibitor, and rifampicin, a known mammalian CYP3A4 inducer. Zebrafish treated with 50 μM dexamethasone for 24 h were used as a positive control. Zebrafish were also treated with no-effect compound α-naphthoflavone as a negative control (Cali et al., 2006), and vehicle control was 0.1% DMSO. At the end of treatment, CYP3A4 activity was quantitated in control and drug-treated zebrafish, as described in Section 10.3. Treatment with 0.1% DMSO for 24 h had no effect on CYP3A4 functional activity (data not shown), indicating that vehicle solvent did not affect CYP3A4 function. 0.01, 0.1, 1, 10, and 100 μM concentrations of azamulin inhibited CYPA4 by 0%, 13%, 20%, 30%, and 31%, respectively (Fig. 9.1a). 0.01, 0.1, 1, 10, and 100 μM concentrations of rifampicin induced CYPA4 by 0%, 17%, 37%, 65%, and 99%, respectively (Fig. 9.1b). Specificity of the assay was further supported by insensitivity to α-naphthoflavone, a no-effect negative control compound (data not shown).

9.4.2 Determination of Optimum Assay Conditions

We also assessed effects of zebrafish developmental stage and found that CYP3A4 level was higher at 3dpf than at 2dpf. However, since rifampicin treatment caused a statistically significant effect ($P < 0.001$) in 2 but not in 3dpf zebrafish ($p > 0.05$) (Fig. 9.2), we used 2dpf zebrafish as the optimum stage for performing the CYP3A4 assay.

To develop a reproducible assay, we also assessed the relationship between number of zebrafish per microwell and luminescence intensity. Three dpf zebrafish (one, two, and three zebrafish per well) were deposited into 96-well plates and incubated in substrate for 30 min. Untreated zebrafish without substrate were used as background controls. After subtracting nonspecific background, a linear relationship between chemiluminescent signal and number of zebrafish per well was observed (data not shown). These data further confirmed specificity and sensitivity of the microplate-based whole zebrafish CYP3A4 assay.

9.4.3 Validation of Whole Zebrafish CYP Microplate Assay

To further validate use of the microplate-based whole zebrafish CYP3A4 functional assay for assessing drug metabolism and safety, using the optimum zebrafish stage and assay procedures, we assessed five additional human CYP3A4 inhibitors, disopyramide, erythromycin, fluvoxamine, omeprazole, and cimetidine, and six additional human CYP3A4 inducers, carbamazepine, hydrocortisone, prednisone, pregnenolone-16α-carbonitrile, lovastatin, and phenytoin (Gomez-Lechon et al., 2003; Daly, 2004; Luo et al., 2004; Cali et al., 2006). We selected these drugs because CYP effects have been evaluated either in mammalian models or in human clinical trials.

Drug treatment and the zebrafish CYP3A4 assay were performed as described above. Zebrafish treated with 50 μM concentration of dexamethasone for 24 h were

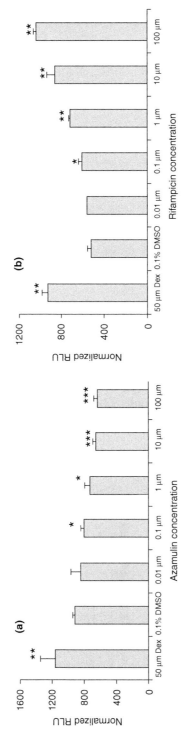

Figure 9.1 CYP3A4 inhibition and induction using microplate-based whole zebrafish CYP3A4 assay. Two dpf zebrafish were treated with azamulin or rifampicin for 24 h and zebrafish CYP3A4 activity was measured using a CYP3A4 chemiluminogenic substrate. Zebrafish treated with 50 μM dexamethasone (Dex) were used as a positive control. Data are expressed as mean ± SD. $^*p < 0.05$, $^{**}p < 0.01$, and $^{***}p < 0.001$ compared with 0.1% DMSO control. RLU = relative luminescence units.

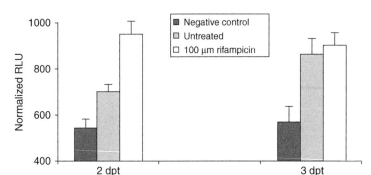

Figure 9.2 Effect of zebrafish stage on whole zebrafish microplate CYP3A4 assay. Two and three dpf zebrafish were treated with $100\,\mu M$ rifampicin for 24 h and zebrafish CYP3A4 activity was measured using a CYP3A4 chemiluminogenic substrate. Untreated 2 and 3dpf zebrafish served as negative controls. Untreated zebrafish without CYP3A4 chemiluminogenic substrate were used as a negative control for CYP3A4 assay. Data are expressed as mean $\pm$ SD. $^{*}p < 0.001$ compared with untreated zebrafish in the same stage. RLU = relative luminescence units.

used as a positive control. Zebrafish treated with vehicle (0.1% DMSO) served as a negative control.

All five human CYP3A4 inhibitors suppressed CYP3A4 functional activity in zebrafish in a concentration-dependent manner ($p < 0.05$–0.001, compared to vehicle controls). For test concentrations ranging from 0.01 to $100\,\mu M$, percent inhibition of zebrafish CYP3A4 functional activity compared to 0.1% DMSO control was 18–42% for disopyramide, 14–42% for erythromycin, 18–53% for fluvoxamine, 24–69% for omeprazole, and 18–53% for cimetidine (Fig. 9.3).

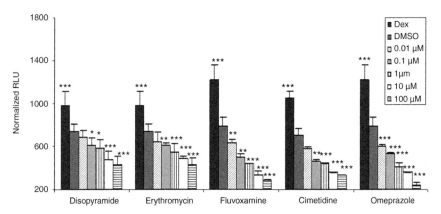

Figure 9.3 CYP3A4 inhibition using microplate-based whole zebrafish CYP3A4 assay. Two dpf zebrafish were treated with CYP3A4 inhibitors for 24 h and zebrafish CYP3A4 activities were measured using a CYP3A4 chemiluminogenic substrate. Zebrafish treated with 0.1% DMSO (vehicle) were used as a negative control and zebrafish treated with $50\,\mu M$ dexamethasone (Dex) served as a positive control. Data are expressed as mean $\pm$ SD. $^{*}p < 0.05$, $^{**}p < 0.01$, and $^{***}p < 0.001$ compared with 0.1% DMSO control.

Table 9.1 CYP3A4 Inhibition in Zebrafish: IC_{50} Values

CYP3A4 inhibitors	CYP3A4 inhibition IC_{50} values in zebrafish (μM)
Azamulin	187.3
Disopyramide	123.1
Erythromycin	128.9
Fluvoxamine	42.7
Omeprazole	10.2
Cimetidine	80.2

IC_{50} values (the concentration at which 50% CYP3A4 inhibition is observed) are summarized in Table 9.1.

Four of the six human CYP3A4 inducers (prednisone, hydrocortisone, carbamazepine, and pregnenolone-16α-carbonitrile) increased zebrafish CYP3A4 functional activity. Percent increase in induction compared to 0.1% DMSO control was 111–126% for carbamazepine, 119–139% for hydrocortisone, 122% for prednisone, and 123–135% for pregnenolone-16α-carbonitrile; human CYP3A4 inducers lovastatin and phenytoin did not affect zebrafish CYP3A4 functional activity (Fig. 9.4). EC_{50} values (the concentration at which 50% CYP3A4 induction was observed) are summarized in Table 9.2.

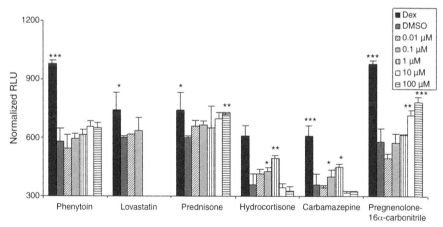

Figure 9.4 CYP3A4 induction using microplate-based whole zebrafish CYP3A4 assay. Two dpf zebrafish were treated with CYP3A4 inducers for 24 h and zebrafish CYP3A4 activities were measured using a CYP3A4 chemiluminogenic substrate. Zebrafish treated with 0.1% DMSO (vehicle) were used as a negative control and zebrafish treated with 50 μM dexamethasone (Dex) served as a positive control. Data are expressed as mean ± SD. $^{*}p < 0.05$, $^{**}p < 0.01$, and $^{***}p < 0.001$ compared with 0.1% DMSO control. Phenytoin and lovastatin did not cause significant effects on CYP3A4 activity in zebrafish.

Table 9.2   CYP3A4 Induction in Zebrafish: EC_{50} Values

CYP3A4 inducers	CYP3A4 induction EC_{50} values in zebrafish (μM)
Rifampicin	8.4
Phenytoin	No induction
Lovastatin (mevinolin)	No induction
Prednisone	225.5
Hydrocortisone	NA
Carbamazepine	NA
Pregnenolone-16α-carbonitrile	126.2

Table 9.3   Comparison of CYP3A4 Inhibitors and Inducers in Zebrafish and Mammals

Compound	Pharmaceutical function in mammals	CYP3A4 effects in zebrafish	CYP3A4 effects in mammals	Correct prediction
Azamulin	Highly selective human CYP3A4 inhibitor	Inhibition	Inhibition	Yes
Disopyramide	Class Ia antiarrhythmic	Inhibition	Inhibition	Yes
Erythromycin	Antibiotic	Inhibition	Inhibition	Yes
Fluvoxamine	Antidepressant	Inhibition	Inhibition	Yes
Omeprazole	Antacid medication	Inhibition	Inhibition	Yes
Cimetidine	Antacid medication	Inhibition	Inhibition	Yes
Rifampicin	Antibiotic prescribed to treat *Mycobacterium* infection	Induction	Induction	Yes
Carbamazepine	Anticonvulsant and mood stabilizing drug	Induction	Induction	Yes
Hydrocortisone	Corticosteroid similar to a naturally produced hormone	Induction	Induction	Yes
Prednisone	Synthetic corticosteroid	Induction	Induction	Yes
Lovastatin	Statin drug class used for lowering cholesterol	None	Induction	No
Pregnenolone-16α-carbonitrile	Activates mouse but not human pregnane X receptor (PXR) and inhibits rodent liver fibrogenesis	Induction	Induction	Yes
Phenytoin	Anticonvulsant drug useful for treating epilepsy	None	Induction	No
α-Naphthoflavone	A potent antiplatelet flavonoid	None	None	Yes

9.4.4 Comparison of CYP3A4 Functional Activity in Zebrafish and Mammals

To further validate use of the microplate-based whole zebrafish CYP3A4 functional assay for assessing drug metabolism and safety, we compared results in zebrafish with results in mammals. For all CYP inhibitors and inducers tested in zebrafish, effects in mammalian models have been investigated or reported in clinical trials. As shown in Table 9.3, results from our whole zebrafish microplate CYP3A4 functional activity assay correlated well with results in mammals. Overall prediction success rate for CYP3A4 inhibition and induction in zebrafish was 86%: 100% for inhibition and 71% for induction. According to criteria for assessing the predictive value of a new assay developed by the European Center for the Validation of Alternative Methods (ECVAM) (Genschow et al., 2002), the assay was ranked "good" (>75%) for identifying inducers and "excellent" (>85%) for identifying inhibitors. These data suggest that zebrafish exhibit comparable CYP drug profiles as mammals and that the whole zebrafish CYP microplate assay represents a predictive, reproducible animal model for assessing drug metabolism and safety.

9.5 CONCLUSIONS

Since single zebrafish embryos and larvae can be maintained in fluid volumes as small as 100 μL for the first week of development, physiologically intact animals can be maintained in individual microtiter wells, an advantage that no other vertebrate model provides. In addition to the microplate-based whole zebrafish CYP assay, we have used similar principles to develop quantitative assays for angiogenesis, Src kinase, TUNEL, caspase, ROS, and cancer cell xenotransplant. Accumulating data suggest that *in vivo* zebrafish microplate assays are highly sensitive, reproducible, and relatively high throughput.

ACKNOWLEDGMENT

This research was supported by a grant from the National Institutes of Health: 2R44GM083366.

REFERENCES

BERTILSSON G, HEIDRICH J, SVENSSON K, ASMAN M, JENDEBERG L, SYDOW-BACKMAN M, OHLSSON R, POSTLIND H, BLOMQUIST P, and BERKENSTAM A (1998). Identification of a human nuclear receptor defines a new signaling pathway for CYP3A induction. Proc Natl Acad Sci USA 95(21): 12208–12213.

BRESOLIN T, DE FREITAS REBELO M, and CELSO DIAS BAINY A (2005). Expression of PXR, CYP3A and MDR1 genes in liver of zebrafish. Comp Biochem Physiol C Toxicol Pharmacol 140(3–4): 403–407.

CALI J, MA D, SOBOL M, SIMPSON DJ, and FRACKMAN S (2006). Luminogenic cytochrome P450 assays. Expert Opin Drug Metab Toxicol 2: 629–645.

CARNEY SA, PETERSON RE, and HEIDEMAN W (2004). 2,3,7,8-Tetrachlorodibenzo-*p*-dioxin activation of the aryl hydrocarbon receptor/aryl hydrocarbon receptor nuclear translocator pathway causes developmental toxicity through a CYP1A-independent mechanism in zebrafish. Mol Pharmacol 66(3): 512–521.

COLLODI P, MIRANDA CL, ZHAO X, BUHLER DR, and BARNES DW (1994). Induction of zebrafish (*Brachydanio rerio*) P450 *in vivo* and in cell culture. Xenobiotica 24(6): 487–493.

CORLEY-SMITH GE, SU HT, WANG-BUHLER JL, TSENG HP, and HU CH (2006). CYP3C1, the first member of a new cytochrome P450 subfamily found in zebrafish (*Danio rerio*). Biochem Biophys Res Commun 340(4): 1039–1046.

DALY AK (2004). Pharmacogenetics of the cytochromes P450. Curr Top Med Chem 4(16): 1733–1744.

DAVIDSON MH (2000). Does differing metabolism by cytochrome P450 have clinical importance? Curr Atheroscler Rep 2(1): 14–19.

DONG W, TERAOKA H, KONDO S, and HIRAGA T (2001). 2,3,7,8-Tetrachlorodibenzo-*p*-dioxin induces apoptosis in the dorsal midbrain of zebrafish embryos by activation of arylhydrocarbon receptor. Neurosci Lett 303(3): 169–172.

GENSCHOW E, SPIELMANN H, SCHOLZ G, SEILER A, and BROWN N (2002). The ECVAM international validation study on *in vitro* embryotoxicity tests: results of the definitive phase and evaluation of prediction models. European Centre for the Validation of Alternative Methods. Altern Lab Anim 30(2): 151–176.

GOMEZ-LECHON MJ, DONATO MT, CASTELL JV, and JOVER R (2003). Human hepatocytes as a tool for studying toxicity and drug metabolism. Curr Drug Metab 4(4): 292–312.

GONZALEZ FJ and YU AM (2006). Cytochrome P450 and xenobiotic receptor humanized mice. Annu Rev Pharmacol Toxicol 46: 41–64.

GU X, XU F, WANG X, GAO X, and ZHAO Q (2005). Molecular cloning and expression of a novel CYP26 gene (cyp26d1) during zebrafish early development. Gene Expr Patterns 5(6): 733–739.

LUO G, GUENTHNER T, GAN LS, and HUMPHREYS WG (2004). CYP3A4 induction by xenobiotics: biochemistry, experimental methods and impact on drug discovery and development. Curr Drug Metab 5(6): 483–505.

MIRANDA CL, COLLODI P, ZHAO X, BARNES DW, and BUHLER DR (1993). Regulation of cytochrome P450 expression in a novel liver cell line from zebrafish (*Brachydanio rerio*). Arch Biochem Biophys 305(2): 320–327.

PARNG C, SEMINO C, ZHANG C, and MCGRATH P (2002). Gene expression profiling in zebrafish after drug treatment. Toxicologist 66: LB75.

RUBINSTEIN AL (2006). Zebrafish assays for drug toxicity screening. Expert Opin Drug Metab Toxicol 2(2): 231–240.

SEMINO C, ZHANG C, PARNG C, and MCGRATH P (2002). Zebrafish as an animal model to assess drug toxicology. Toxicologist 66: LB59.

SERBEDZIJA G, SEMINO C, FROST D, and MCGRATH P (2003). Methods of screening agents for activity using teleosts. U.S. Patent 6,656,449.

TRANT JM, GAVASSO S, ACKERS J, CHUNG BC, and PLACE AR (2001). Developmental expression of cytochrome P450 aromatase genes (CYP19a and CYP19b) in zebrafish fry (*Danio rerio*). J Exp Zool 290(5): 475–483.

TSENG HP, HSEU TH, BUHLER DR, WANG WD, and HU CH (2005). Constitutive and xenobiotics-induced expression of a novel CYP3A gene from zebrafish larva. Toxicol Appl Pharmacol 205(3): 247–258.

VERMEIR M, ANNAERT P, MAMIDI RN, ROYMANS D, MEULDERMANS W, and MANNENS G (2005). Cell-based models to study hepatic drug metabolism and enzyme induction in humans. Expert Opin Drug Metab Toxicol 1(1): 75–90.

WANG L, YAO J, CHEN L, CHEN J, XUE J, and JIA W (2007). Expression and possible functional roles of cytochromes P450 2J1 (zfCyp 2J1) in zebrafish. Biochem Biophys Res Commun 352(4): 850–855.

WANG-BUHLER JL, LEE SJ, CHUNG WG, STEVENS JF, and TSENG HP (2005). CYP2K6 from zebrafish (*Danio rerio*): cloning, mapping, developmental/tissue expression, and aflatoxin B1 activation by baculovirus expressed enzyme. Comp Biochem Physiol C Toxicol Pharmacol 140(2): 207–219.

WESTERFIELD M (1993). The Zebrafish Book: A Guide for the Laboratory Use of Zebrafish (*Danio rerio*). University of Oregon Press, Eugene, OR.

WIEGAND C, PFLUGMACHER S, GIESE M, FRANK H, and STEINBERG C (2000). Uptake, toxicity, and effects on detoxication enzymes of atrazine and trifluoroacetate in embryos of zebrafish. Ecotoxicol Environ Saf 45(2): 122–131.

WIENKERS LC (2001). Problems associated with *in vitro* assessment of drug inhibition of CYP3A4 and other P-450 enzymes and its impact on drug discovery. J Pharmacol Toxicol Methods 45(1): 79–84.

YUEH MF, KAWAHARA M, and RAUCY J (2005). High volume bioassays to assess CYP3A4-mediated drug interactions: induction and inhibition in a single cell line. Drug Metab Dispos 33(1): 38–48.

ZHANG W, PURCHIO AF, CHEN K, WU J, and LU L (2003). A transgenic mouse model with a luciferase reporter for studying *in vivo* transcriptional regulation of the human CYP3A4 gene. Drug Metab Dispos 31(8): 1054–1064.

ZHANG W, PURCHIO AF, COFFEE R, and WEST DB (2004). Differential regulation of the human CYP3A4 promoter in transgenic mice and rats. Drug Metab Dispos 32(2): 163–167.

Chapter 10

Methods for Assessing Neurotoxicity in Zebrafish

Chunqi Li, Wen Lin Seng, Demian Park, and Patricia McGrath

Phylonix, Cambridge, MA, USA

10.1 INTRODUCTION

Neurotoxicity, defined as an adverse effect on the structure or function of the nervous system, can result from exposure to drugs used for chemotherapy, radiation treatment, and organ transplantation as well as from food additives and environmental toxicants. Neurotoxicity is the second leading cause (after cardiovascular toxicity) of drug withdrawals. Examples of costly, high-profile drug withdrawals due to neurotoxicity include diamthazole, vinyl chloride aerosol, and clioquinol (Wysowski and Swartz, 2005). Neurotoxicity profiles of numerous approved drugs are incomplete and many have been shown to cause neurotoxic side effects. For example, chloramphenicol (an antibiotic) and ethambutol and isoniazid (antituberculosis drugs) can cause optic neuritis (Chang et al., 1966; Noguera-Pons et al., 2005). Phenytoin (an antiepileptic drug), fluorouracil and cytarabine (chemotherapy agents), and aminoglycosides can induce cerebellar syndromes such as ataxia, dysarthria, and nystagmus (Sylvester et al., 1987; Macdonald, 1991). Hormone replacement therapy (HRT) and estrogen combined oral contraceptives (COCs) are known to induce stroke and intracranial venous thrombosis (Bushnell et al., 2001; Nightingale and Farmer, 2004). Antipsychotic drugs such as prochlorperazine can induce Parkinson's-like symptoms (Nath et al., 2000; Catalano et al., 2005). Chemotherapy and anti-glycolipid antibodies can induce peripheral neuropathy (Willison and Yuki, 2002; Cavaletti and Marmiroli, 2004). In addition, the onset of demyelinating disease relapse has been associated with vaccines (Kaplanski et al., 1995; DeStefano et al., 2003). Many other drug-induced neurological complications such as cognitive impairment, headache, and neuromuscular disorders have also been reported (Grosset and Grosset, 2004).

Zebrafish: Methods for Assessing Drug Safety and Toxicity, Edited by Patricia McGrath.
© 2012 John Wiley & Sons, Inc. Published 2012 by John Wiley & Sons, Inc.

10.2 LIMITATIONS OF CURRENT NEUROTOXICITY TESTING

Approaches for assessing neurotoxicity are categorized into three general groups: behavioral, morphological (neurohistopathology), and biochemical (measurements of altered cellular metabolism and function). Current preclinical neurotoxicity testing generally relies on detection of behavioral abnormalities and/or the appearance of overt histopathological lesions in nerve tissue. Animal behavioral studies are most effective in detecting neurotoxicity that adversely affects well-defined and easily detectable parameters such as survival, motor function, aggression, feeding, grooming, and reproductive and maternal behavior. However, it is likely that a majority of the effects on the central nervous system are silent and produce changes in function that result in subtle alterations in parameters such as emotion, cognition, temperament, or mood that cannot be adequately evaluated in animals and may not be easily identifiable in human studies. Histological methods are effective in detecting neurotoxicity only when lesions in the nervous system are extensive and can be detected by immunohistochemical staining methods. The number of sections that can be analyzed and the availability of skilled neuropathologists are limiting factors. The application of biochemical markers for neurotoxicity is an ongoing area of research within both government agencies and the academic community. Several specific biochemical markers (e.g., changes in enzyme activity, protein phosphorylation) have been examined but have proven useful for detecting only specific types of neurotoxicity. Other biochemical assessments that correlate brain functions with metabolism have not been rigorously tested. More recently, a variety of "-omic" technologies are increasingly applied for preclinical safety assessment; however, these approaches have not yet been implemented in neurotoxicity safety evaluations (O'Callaghan and Sriram, 2005). Development of rapid assay methods and new animal models to predict neurotoxicity are urgently needed.

Zebrafish is exceptionally well suited for neurotoxicity studies that combine cellular, molecular, and genetic approaches. Because the embryo is transparent and develops rapidly, development of specific neurons and axon tracts can be visualized in live embryos using differential interference contrast (DIC) microscopy or by injecting live dyes (Kuwada and Bernhardt, 1990). Specific types of neurons can be visualized in whole, fixed embryos by immunohistochemistry or *in situ* hybridization (Chandrasekhar et al., 1997; Moens and Fritz, 1999). Motor neuron activity can be monitored *in vivo* by calcium imaging and patch clamp recording (Drapeau et al., 2002). Function of individual neurons can be elucidated by specific neural lesion using toxic dye injection (Gahtan and O'Malley, 2001) or laser ablation (Fetcho and Liu, 1998). In addition, mutants exhibiting observable morphological phenotypes or behaviors have been isolated from large-scale screens and have been useful in understanding early CNS patterning (Jiang et al., 1996; Schier et al., 1996).

10.3 ASSESSING NEUROTOXICITY IN ZEBRAFISH

Recently, zebrafish has been shown to be a useful animal model for assessing compound-induced neurotoxicity. During the first 2 weeks of development, zebrafish are transparent, providing excellent accessibility for visualization of vital dyes, fluorescent tracers, antibodies, and riboprobes in live and whole mount fixed specimens. Zebrafish have an early population of neurons, called primary neurons, which are part of a relatively simple nervous system that differentiates in order to coordinate larval movement (Westerfield et al., 1986; Kimmel et al., 1988). The distribution and projection patterns of specific primary neurons have been described in detail (Eisen, 1991). By 24 h post fertilization (hpf), these primary zebrafish neurons differentiate and establish many of their projections. During this period, the relatively large neuronal cell bodies can be identified *in vivo* using Normarski optics. By 48hpf, zebrafish brain ventricles have formed (Jiang et al., 1996; Schier et al., 1996). Zebrafish body length increases from 1 mm on 1dpf (day post fertilization) to 5 mm by 6dpf and the animals do not have skulls. Whole animal staining can be performed to examine the entire nervous system in extensive detail. Moreover, zebrafish motor behavior develops in a predictable sequence (Drapeau et al., 2002), supporting assessment of drug effects on locomotion. Classical neurotoxins tested in zebrafish include (1) dopaminergic neurotoxins such as MPTP (1-methyl-4-phenyl-1,2,3, 6-tetrahydropyridine), 6-hydroxydopamine (6-OHDA), rotenone, and paraquat; (2) non-NMDA-type glutamate receptor (AMPA) agonists or antagonists such as domoic acid, 6-cyano-7-nitroquinoxaline-2,3-dione, alpha-latrotoxin, and picrotoxin; (3) nicotinic acetylcholine receptor (nAChR) antagonists such as bungarotoxins and cobratoxins or acetylcholinesterase (AChE) inhibitors; and (4) NMDA receptor antagonist, DL-2-amino-5-phosphonovalerate (AP-5) (Legendre, 1997; Ali et al., 2000; Hatta et al., 2001; Rigo et al., 2003).

10.3.1 Compound Effects on Motor Neurons and Neuronal Proliferation

In vertebrates, motor neurons are defined as neurons located in the central nervous system that project their axons outside the CNS and directly or indirectly control muscles. Drug-induced neurotoxicity on motor neurons and neuronal proliferation has been extensively reported in mammalian models; for example, ethanol has been shown to induce motor neuron death and inhibit neuronal proliferation. Ethanol has also been found to affect the brain and motor functions in humans. To assess compound effects on morphology of motor neurons and pattern of neuronal proliferation, 5hpf zebrafish were exposed to 2.5% ethanol for 1 h and zebrafish were fixed at 2dpf for whole mount immunostaining using Znp-1, a mouse monoclonal antibody specific to zebrafish primary motor neurons or a mouse monoclonal antibody against PCNA (proliferating cell nuclear antigen) (Parng et al., 2007). Primary motor neurons in untreated, wild-type zebrafish exhibited an organized, stereotypical vertical pattern. After ethanol treatment, primary motor neuron staining in the somite region was absent (Fig. 10.1a and b) and the pattern of neuronal proliferation was abnormal (Fig. 10.1c and d). These results suggest that zebrafish is a suitable animal model for visually assessing compound effects on motor neurons and neuron proliferation (Parng et al., 2007).

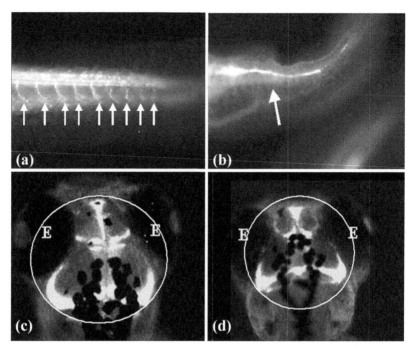

Figure 10.1 Effects of ethanol on zebrafish embryos. Motor neuron examination (a and b): 48hpf zebrafish were stained with Znp-1, a motor neuron-specific antibody. In untreated controls, the motor neurons are highly organized along the somites (arrows). However, after treatment with 2.5% ethanol, motor neuron loss was observed. Assessment of neuron proliferation (c and d): neuron proliferation was examined by immunostaining with an anti-PCNA antibody (white area). The pattern of proliferating neuronal cells in the brain was different and proliferation zone was smaller in ethanol-treated embryos.

10.3.2 Compound Effects on Dopaminergic Neurons

Dopaminergic neurons of the midbrain are the main source of dopamine (DA) in the mammalian central nervous system and loss of DA is associated with Parkinson's disease (PD), one of the most prominent human neurological disorders. Dopaminergic neurons, which are present in low numbers, are found in a "harsh" region of the brain, the substantia nigra pars compacta, which is DA-rich and contains both redox available neuromelanin and high iron content. Dopaminergic neurons play an important role in the control of multiple brain functions including voluntary movement and a broad array of behavioral processes such as mood, reward, addiction, and stress. To assess compound effects on dopaminergic neurons, 2dpf zebrafish were treated with 250 μM 6-OHDA, a neurotoxin that destroys catecholaminergic terminals, for 3 days and then processed for immunostaining using an anti-tyrosine hydroxylase (TH) antibody. Although this antibody stains catecholaminergic neurons, it has been demonstrated that all tyrosine hydroxylase positive neurons in zebrafish diencephalon (hypothalamus, posterior tuberculum, ventral thalamus, and pretectum) are DA neurons

(a)   **(b)**

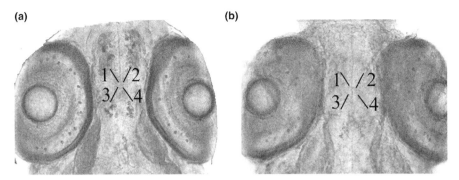

Figure 10.2 Effects of 6-hydroxydopamine in zebrafish. Two dpf zebrafish were untreated (a) or treated with 6-hydroxydopamine (b) and stained with anti-tyrosine hydroxylase antibody at 5dpf. TH positive cells were present in the diencephalon, eyes, and hindbrain (circled). The total number of TH positive DA neurons in the populations 1 and 2 was 44 ± 4 for wild-type zebrafish (a) and 3 ± 4 for 6-OHDA-treated zebrafish (b) ($P < 0.0001$) indicating occurrence of dopaminergic neuron neurotoxicity.

(Rink and Wullimann, 2001). For quantitative assessment, TH-immunoreactive DA neurons were counted by microscopy ($n = 10$). Statistical analysis was performed to determine significance. Distinct populations of dopaminergic neurons were observed in the diencephalons of wild-type zebrafish (Fig. 10.2a). After 6-OHDA treatment, the number of dopaminergic neurons significantly decreased (Fig. 10.2b); total number of TH positive DA neurons in populations 1 and 2 was 44 ± 4 for wild-type zebrafish and 3 ± 4 for 6-OHDA-treated zebrafish ($P < 0.0001$). These results indicate 6-OHDA-induced dopaminergic neurotoxicity (Parng et al., 2007). In mammals, 6-OHDA has been shown to induce oxidative stress and results in neuronal death (Elkon et al., 2004). Immunostaining using an anti-nitrotyrosine antibody was performed and nitrotyrosine immunoreactivity was observed in the brain of 6-OHDA-treated zebrafish but not in untreated controls, suggesting that oxidative stress in response to 6-OHDA treatment occurs in zebrafish (Parng et al., 2007).

10.3.3 Compound Effects on Myelin

Myelin is a dielectric (electrically insulating) material that forms a layer, the myelin sheath, usually specifically around axons on neurons. Myelin is essential for the proper functioning of the nervous system. Schwann cells supply the myelin for peripheral neurons, whereas oligodendrocytes, specifically of the interfascicular type, supply it to those of the central nervous system. Myelin is considered a defining characteristic of vertebrates, but it has also arisen by parallel evolution in some invertebrates. The myelin sheath has been shown to be affected by many compounds. To assess compound effects on the integrity of the myelin sheath in zebrafish, expression of myelin basic protein (MBP) was quantified after compound treatment and whole mount immunostaining. Acrylamide, which has been shown to cause demyelination in mammals (Harry et al., 1992; LoPachin et al., 1992, 1993;

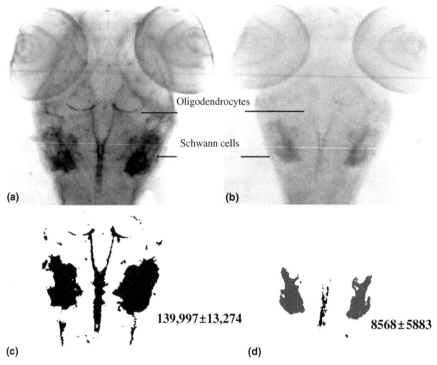

Figure 10.3 Myelin sheath assessment. Three dpf zebrafish were treated with vehicle control
(a and c) or acrylamide (b and d) for 24 h and *in situ* hybridization was performed to examine MBP
expression (a and b). Untreated control zebrafish show MBP expression in oligodendrocytes in the brain
and in Schwann cells in the craniofacial nerves (labeled) (a). Acrylamide-treated zebrafish show
significant loss of MBP expression in both the brain and craniofacial nerves. The average staining signal
was 139,997 ± 13,274 for untreated zebrafish and 8568 ± 5883 for acrylamide-treated zebrafish.
(c and d). Luxol blue staining was performed to examine the integrity of the myelin sheath; light gray:
neurophils; black: myelin; gray: nerve tissues. Control-treated zebrafish exhibited blue myelin staining
in the forebrain and midbrain regions (c). Acrylamide treatment caused loss of myelin in the forebrain
and midbrain regions (d).

Lehning et al., 1998), was used to generate lesions in zebrafish myelin sheath. MBP
expression in oligodendrocytes and Schwann cells decreased after acrylamide
treatment (Fig. 10.3). Morphometric analysis demonstrated that average fluorescence
intensity was 139,997 ± 13,274 for untreated control group and 8568 ± 5883 for the
acrylamide-treated group ($N = 10$); fluorescence intensity for acrylamide-treated
group (6%) significantly decreased compared to untreated control (100%,
$P < 0.0001$). These results imply that acrylamide either induced loss of MBP
producing cells or reduced production of MBP transcripts. To further examine effects
of acrylamide on the structure of the myelin sheath, Luxol blue staining was
performed: myelinated fibers stained blue (DeStefano et al., 2003), neurophils stained
pink (light gray), and nerve cells stained purple (gray). The level of myelin sheath
staining was significantly less in the brain and spinal cord of acrylamide-treated

zebrafish (Parng et al., 2007), supporting use of zebrafish as a model organism for assessing demyelination and probably remyelination.

10.3.4 Compound-Induced Brain-Specific Apoptosis

Although apoptosis occurs naturally during development of the nervous system, increased apoptosis in the mature nervous system is deleterious and a hallmark of many neurodegenerative diseases. Apoptosis is also a common cellular response to exposure to chemical toxins (Corcoran et al., 1994). Apoptosis is induced in a number of mammalian cell lines, including neuronal cell lines (Ahmadi et al., 2003; Caughlan et al., 2004) and mouse embryos after exposure to pesticides (Greenlee et al., 2004). Organophosphate pesticides induce apoptosis in cultured rat cortical cells (Kim et al., 2004). Conventional assay formats include absorbance, fluorescence, luminescence for *in vitro* detection of caspase enzymatic activity, intracellular oxidation, mitochondrial permeability, or double stranded DNA breaks. However, these *in vitro* approaches have a common shortfall: none of the assays can predict if compounds will work under *in vivo* physiological conditions.

In recent investigations, using both fluorescent acridine orange (AO) live staining (Parng et al., 2004; Ton et al., 2006) and terminal deoxynucleotidyl transferase dUTP nick end labeling (TUNEL), zebrafish treated with L-2-hydroxylglutaric acid (LGA) exhibited brain-specific apoptosis (Fig. 10.4) (Serbedzija et al., 2001; Parng et al., 2006; McGrath et al., 2010). Accumulation of LGA in cerebrospinal and other body fluids has been observed in patients with hydroxyglutaric aciduria (OHGA) (Hoffmann et al., 1993). In addition, LGA has been shown to promote oxidative stress and

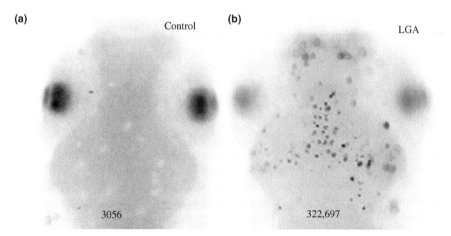

Figure 10.4 LGA-induced brain apoptosis. After TUNEL staining, images were captured, inverted, and used for morphometric analysis. Apoptotic cells display a punctate staining pattern in the brain. Fluorescent signals from the brain region are quantified using ImageJ software (NIH). Total fluorescence was calculated as fluorescence intensity × staining area. Compared to control (a), LGA-treated zebrafish (b) exhibited increased apoptosis (3056 versus 322,697).

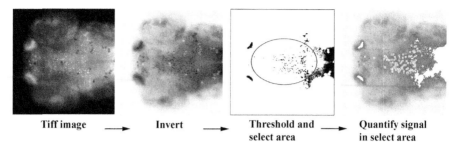

| Tiff image | $\longrightarrow$ | Invert | $\longrightarrow$ | Threshold and select area | $\longrightarrow$ | Quantify signal in select area |

Figure 10.5 Quantification of apoptosis in the brain by TUNEL staining followed by morphometric image analysis. Dorsal views of stained embryos were photographed, inverted, and analyzed using ImageJ software (NIH). Fluorescence images of region of interest (ROI) in each animal for each treatment condition were captured using the same gain and exposure time. Then, a constant threshold for acceptable fluorescence signal was applied to brain images from each zebrafish. This threshold was set automatically by the software and adjusted manually to take into account artifacts. Then, using the automated measurement feature of the software, total fluorescence intensity was measured from threshold-processed fluorescence images. Fluorescence intensity in ROI was then quantitated (dorsal view; anterior, top).

neuronal apoptosis in the rodent brain (Latini et al., 2003a, 2003b). In order to increase screening throughput, brain-specific apoptosis can be assessed using quantitative morphometric image analysis. Using the same gain and exposure time, fluorescence images of region of interest (ROI) in each animal were acquired. Then, a constant threshold for acceptable fluorescence signal was applied to brain images. This threshold was set automatically by the software and adjusted manually to take into account artifacts. Next, using the automated measurement feature of the software, total fluorescence intensity was measured from threshold-processed fluorescence images. Fluorescent signals (Fl) were quantified as Fl = staining area × staining intensity of apoptotic cells by particle analysis. Images of lateral sides of each animal were obtained using the same exposure time and fluorescent gain (anterior, left; posterior, right; dorsal, top) (Fig. 10.5). Neurotoxicity was calculated and expressed as percent increase in apoptosis in zebrafish brain (Parng et al., 2004, 2006). Using the same methods and procedures, effects of additional mammalian neurotoxins, including taxol, acrylamide, and c-Jun kinase (JNK) inhibitor (SP600125), on brain apoptosis were examined; all compounds significantly increased brain apoptosis in zebrafish. These results indicate that zebrafish is susceptible to apoptosis induced by mammalian neurotoxicants.

10.3.5 Blood–Brain Barrier in Zebrafish

The blood–brain barrier (BBB) is a specialized system of capillary endothelial cells that protects the brain from harmful substances in the blood stream while supplying the brain with nutrients that are required to maintain essential physiological functions. The BBB consists of a complex cellular system comprised of specialized endothelial cells, a large number of pericytes embedded in the basal membrane, perivascular macrophages, and astrocyte endfeet (Gloor et al., 2001; Ballabh et al., 2004). Complex tight junctions (TJs) between endothelial cells act as a barrier to the passage of macromolecules (Engelhardt, 2003). It is known that some small molecules can cross the BBB by passive diffusion while some are dependent on transport systems that cross the BBB via carrier-mediated transport, receptor-mediated transcytosis, or adsorption-mediated transcytosis (Greig et al., 1988; Bobo et al., 1994; Pardridge, 2003). Three groups of transporters have been characterized in mammals: (1) active efflux transporter (AET), which includes Pgp, a major efflux transporter, (2) carrier-mediated transporter (CMT), and (3) receptor-mediated transporter (RMT). These BBB transporters are localized to the luminal endothelial membrane (on the blood side of the capillaries) and abluminal endothelial membrane (on the brain side of the capillaries) and regulate drug influx and efflux. Moreover, proteins and polypeptides, which represent a promising category of potential drugs for treating various CNS disorders, cannot cross the BBB. It is widely believed that in order to cross the BBB, a small molecule must have the following characteristics: (1) molecular weight (MW) ≤ 400 Da, (2) lipid soluble, and (3) not a substrate for a BBB active efflux transporter. Conventional methods for assessing compound BBB permeability include both *in silico*, cell-free and cell culture system and mammalian models. These methods often involve dissection, sectioning, and histological analyses that can be less reliable and less efficient.

In zebrafish, initial neurulation starts in the late stage of gastrulation (9hpf) and by 30hpf, four separate brain ventricles have formed and are visible throughout development (Wilson et al., 2002), providing excellent accessibility for assessing drug permeation and molecular interactions *in vivo*. BBB is present in all vertebrates, including zebrafish (Cserr and Bundgaard, 1984), and during development, zebrafish exhibit a complex angiogenic network in the brain (Lawson and Weinstein, 2002). An extensive capillary network that exhibits angiogenic vessels has developed by 3dpf and continues remodeling throughout development (Lawson and Weinstein, 2002). Zebrafish exhibit comparable drug metabolism as mammals (Langheinrich et al., 2003; Parng, 2005; Parng et al., 2006). To assess BBB formation during embryogenesis, development of tight junctions in day 3 zebrafish was recently confirmed using ZO-1 antibody staining (Panizzi et al., 2007; McGrath et al., 2010). Evans blue dye permeation at different stages of development was also examined and blockage of dye diffusion from the vessels to the brain in day 3 animals was detected, indicating that BBB functions are present in 3dpf zebrafish (Fig. 10.6). Because the brain structure is transparent in zebrafish during early development (days 0–22), it is possible to study drug permeation and the molecular machinery underlying BBB formation and disruption. Drug or dye trafficking between the blood and the brain can be easily assessed without complicated histology. Although the BBB may not fully mature until a later stage, a functioning BBB in the transparent embryo brain can still serve as an excellent model for studying drug permeation and efficacy of CNS drugs. As additional data supporting the utility of zebrafish for screening drugs that permeate the BBB, we screened 14 well-characterized neuroprotectants using a drug-induced zebrafish brain-specific apoptosis model. In this study, 11 of the 11 compounds that showed protection in other animal models caused significant protection in zebrafish (Parng, 2005; Parng et al., 2006). Furthermore, three BBB impermeable

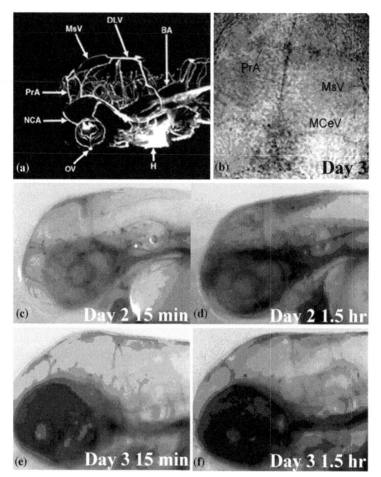

Figure 10.6 Confirmation of BBB formation in zebrafish by microangiography, antibody staining, and dye injection. (a) After injecting dextran–rhodamine beads, microangiography was used to image the entire brain vasculature in 3dpf zebrafish; the major brain ventricles are completely formed, angiogenic vessels are well formed, and the blood vessel network, which transports nutrients to brain cells, can be visualized in the brain. (b) Tight junctions are present during early development. Whole mount ZO-1, a tight junction component, antibody immunostaining showed that tight junctions form in the brain (brown spots) adjacent to the mesencephalic vein (MsV), middle cerebral vein (MCeV), and prosencephalic artery (PrA) (arrows). BA, basilar artery; DLV, dorsal longitudinal anastomotic vessel; NCA, nasal ciliary artery; OV, optic vein; H, heart. (c–f) Conventional Evans blue dye injection method was used to confirm the presence of BBB in zebrafish. Evans blue dye was microinjected into the peripheral blood vessels in 2 (c and d) and 3dpf (e and f) zebrafish. Distribution of the dye was assessed in the brain at 15 min (c and e) and 1.5 h (d and f) after microinjection. In 2dpf zebrafish (c and d), Evans blue dye was observed in the brain blood vessels (arrow) at 15 min (c); however, permeation of dye to the midbrain and hindbrain regions was observed at 1.5 h (d). In contrast, in 3dpf zebrafish (e and f), after injection, Evans blue dye was retained in the brain blood vessels at both 15 min (e) and 1.5 h (f) and was not found in brain tissue, indicating the presence of BBB in 3dpf zebrafish. (See the color version of this figure in Color Plates section.)

compounds caused no significant effects, supporting the utility of zebrafish for identifying potential neuroprotectants (Parng, 2005; Parng et al., 2006).

10.3.6 Compound-Induced Effects on Motility

One of the greatest challenges in developing methods for assessing neurotoxicity is
associating neuromorphological, neurochemical, and neurophysiological alterations
with behavioral changes, frequently assessed as abnormal movement (NRC, 1992).
In zebrafish larvae, patterns of motility or locomotion are stage specific and
behavioral abnormalities can be readily distinguished (Granato and Nusslein-Volhard, 1996). Embryonic motor behavior develops sequentially and consists of an early
period of transient spontaneous coiling contractions, followed by the emergence of
twitching in response to touch, and later, by the ability to swim (Drapeau et al., 2002).
By 4dpf, embryos are free swimming and change direction spontaneously with
characteristic speed and distance intervals. Zebrafish also exhibit basic behavior
including memory, nonassociative learning, conditioned responses, and social behavior such as schooling (Burgess and Granato, 2007; Best et al., 2008).

Several studies have shown that zebrafish motility can be tracked using a
computer-driven motion detector (ViewPoint Life Sciences, Lyons, France) (Emran
et al., 2007; McGrath and Li, 2008; Winter et al., 2008). Using this automated system,
zebrafish movement is captured electronically with a high-resolution black and white
camera (Fig. 10.7a). The number of movements in a given time period, duration, and
distance traveled can be assessed (Fig. 10.7b). Investigators (Baraban et al., 2005;
Prober et al., 2006; Berghmans et al., 2007; McGrath and Li, 2008; Winter et al., 2008)
have demonstrated the usefulness of this assay format for assessing compound-
induced neurotoxicity, including seizures. Similar to results in mammals,
pentylenetetrazole (PTZ), a GABA antagonist known to induce convulsions in
humans, has been shown to induce seizures in zebrafish and the behavioral,
electrophysiological, and molecular changes in PTZ-treated zebrafish were comparable to effects observed in a rodent seizure model. In recent experiments, using the
VideoTrack System, effect of treatment with PTZ on zebrafish swim movement was
assessed. Zebrafish were placed in 24-well microplates, one animal per well.
Acquisition software was programmed to separate swim movements into three
distinct color-coded patterns based on speed: inactivity/low-speed movement (<4
mm/s), normal movement (4–20 mm/s), and high-speed movement (>20 mm/s).
Zebrafish exposed to PTZ exhibited a dramatic increase in high-speed swim movement, compared to carrier control-treated zebrafish that primarily exhibited normal
movement during 60 min of motility recording. Total distance traveled at high speed
by combining speed and duration of high-speed movement was analyzed at 1 min
intervals. The average distance traveled (mm) by 10 zebrafish in 1 min intervals
increased dramatically and peaked ~7 min after PTZ treatment; increased distance
traveled at high speed continued for the duration of the 60 min recording period.
Although untreated zebrafish also exhibited high-speed swim movement, this represented a small percentage of total swim movement and did not increase.

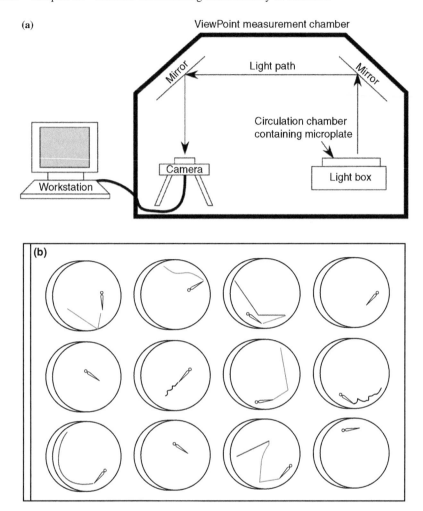

Figure 10.7 (a) Diagram of ViewPoint measurement chamber specifically designed for zebrafish motility research. The chamber contains a camera, light box, and a chamber that can circulate water to hold microplates containing zebrafish. The size of the box is minimized by use of mirrors to increase the light path length. (b) Diagram of zebrafish motion tracking results. Diagram shows a portion of a 96-well microplate, one zebrafish per well. VideoTrack software analyzes length, speed, and direction of movement. Speed of movement is color coded: black, slow; green, medium; red, fast. (See the color version of this figure in Color Plates section.)

10.3.7 Compound-Induced Developmental Neurotoxicity

Developmental neurotoxicity involves alterations in behavior, neurohistology, neurochemistry, neurophysiology, or gross dysmorphology due to exposure during development. The developing embryonic brain is more sensitive to perturbation by toxins due to the disruption of sensitive processes that occur only during development, such as differentiation of specific cell type, and proliferation and migration of nascent neurons (Barone et al., 2000). In addition, the developing brain is more exposed to blood-borne toxins prior to formation of the BBB. Therefore, critical assessment of developmental neurotoxicity requires use of an embryonic model. The U.S. Environmental Protection Agency (U.S. EPA) and the Organization for Economic Cooperation and Development (OECD) guidelines for developmental toxicity testing contain protocols using a rat embryonic model (U.S. EPA, 1998; Mileson and Ferenc, 2001; OECD, 2003). This model is well developed and includes several developmental, histological, and behavioral end points. Developmental landmarks include size, time of vaginal opening, and time of preputial separation. Histological methods include assessment of brain weight, histopathology, and qualitative and semiquantitative morphometric analyses. Behavioral end points include measurement of motor activity, auditory startle, and learning and memory. The rat model allows comprehensive and predictive testing; however, these studies require a minimum of 60 days per test and are experimentally complex, which hinders testing.

In a recent study, seven well-characterized compounds were selected to validate zebrafish as a developmental neuorotoxicity model (Table 10.1) (Ton et al., 2006). Embryos were exposed by semistatic immersion from 6hpf to 96 or 120hpf; teratogenicity was assessed using a modified method previously reported (Ton et al., 2006). Atrazine, dichlorodiphenyltrichloroethane (DDT), and 2,3,7,8-tetrachlorodibenzo-*p*-dioxin (TCDD) were found to be primarily teratogenic and not specifically neurotoxic. 2,4-Dichlorophenoxyacetic acid (2,4-D), dieldrin, and nonylphenol showed specific neurotoxicity; dieldrin and nonylphenol were specifically toxic to catecholaminergic neurons. Exposure to 100 µM dieldrin also caused reduced axon tracts in the brain (Fig. 10.8). Although not teratogenic, malathion showed some nonspecific toxicity. Induction of brain apoptosis or necrosis was confirmed as an indicator of developmental neurotoxicity and an effect on motor neurons in the caudal third of the embryo was shown to correlate with expected defects in motility. In another developmental neurotoxicity project funded by the U.S. National Science Foundation, 200 environmental contaminants were screened for effects on brain apoptosis, axon tract and motor neuron formation, and catecholaminergic neurons in developing zebrafish and several compounds that showed effects on one or more of the neurotoxicity parameters were identified and some compounds showed broad toxicity

Table 10.1 Comparison of Teratogenicity in Mammals and Zebrafish

Compound	Mammals	Zebrafish (96hpf)
2,4-D	No (rats, rabbits), slight (mouse)	Slight
Atrazine	No (mouse)	Yes
DDT	Yes (mouse)	Yes
Dieldrin	No	No
Malathion	No (rat, rabbit)	No
Nonylphenol	No (rat)	No
TCDD	High (rat, mouse, hamster)	High

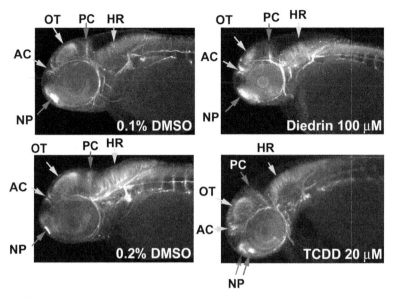

Figure 10.8   Visual assessment of axon tracts in the brain of control and compound-exposed embryos at 48hpf. AC, anterior commissure; HR, hindbrain region; NP, nasal placode; OT, optic tectum; PC, posterior commissure. Anterior, left and dorsal, up.

across all neuroanatomical end points (Ton et al., 2006). Overall, these results showed strong correlation with mammalian data and suggest that zebrafish is a predictive animal model for developmental neurotoxicity screening.

10.4 SUMMARY

Zebrafish have been shown to be a predictive animal model for assessing neurotoxicity. The ability to visually examine several distinct end points and to elucidate the mechanisms of toxicity *in vivo* is a significant advantage of zebrafish as a model for assessing neurotoxicity morphologically. Many qualitative and quantitative toxicity end points can be rapidly examined in the intact zebrafish nervous system without the artifacts that can result from dissecting organs. Genetic tools, including mutants and transgenic and gene knockdown animals, can be used to assist in developing comprehensive morphological and behavioral studies. Technology and instrumentation including microplate readers, digital image systems, and fluorescence-activated cell sorting have been adapted for quantitative analysis of drug effects in zebrafish, significantly increasing assay reliability and efficiency. Quantitative studies including 3D computer analysis for detecting gene expression, automated image analysis for quantitative neuronal phenotyping, and morphometric analysis for quantifying hair cells have been reported (Ton and Parng, 2005). Although many qualitative end points for predicting neurotoxicity in mammals can be assessed in zebrafish, direct comparison between results in zebrafish and results in mammals requires further validation. Several significant challenges remain in order to establish zebrafish as a predictive animal model for neurotoxicity, including determining (1) sensitivity and susceptibility of zebrafish to specific compounds and general compound classes compared to mammalian models and humans; (2) comparability of ADME in zebrafish using delivery in fish water to ADME in humans using other routes of administration; and (3) role of BBB, shown to be present in zebrafish by 3dpf (McGrath and Li, 2008).

ACKNOWLEDGMENTS

This research was supported by grants from National Institutes of Health (2R44CA83256) and National Science Foundation (OII-0548657).

REFERENCES

AHMADI FA, LINSEMAN DA, GRAMMATOPOULOS TN, JONES SM, BOUCHARD RJ, FREED CR, HEIDENREICH KA, and ZAWADA WM (2003). The pesticide rotenone induces caspase-3-mediated apoptosis in ventral mesencephalic dopaminergic neurons. J Neurochem 87(4): 914–921.

ALI DW, BUSS RR, and DRAPEAU P (2000). Properties of miniature glutamatergic EPSCs in neurons of the locomotor regions of the developing zebrafish. J Neurophysiol 83(1): 181–191.

BALLABH P, BRAUN A, and NEDERGAARD M (2004). The blood–brain barrier: an overview—structure, regulation, and clinical implications. Neurobiol Dis 16(1): 1–13.

BARABAN SC, TAYLOR MR, CASTRO PA, and BAIER H (2005). Pentylenetetrazole induced changes in zebrafish behavior, neural activity and c-fos expression. Neuroscience 131(3): 759–768.

BARONE S. JR., DAS KP, LASSITER TL, and WHITE LD (2000). Vulnerable processes of nervous system development: a review of markers and methods. Neurotoxicology 21(1–2): 15–36.

BERGHMANS S, HUNT J, ROACH A, and GOLDSMITH P (2007). Zebrafish offer the potential for a primary screen to identify a wide variety of potential anticonvulsants. Epilepsy Res 75(1): 18–28.

BEST JD, BERGHMANS S, HUNT JJ, CLARKE SC, FLEMING A, GOLDSMITH P, and ROACH AG (2008). Non-associative learning in larval zebrafish. Neuropsychopharmacology 33(5): 1206–1215.

BOBO RH, LASKE DW, AKBASAK A, MORRISON PF, DEDRICK RL, and OLDFIELD EH (1994). Convection-enhanced delivery of macromolecules in the brain. Proc Natl Acad Sci USA 91(6): 2076–2080.

BURGESS HA and GRANATO M (2007). Modulation of locomotor activity in larval zebrafish during light adaptation. J Exp Biol 210 (Pt 14): 2526–2539.

BUSHNELL CD, SAMSA GP, and GOLDSTEIN LB (2001). Hormone replacement therapy and ischemic stroke severity in women: a case-control study. Neurology 56(10): 1304–1307.

CATALANO G, GRACE JW, CATALANO MC, MORALES MJ, and CRUSE LM (2005). Acute akathisia associated with quetiapine use. Psychosomatics 46(4): 291–301.

CAUGHLAN A, NEWHOUSE K, NAMGUNG U, and XIA Z (2004). Chlorpyrifos induces apoptosis in rat cortical neurons that is regulated by a balance between p38 and ERK/JNK MAP kinases. Toxicol Sci 78(1): 125–134.

CAVALETTI G and MARMIROLI P (2004). Chemotherapy-induced peripheral neurotoxicity. Expert Opin Drug Saf 3(6): 535–546.

CHANDRASEKHAR A, MOENS CB, WARREN J.T. JR., KIMMEL CB, and KUWADA JY (1997). Development of branchiomotor neurons in zebrafish. Development 124(13): 2633–2644.

CHANG N, GILES CL, and GREGG RH (1966). Optic neuritis and chloramphenicol. Am J Dis Child 112(1): 46–48.

CORCORAN GB, FIX L, JONES DP, MOSLEN MT, NICOTERA P, OBERHAMMER FA, and BUTTYAN R (1994). Apoptosis: molecular control point in toxicity. Toxicol Appl Pharmacol 128(2): 169–181.

CSERR HF and BUNDGAARD M (1984). Blood–brain interfaces in vertebrates: a comparative approach. Am J Physiol 246 (3 Pt 2): R277–R288.

DESTEFANO F, VERSTRAETEN T, JACKSON LA, OKORO CA, BENSON P, BLACK SB, SHINEFIELD HR, MULLOOLY JP, LIKOSKY W, and CHEN RT (2003). Vaccinations and risk of central nervous system demyelinating diseases in adults. Arch Neurol 60(4): 504–509.

DRAPEAU P, SAINT-AMANT L, BUSS RR, CHONG M, MCDEARMID JR, and BRUSTEIN E (2002). Development of the locomotor network in zebrafish. Prog Neurobiol 68(2): 85–111.

EISEN JS (1991). Motoneuronal development in the embryonic zebrafish. Development (Suppl 2): 141–147.

ELKON H, MELAMED E, and OFFEN D (2004). Oxidative stress, induced by 6-hydroxydopamine, reduces proteasome activities in PC12 cells: implications for the pathogenesis of Parkinson's disease. J Mol Neurosci 24(3): 387–400.

EMRAN F, RIHEL J, ADOLPH AR, WONG KY, KRAVES S, and DOWLING JE (2007). OFF ganglion cells cannot drive the optokinetic reflex in zebrafish. Proc Natl Acad Sci USA 104(48): 19126–19131.

ENGELHARDT B (2003). Development of the blood–brain barrier. Cell Tissue Res 314(1): 119–129.

FETCHO JR and LIU KS (1998). Zebrafish as a model system for studying neuronal circuits and behavior. Ann NY Acad Sci 860: 333–345.

GAHTAN E and O'MALLEY DM (2001). Rapid lesioning of large numbers of identified vertebrate neurons: applications in zebrafish. J Neurosci Methods 108(1): 97–110.

GLOOR SM, WACHTEL M, BOLLIGER MF, ISHIHARA H, LANDMANN R, and FREI K (2001). Molecular and cellular permeability control at the blood–brain barrier. Brain Res Brain Res Rev 36(2–3): 258–264.

GRANATO M and NUSSLEIN-VOLHARD C (1996). Fishing for genes controlling development. Curr Opin Genet Dev 6(4): 461–468.

GREENLEE AR, ELLIS TM, and BERG RL (2004). Low-dose agrochemicals and lawn-care pesticides induce developmental toxicity in murine preimplantation embryos. Environ Health Perspect 112(6): 703–709.

GREIG NH, FREDERICKS WR, HOLLOWAY HW, SONCRANT TT, and RAPOPORT SI (1988). Delivery of human interferon-alpha to brain by transient osmotic blood–brain barrier modification in the rat. J Pharmacol Exp Ther 245(2): 581–586.

GROSSET KA and GROSSET DG (2004). Prescribed drugs and neurological complications. J Neurol Neurosurg Psychiatry 75 (Suppl 3): iii2–iii8.

HARRY GJ, MORELL P, and BOULDIN TW (1992). Acrylamide exposure preferentially impairs axonal transport of glycoproteins in myelinated axons. J Neurosci Res 31(3): 554–560.

HATTA K, ANKRI N, FABER DS, and KORN H (2001). Slow inhibitory potentials in the teleost Mauthner cell. Neuroscience 103(2): 561–579.

HOFFMANN GF, MEIER-AUGENSTEIN W, STOCKLER S, SURTEES R, RATING D, and NYHAN WL (1993). Physiology and pathophysiology of organic acids in cerebrospinal fluid. J Inherit Metab Dis 16(4): 648–669.

JIANG YJ, BRAND M, HEISENBERG CP, BEUCHLE D, FURUTANI-SEIKI M, KELSH RN, WARGA RM, GRANATO M, HAFFTER P, HAMMERSCHMIDT M, KANE DA, MULLINS MC, ODENTHAL J, VAN EEDEN FJ, and NUSSLEIN-VOLHARD C (1996). Mutations affecting neurogenesis and brain morphology in the zebrafish, *Danio rerio*. Development 123: 205–216.

KAPLANSKI G, RETORNAZ F, DURAND J, and SOUBEYRAND J (1995). Central nervous system demyelination after vaccination against hepatitis B and HLA haplotype. J Neurol Neurosurg Psychiatry 58(6): 758–759.

KIM SJ, KIM JE, and MOON IS (2004). Paraquat induces apoptosis of cultured rat cortical cells. Mol Cells 17 (1): 102–107.

KIMMEL CB, SEPICH DS, and TREVARROW B (1988). Development of segmentation in zebrafish. Development 104 (Suppl): 197–207.

KUWADA JY and BERNHARDT RR (1990). Axonal outgrowth by identified neurons in the spinal cord of zebrafish embryos. Exp Neurol 109(1): 29–34.

LANGHEINRICH U, VACUN G, and WAGNER T (2003). Zebrafish embryos express an orthologue of HERG and are sensitive toward a range of QT-prolonging drugs inducing severe arrhythmia. Toxicol Appl Pharmacol 193(3): 370–382.

LATINI A, SCUSSIATO K, ROSA RB, LEIPNITZ G, LLESUY S, BELLO-KLEIN A, DUTRA-FILHO CS, and WAJNER M (2003a). Induction of oxidative stress by L-2-hydroxyglutaric acid in rat brain. J Neurosci Res 74(1): 103–110.

LATINI A, SCUSSIATO K, ROSA RB, LLESUY S, BELLO-KLEIN A, DUTRA-FILHO CS, and WAJNER M (2003b). D-2-Hydroxyglutaric acid induces oxidative stress in cerebral cortex of young rats. Eur J Neurosci 17(10): 2017–2022.

LAWSON ND and WEINSTEIN BM (2002). Arteries and veins: making a difference with zebrafish. Nat Rev Genet 3(9): 674–682.

LEGENDRE P (1997). Pharmacological evidence for two types of post synaptic glycinergic receptors on the Mauthner cell of 52-h-old zebrafish larvae. J Neurophysiol 77(5): 2400–2415.

LEHNING EJ, PERSAUD A, DYER KR, JORTNER BS, and LOPACHIN RM (1998). Biochemical and morphologic characterization of acrylamide peripheral neuropathy. Toxicol Appl Pharmacol 151(2): 211–221.

LOPACHIN RM, CASTIGLIA CM, LEHNING E, and SAUBERMANN AJ (1993). Effects of acrylamide on subcellular distribution of elements in rat sciatic nerve myelinated axons and Schwann cells. Brain Res 608(2): 238–246.

LOPACHIN RM, CASTIGLIA CM, and SAUBERMANN AJ (1992). Acrylamide disrupts elemental composition and water content of rat tibial nerve. II. Schwann cells and myelin. Toxicol Appl Pharmacol 115(1): 35–43.

MACDONALD DR (1991). Neurologic complications of chemotherapy. Neurol Clin 9(4): 955–967.

MCGRATH P and LI CQ (2008). Zebrafish: a predictive model for assessing drug-induced toxicity. Drug Discov Today 13(9–10): 394–401.

MCGRATH P, PARNG C, and SERBEDZIJA G (2010). Methods of screening agents using teleosts. U.S. Patent 7,767,880.

MILESON BE and FERENC SA (2001). Methods to identify and characterize developmental neurotoxicity for human health risk assessment: overview. Environ Health Perspect 109 (Suppl 1): 77–78.

MOENS CB and FRITZ A (1999). Techniques in neural development. Methods Cell Biol 59: 253–272.

NATH A, ANDERSON C, JONES M, MARAGOS W, BOOZE R, MACTUTUS C, BELL J, HAUSER KF, and MATTSON M (2000). Neurotoxicity and dysfunction of dopaminergic systems associated with AIDS dementia. J Psychopharmacol 14(3): 222–227.

NIGHTINGALE AL and FARMER RD (2004). Ischemic stroke in young women: a nested case-control study using the UK General Practice Research Database. Stroke 35(7): 1574–1578.

NOGUERA-PONS R, BORRAS-BLASCO J, ROMERO-CRESPO I, ANTON-TORRES R, NAVARRO-RUIZ A, and GONZALEZ-FERRANDEZ JA (2005). Optic neuritis with concurrent etanercept and isoniazid therapy. Ann Pharmacother 39(12): 2131–2135.

NRC (1992). Committee on Neurotoxicology and Models for Assessing Risk. National Academy Press, Washington, DC.

O'CALLAGHAN JP and SRIRAM K (2005). Glial fibrillary acidic protein and related glial proteins as biomarkers of neurotoxicity. Expert Opin Drug Saf 4(3): 433–442.

OECD (2003). OECD Guideline for the Testing of Chemicals: Proposal for a New Guideline 426. Developmental Neurotoxicity Study. pp. 1–20.

PANIZZI JR, JESSEN JR, DRUMMOND IA, and SOLNICA-KREZEL L (2007). New functions for a vertebrate Rho guanine nucleotide exchange factor in ciliated epithelia. Development 134(5): 921–931.

PARDRIDGE WM (2003). Blood–brain barrier drug targeting: the future of brain drug development. Mol Interv 3(2): 90–105.

PARNG C (2005). In vivo zebrafish assays for toxicity testing. Curr Opin Drug Discov Dev 8(1): 100–106.

PARNG C, ANDERSON N, TON C, and MCGRATH P (2004). Zebrafish apoptosis assays for drug discovery. Methods Cell Biol 76: 75–85.

PARNG C, ROY NM, TON C, LIN Y, and MCGRATH P (2007). Neurotoxicity assessment using zebrafish. J Pharmacol Toxicol Methods 55(1): 103–112.

PARNG C, TON C, LIN YX, ROY NM, and MCGRATH P (2006). A zebrafish assay for identifying neuroprotectants in vivo. Neurotoxicol Teratol 28(4): 509–516.

PROBER DA, RIHEL J, ONAH AA, SUNG RJ, and SCHIER AF (2006). Hypocretin/orexin overexpression induces an insomnia-like phenotype in zebrafish. J Neurosci 26(51): 13400–13410.

RIGO JM, BADIU CI, and LEGENDRE P (2003). Heterogeneity of post synaptic receptor occupancy fluctuations among glycinergic inhibitory synapses in the zebrafish hindbrain. J Physiol 553 (Pt 3): 819–832.

RINK E and WULLIMANN MF (2001). The teleostean (zebrafish) dopaminergic system ascending to the subpallium (striatum) is located in the basal diencephalon (posterior tuberculum). Brain Res 889(1–2): 316–330.

SCHIER AF, NEUHAUSS SC, HARVEY M, MALICKI J, SOLNICA-KREZEL L, STAINIER DY, ZWARTKRUIS F, ABDELILAH S, STEMPLE DL, RANGINI Z, YANG H, and DRIEVER W (1996). Mutations affecting the development of the embryonic zebrafish brain. Development 123: 165–178.

SERBEDZIJA G, SEMINO C, and FROST D (2001). Methods of screening agents for activity using teleosts. U.S. Patent 6,299,858.

SYLVESTER RK, FISHER AJ, and LOBELL M (1987). Cytarabine-induced cerebellar syndrome: case report and literature review. Drug Intell Clin Pharm 21(2): 177–180.

TON C, LIN Y, and WILLETT C (2006). Zebrafish as a model for developmental neurotoxicity testing. Birth Defects Res A Clin Mol Teratol 76(7): 553–567.

TON C and PARNG C (2005). The use of zebrafish for assessing ototoxic and otoprotective agents. Hear Res 208(1–2): 79–88.

U.S. EPA (1998). Health Effects Test Guidelines. OPPTS 870.6300. Developmental Neurotoxicity Study. pp. 1–12.

WESTERFIELD M, MCMURRAY JV, and EISEN JS (1986). Identified motoneurons and their innervation of axial muscles in the zebrafish. J Neurosci 6(8): 2267–2277.

WILLISON HJ and YUKI N (2002). Peripheral neuropathies and anti-glycolipid antibodies. Brain 125 (Pt 12): 2591–2625.

WILSON SW, BRAND M, and EISEN JS (2002). Patterning the zebrafish central nervous system. Results Probl Cell Differ 40: 181–215.

WINTER MJ, REDFERN WS, HAYFIELD AJ, OWEN SF, VALENTIN JP, and HUTCHINSON TH (2008). Validation of a larval zebrafish locomotor assay for assessing the seizure liability of early-stage development drugs. J Pharmacol Toxicol Methods 57(3): 176–187.

WYSOWSKI DK and SWARTZ L (2005). Adverse drug event surveillance and drug withdrawals in the United States, 1969–2002: the importance of reporting suspected reactions. Arch Intern Med 165(12): 1363–1369.

Chapter 11

Zebrafish: A Predictive Model for Assessing Cancer Drug-Induced Organ Toxicity

Louis D'Amico[1], Chunqi Li[1], Elizabeth Glaze[2], Myrtle Davis[2], and Wen Lin Seng[1]

[1]*Phylonix, Cambridge, MA, USA*
[2]*NCI, NIH, Bethesda, MD, USA*

11.1 INTRODUCTION

In this study, we visually assessed acute zebrafish organ toxicity of five cancer drug candidates: 17-DMAG, HDAC inhibitor X, paclitaxel, SMA-838, and zebularine. To assess acute toxicity, we initially determined the lethal compound concentration after treatment for 24 h. Zebrafish were then treated for 24 h with six drug concentrations $\leq LC_{10}$ and effects on heart, central nervous system, liver, and kidney morphology, function, and organ-specific cell death were assessed. Based on results for assessing organ morphology and function, we estimated the no observable effect concentration (NOEC). Next, we assessed reversibility of toxic effects. After treatment for 24 h, drugs were removed by washing zebrafish in fresh fish water and incubating for an additional 48 h. Organ morphology and function were then reassessed. Finally, results in zebrafish showed good correlation with available results in mammals and humans. These results show that drugs can be added directly to fish water and effects on major organs can be visually assessed in this transparent animal.

Zebrafish: Methods for Assessing Drug Safety and Toxicity, Edited by Patricia McGrath.
© 2012 John Wiley & Sons, Inc. Published 2012 by John Wiley & Sons, Inc.

11.2 MATERIALS AND METHODS

11.2.1 Zebrafish Breeding and Handling

Embryos were generated by natural pairwise mating as described in *The Zebrafish Book* (Westerfield, 1993). Embryos were maintained at 28°C in fish water (200 mg Instant Ocean Salt (Aquarium Systems, Mentor, OH) per liter of deionized water; pH 6.6–7.0, maintained with 2.5 mg/L Jungle pH Stabilizer (Jungle Laboratories Corporation, Cibolo, TX); conductivity 670–760 μS). Embryos were cleaned (dead and unfertilized embryos removed) and staged at 4 h post fertilization (hpf). Because embryos receive nourishment from an attached yolk sac, no feeding is required for 7 days post fertilization (dpf).

11.2.2 Compounds

Five test compounds (Table 11.1) were supplied blinded by the Developmental Therapeutics Program, National Cancer Institute (SAIC-Frederick, Frederick, MD). Compounds were dissolved in dimethyl sulfoxide (DMSO) (Sigma, St. Louis, MO) or fish water following Study Sponsor's instructions. Master stock solutions were aliquoted and stored at −20°C until the day of experiments when substock solutions were prepared at designated concentrations. Table 11.2 shows compound concentrations and solvents used for master stock solutions.

11.2.3 Compound Treatment

Zebrafish organs form at different stages. To ensure comparability for assessing drug-induced lethality and acute toxicity, central nervous system and heart were assessed after treating 2dpf zebrafish for 24 h and liver and kidney were assessed after treating 4dpf zebrafish for 24 h. For each compound concentration, 10 zebrafish were distributed into six-well plates containing 3 mL sterile fish water. For compounds

Table 11.1 Mechanism of Action and Potential Therapeutic Indication for Five Test Compounds

Compound	Mechanism of action	Indication
17-DMAG	Binds HSP90 and destabilizes HSP90–oncoprotein complex	Lymphoma
HDAC inhibitor X	HDAC inhibitor	Lymphoma
Paclitaxel	Promotes assembly and prevents depolymerization of microtubules	Breast cancer
SMA-838	Transcriptional inhibitor—upregulates p53 and downregulates p21 and survivin	Leukemia
Zebularine	DNA methyltransferase inhibitor	Leukemia

Table 11.2 Final Compound Concentrations for Master Stock Solutions

Compound	MW	Concentration (mM)	Solvent
17-DMAG	653	10.8	Fish water
HDAC inhibitor X	540.7	46.2	DMSO
Paclitaxel	853.9	50.0	DMSO
SMA-838	375.8	55.6	Fish water
Zebularine	228	500.0	Fish water

dissolved in DMSO, 3 µL of compound substock solution was added directly to fish water and final DMSO concentration was 0.4%. For stock solutions prepared in fish water, 12 µL DMSO was added and final concentration was 0.4%. During early embryogenesis, a protective chorion membrane, which may interfere with compound uptake, is present. Therefore, to facilitate drug delivery, 2dpf zebrafish were dechorionated using a mild protease solution (0.5 mg/mL for ~1.5 min at room temperature). Two or four dpf zebrafish were then incubated with compounds for 24 h.

For each experiment, untreated zebrafish were used as assay control and 0.4% DMSO-treated zebrafish were used as carrier control. If >10% of embryos in the untreated control group were dead or malformed, the experiment was considered invalid and aborted.

11.2.4 Lethality Assessment

To assess lethality, 10 concentrations were initially tested: 0.01, 0.5, 0.1, 0.5, 1, 5, 10, 50, 100, and 400 µM. To establish lethality curves, after treatment for 24 h, surviving zebrafish were counted to determine number of dead zebrafish; dead zebrafish can disintegrate, impeding counting. Experiments were performed three times. LC_{50} and LC_{10} were calculated using logistic regression analysis (JMP software, v7.0, The SAS Institute, Cary, NC). If no lethality was observed up to 400 µM, additional concentrations up to 10,000 µM were assessed.

11.2.5 Assessment of Organ Morphology/Function Toxicity After Treatment for 24 h

Six concentrations $\leq LC_{10}$, determined in lethality studies, were used to assess organ-specific toxicity. Compound treatment was performed as described above. Thirty micromolar celecoxib (Celebrex) was used as a positive control to assess cardiotoxicity; 7.5 and 10 µM brefeldin A were used as positive controls to assess CNS and liver toxicity, respectively, and 3% ethanol was used as a positive control to assess kidney toxicity. Controls were assessed concurrently with test compounds. After heart rate and circulation were assessed zebrafish were anesthetized with MESAB (0.5 mM 3-aminobenzoic acid ethyl ester, 2 mM Na_2HPO_4) and visually assessed using a Zeiss

Stemi 2000C stereomicroscope (Zeiss, Thornwood, NY) equipped with a SPOT Insight digital camera (MVI, Avon, MA). Three percent methylcellulose (Sigma, St. Louis, MO) was used to immobilize zebrafish to facilitate imaging.

11.2.6 End Points for Organ Morphology/Function Toxicity

Heart: Abnormal heart rate: tachycardia (increased heart rate), bradycardia (decreased heart rate), arrhythmia; abnormal circulation, pericardial edema, abnormal heart chamber morphology, and stagnant blood flow, which can lead to thrombosis.
CNS: Misshapen brain, brain hemorrhage, or degenerated brain tissue, which is opaque, brown/gray in color.
Liver: Liver size and color. Degenerated liver tissue is opaque, brown/gray. Normal liver is transparent and clear/light brown, and blood flow can be visualized.
Kidney: Fluid accumulation around the kidney, cyst formation, or trunk edema.

11.2.7 Assessment of Cell Death After 24 h Treatment

The same compound treatment strategy described in Section 11.2.3 was used for the cell death assay. Cell death in target organs was assessed using fluorescent acridine orange (AO) dye that penetrates dead cells and labels nucleotides. Zebrafish samples were stained with AO (1 mg/mL) for 1 h, and then washed to remove excess dye. After washing, stained zebrafish were anesthetized with MESAB and dead cells in heart, CNS, liver, and kidney were visually assessed using a fluorescence microscope (Zeiss M2Bio fluorescence microscope) equipped with a rhodamine (red) cube, a FITC (green) filter, and a chilled CCD camera (AxioCam MRm, Zeiss, NY).

11.2.8 Assessment of Reversibility of Morphology/ Function Organ Toxicity

In order to determine if compound-induced acute toxicity was reversible, after treatment for 24 h, compounds were removed by washing and incubating zebrafish in fresh fish water for an additional 48 h, and organ toxicity was reassessed.

11.2.9 NOEC Determination

To determine the concentration that did not induce observable toxicity, for each condition, zebrafish exhibiting toxicity in any of the four organs were counted and expressed as percent incidence. Percent incidence and concentration were then correlated and NOEC was estimated for each drug.

11.3 RESULTS

In initial studies, we determined the lethal concentration of each compound. Next we assessed acute zebrafish organ toxicity of five cancer drug candidates: 17-DMAG, HDAC inhibitor X, paclitaxel, SMA-838, and zebularine. Zebrafish were treated for 24 h with six drug concentrations $\leq LC_{10}$ and effects on heart, central nervous system, liver, and kidney morphology, function, and organ-specific cell death were assessed. Based on these results, we estimated the no observable effect concentration. Next, we assessed reversibility of toxic effects. After treatment for 24 h, drugs were removed by washing zebrafish in fresh fish water and incubating for an additional 48 h. Organ morphology and function were then reassessed. Finally, we compared results in zebrafish with available results in mammalian models and humans.

11.3.1 LC_{10} and LC_{50} Determination

Two and four dpf zebrafish were initially treated for 24 h with compound concentrations ranging from 0.01 to 400 µM. Triplicate experiments were performed using 10 zebrafish per condition. No lethality was observed for any compound at any concentration at either stage. However, paclitaxel precipitated from the fish water and 2 µM was estimated as the highest soluble concentration. During incubation in aqueous solution, SMA-838 changed color to dark orange from light yellow, indicating possible disintegration. Additional 17-DMAG, SMA-838, and zebularine concentrations, including 800, 1200, 1600, 2000, 3000, 4000, 5000, and 10,000 µM, were then assessed in single experiments, due to limited compound availability. For higher drug concentrations, final DMSO concentration was increased to 1.0%. Percent lethality was calculated for each condition and LC_{10} and LC_{50} for 17-DMAG, SMA-838, and zebularine were then estimated using logistic regression analysis (JMP Software, v7.0). Results showed that for 17-DMAG, LC_{10} and LC_{50} at 2dpf were 1664 and 3141 µM, respectively, and at 4dpf were 2000 and 2080 µM, respectively. For HDAC inhibitor X, no lethality was detected at concentrations up to 400 µM. For paclitaxel, no lethality was detected up to the highest soluble concentration, 2 µM. For SMA-838, at 2dpf, no lethality was detected up to 5000 µM, and at 4dpf, LC_{10} and LC_{50} were 2000 and 2528 µM, respectively. For zebularine, no lethality was detected up to 10,000 µM at either 2 or 4dpf stage.

To assess toxicity, six concentrations $\leq LC_{10}$ were selected for each compound (Table 11.3). Since zebularine was not lethal, and no abnormalities were observed at concentrations up to 10,000 µM, per Study Sponsor's instructions, organ-specific toxicity was not assessed.

11.3.2 Morphology/Function Organ Toxicity After Compound Treatment for 24 h

Organ morphology and function were assessed after compound treatment for 24 h at 2 or 4dpf stage. For each experiment, in addition to compound-treated zebrafish,

Table 11.3   Concentrations Used to Assess Organ Toxicity

Compound	Concentrations (µM)
17-DMAG	10, 50, 100, 500, 1000, 1500
HDAC inhibitor X	10, 50, 100, 200, 300, 400
Paclitaxel[a]	0.01, 0.05, 0.1, 0.5, 1.0, 2.0
SMA-838	300, 600, 900, 1250, 1500, 2000
Zebularine	Not tested per Study Sponsor's instructions

[a] Since 2 µM was the highest soluble paclitaxel concentration, higher concentrations were not tested.

untreated, carrier-treated, and positive compound-treated zebrafish were assessed concurrently. Results for assessing toxic effects on heart, CNS, liver, and kidney are shown in Fig. 11.2.

In order to reduce repetition, images of control-treated zebrafish are shown separately (Fig. 11.1). No difference was observed for untreated zebrafish (fish water), used as assay control, and 0.4% DMSO-treated zebrafish, used as carrier control, indicating 0.4% DMSO did not cause adverse effects. Celebrex (30 µM), a positive control for cardiotoxicity, caused slow heart rate, arrhythmia, and pericardial edema. 7.5 and 10 µM brefeldin A were used as positive controls for assessing CNS and liver toxicity, respectively. Treatment with brefeldin A caused CNS and liver discoloration, indicating tissue degeneration. Ethanol (3%), which was used as a positive control for assessing kidney toxicity, caused trunk edema, indicating potential dysfunction.

Organ toxicity results are shown in Fig. 11.2. 17-DMAG caused toxicity in all four organs: heart, CNS, liver, and kidney. Reduced heart rate and pericardial edema were also observed. Tissue discoloration was observed in brain, indicating degeneration. Small, discolored liver and hemorrhage were observed. Trunk edema indicating potential defective kidney function was observed. HDAC inhibitor X caused slow heart rate, slow circulation, and pericardial edema. However, toxicity was not observed in the other three organs. At concentrations up to 2 µM, the highest soluble concentration, paclitaxel did not cause visible organ toxicity. SMA-838 caused liver toxicity, including small size and discoloration.

Results for visual assessment of organ toxicity in live zebrafish after treatment for 24 h are summarized in Table 11.4.

11.3.3 Cell Death Assessed After 24 h Compound Treatment

Using the same treatment conditions for visually assessing organ toxicity, we next assessed cell death. Two or four dpf zebrafish were treated with each drug for 24 h and cell death in heart, CNS, liver, and kidney was assessed. Compound-induced cell death in each organ was assessed visually by staining with fluorescent acridine

Heart and CNS   **Liver and kidney**

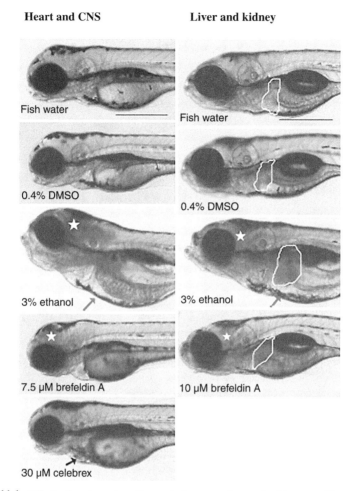

Figure 11.1 Effects of carrier and positive control compounds on zebrafish organs. Untreated (fish water), and carrier- and positive control-treated 2 or 4dpf zebrafish after 24 h exposure at 8× magnification. (Left panels, Heart and CNS) 0.4% DMSO did not affect zebrafish organs. Three % ethanol caused CNS discoloration (white star) and trunk edema (red arrow). 7.5 μM brefeldin A caused CNS tissue discoloration (white star). Thirty μm Celebrex caused pericardial edema (black arrow). Scale bar = 0.5 mm. (Right panels, Liver and kidney) 0.4% DMSO did not affect zebrafish organs. Three % ethanol caused CNS discoloration (white star), trunk edema (red arrow), and liver discoloration (yellow outline). Ten μm brefeldin A caused CNS discoloration (white star) and liver malformation and discoloration (yellow outline). Scale bar = 0.5 mm. (See the color version of this figure in Color Plates section.)

orange and results are shown in Fig. 11.3. After treatment with 17-DMAG, cell death was observed in heart, CNS, and liver. After treatment with HDAC inhibitor X, cell death was observed in CNS and liver. After treatment with paclitaxel, cell death was observed in CNS. After treatment with SMA-838, cell death was observed in CNS and liver.

Heart and CNS   **Liver and kidney**

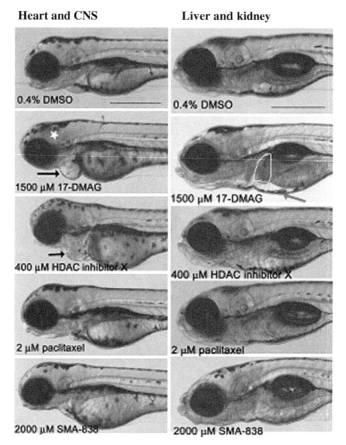

Figure 11.2   Compound-induced toxicity in zebrafish heart, CNS, liver, and kidney after 24 h treatment. Control- and compound-treated zebrafish at 8× magnification. Heart and CNS (left panels 3dpf): 17-DMAG caused CNS toxicity (white star) and pericardial edema (black arrow). HDAC inhibitor X caused pericardial edema (black arrow). Paclitaxel and SMA-838 caused no toxicity in heart or CNS. Liver and kidney (right panels 5dpf): 17-DMAG caused liver hemorrhage (yellow outline) and trunk edema (red arrow). HDAC inhibitor X and paclitaxel caused no discernable effects. Seventy percent of SMA-838-treated zebrafish appeared normal. Scale bar = 0.5 mm. (See the color version of this figure in Color Plates section.)

Table 11.4   Organ Toxicity After 24 h Treatment (Concentration)

Compound	Cardiotoxicity	CNS toxicity	Liver toxicity	Kidney toxicity
17-DMAG	+ (1500 μM)	+ (1500 μM)	+ (1500 μM)	+ (1500 μM)
HDAC inhibitor X	+ (400 μM)	−	−	−
Paclitaxel[a]	−	−	−	−
SMA-838	−	−	+ (2000 μM)	−

"+": toxicity was observed; "−", no toxicity was observed. The value given in parentheses indicates concentration at which toxicity was initially observed, which was also the highest concentration tested for each compound.

[a] Tested up to 2 μM, the highest soluble concentration.

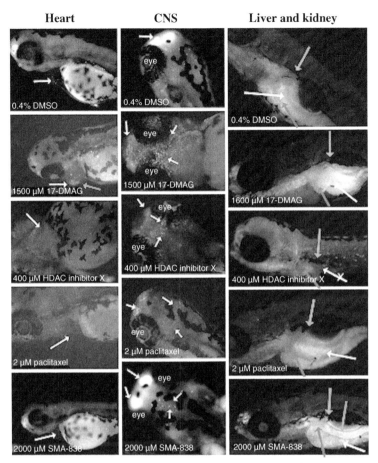

Figure 11.3 Cell death assessed in heart and CNS and liver and kidney in 3 and 5dpf zebrafish, respectively. Note: (1) autofluorescence, particularly in the yolk (white arrows in right panels) and nasal placodes (white arrows in middle panels), was observed during exposure fluorescence light. (2) To highlight the staining pattern observed in different organs, magnification varied from 4× to 10×. (3) Heart effects (left panels): white arrows show heart location. 17-DMAG induced cell death in the heart (red arrow). HDAC inhibitor X, paclitaxel, and SMA-838 did not cause cell death in heart tissue. (4) CNS effects (middle panels): 17-DMAG, HDAC inhibitor X, paclitaxel, and SMA-838 all induced cell death in brain (yellow arrows). (5) Liver effects (right panels): 17-DMAG induced cell death in liver (blue arrows show liver location). HDAC inhibitor X, paclitaxel, and SMA-838 did not induce cell death in liver. However, GI tract staining (red arrow) was observed after treatment with 2000 μM of SMA-838. (6) Kidney effects (right panels, green arrows), cell death was not observed for any test compound. (See the color version of this figure in Color Plates section.)

11.3.4 Evaluation of Reversibility of Toxicity 48 h After Compound Removal

In order to evaluate if compound-induced acute toxicity was reversible, after treatment for 24 h, zebrafish were washed in fresh fish water to remove compounds and incubated for an additional 48 h. For 17-DMAG, although no lethality was observed after treatment with any concentration for 24 h, at concentrations $\geq 500\,\mu M$, 80–100% lethality was observed 48 h after compound removal. Organ morphology and function were then reassessed visually in the surviving animals (Fig. 11.4). Although pericardial edema was still present, degenerated brain tissue was not observed. In addition, liver toxicity and defective kidney function were still present 48 h after compound removal. For HDAC inhibitor X, although no lethality was observed after treatment with any concentration for 24 h, 48 h after removing 400 μM concentration, 30% lethality was observed. In addition, 48 h after removing HDAC inhibitor X, pericardial edema and CNS, liver, and kidney toxicity were visible in surviving animals, indicating delayed drug effects. Forty-eight h after removing paclitaxel, no toxicity was detectable in any organ. Forty-eight h after removing SMA-838, liver toxicity was still detectable. These results suggest that drug-induced toxicity for 17-DMAG, HDAC inhibitor X, and SMA-838 was irreversible. In fact, delayed lethality and toxicity were observed, suggesting that for some drugs treatment longer than 24 h may be required to induce toxicity. Organ toxicity results by compound are summarized below.

11.3.4.1   17-DMAG   Toxicity in all four organs, heart, CNS, liver, and kidney, was observed both after 24 h treatment and 48 h after compound removal, indicating irreversible toxicity. After 24 h treatment, cell death was observed in heart, CNS, and liver, but not in kidney.

11.3.4.2   HDAC Inhibitor X   Heart toxicity was observed after 24 h treatment and 48 h after compound removal, indicating heart toxicity was irreversible. CNS, liver, and kidney toxicity was not observed after 24 h treatment; however, toxicity in these three organs was observed 48 h after compound removal, indicating delayed toxicity. Cell death was observed in CNS, but not in heart, liver, or kidney after 24 h treatment.

11.3.4.3   Paclitaxel   Toxicity was not observed in any organ either after 24 h treatment or 48 h after compound removal. However, low solubility (up to 2 μM) in fish water may have limited assessment of organ toxicity. Cell death was observed in CNS, but not in heart, liver, or kidney after 24 h treatment.

11.3.4.4   SMA-838   Toxicity was not observed in heart, CNS, or kidney, either after 24 h treatment or 48 h after compound removal. However, liver toxicity was observed both after 24 h treatment and 48 h after compound removal, indicating

Heart and CNS   **Liver and kidney**

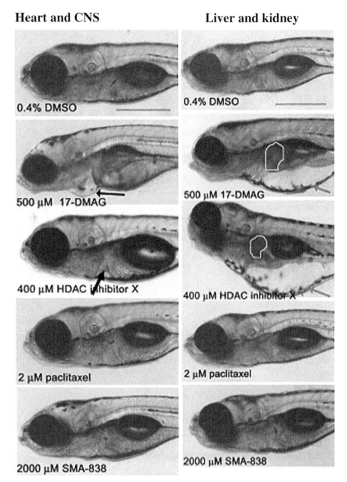

Figure 11.4 Toxicity in zebrafish heart, CNS, liver, and kidney 48 h after compound removal. Heart and CNS (left panels): 48 h after removing 17-DMAG and HDAC inhibitor X, pericardial edema (black arrow) was still visible. No toxicity in heart or CNS was observed 48 h after removing paclitaxel and SMA-838. Liver and kidney (right panels): 48 h after removing 17-DMAG, zebrafish exhibited discolored liver (yellow outlines) and trunk edema (red arrows); 48 h after removing HDAC inhibitor X, liver discoloration was still present, and trunk edema was observed; 48 h after removing paclitaxel, zebrafish appeared normal; 48 h after removing SMA-8383, 40% of animals exhibited discolored liver (yellow outline). Scale bar = 0.5 mm. (See the color version of this figure in Color Plates section.)

effects were irreversible. After 24 h treatment, cell death was observed in CNS and liver, but not in heart or kidney.

11.3.4.5 Zebularine   Based on the absence of toxicity observed during lethality assessment and limited compound availability, organ toxicity and cell death were not evaluated separately.

11.3.5 No Observable Effect Concentration

To determine the concentration that did not cause observable effects, we generated percent incidence curves by determining percent incidence for each drug concentration as described in Section 11.2. Based on these curves, we estimated NOEC as 10 µM (17-DMAG), 10 µM (HDAC inhibitor X), 2 µM (paclitaxel), and 300 µM (SMA-838).

Table 11.5 Comparison of Toxicity of 17-DMAG in Zebrafish, Mammals, and Humans

Toxicity	Zebrafish	Rat	Dog	Human
Heart	Nonreversible pericardial edema and bradycardia at 100 µM Confirmed with cell death assay	No toxic effects observed	No toxic effects observed	Heart block, atrial fib/flutter, atrial dysrhythmia, QTc prolongation, bradycardia
CNS	Nonreversible tissue degeneration at 500 µM Confirmed with cell death assay	No toxic effects observed	No toxic effects observed	Neuropathy
Liver	Nonreversible liver degeneration at 50 µM; hemorrhage at 1500 µM Confirmed with cell death assay	↑ ALT, AST, ALP, TBA at lethal dose (5 mg/kg; $C_{max} = 2$ µM) Liver necrosis at 5 mg/kg	↑ ALT, AST, ALP, GGT at nonlethal dose (0.75 mg/kg; $C_{max} = 0.3$–0.4 µM) Gall bladder and liver necrosis at lethal dose (1.5 mg/kg)	↑ ALT, AST
Kidney	Moderate trunk edema at 100 µM No cell death observed	No toxic effects observed	No toxic effects observed	↑ K, ↓ Na

11.3.6 Comparison of Toxicity in Zebrafish and Mammals

For all four compounds, we then compared results in zebrafish with results for mammalian models. For 17-DMAG and paclitaxel, we also compared results in zebrafish with results available for humans (Tables 11.5–11.8).

11.3.6.1   17-DMAG CNS, heart, liver, and kidney toxicity was observed in zebrafish as well as in human clinical trials after treatment with 17-DMAG. In comparison, in rat and dog studies, liver-specific toxicity was observed (Table 11.5), and no toxicity was reported in other organs.

11.3.6.2   HDAC Inhibitor X Heart, CNS, liver, and kidney toxicity was observed in zebrafish after treatment with HDAC inhibitor X. In dogs, heart and liver toxicity was also identified. However, in mice, only heart toxicity was identified. No human data were available (Table 11.6).

11.3.6.3   Paclitaxel Due to low solubility, paclitaxel was only assessed up to $2 \, \mu M$ concentration, which induced CNS toxicity. Human clinical trial data also

Table 11.6 Comparison of HDAC Inhibitor X Toxicity in Zebrafish and Mammals

Toxicity	Zebrafish	Mouse	Dog
Heart	Nonreversible minor arrhythmia at 50 mM	↑ CK, LDH at $\geq 2 \, \text{mg/kg}$; $C_{max} = ?$)	↑ CK, LDH at 2 mg/kg; $C_{max} = 1\text{--}3 \, \text{mM}$)
	↓ Heart rate, slow circulation, pericardial edema at 400 mM	Inflammation necrosis	Atrial hemorrhage
	No cell death observed		
CNS	Delayed degeneration at 300 mM	No toxic effects observed	No toxic effects observed
	Confirmed with cell death assay		
Liver	Delayed degeneration at 10 mM	No toxic effects observed	↑ ALT, AST, ALP, LDH
	Confirmed with cell death assay		
Kidney	Delayed trunk edema at 50 mM	No toxic effects observed	No toxic effects observed
	No cell death observed		

Table 11.7   Comparison of Paclitaxel Toxicity in Zebrafish, Mammals, and Humans

Toxicity	Zebrafish	Rat	Human
Heart	No toxic effects observed	No toxic effects observed	Rare: atrial fib, dysrhythmia, congestive heart failure, tachycardia
CNS	Cell death observed at 2 mM (highest dose tested)	No toxic effects observed	Common: neuropathy Rare: seizure/grand mal seizure
Liver	No toxic effects observed	No toxic effects observed	No toxic effects observed
Kidney	No toxic effects observed	No toxic effects observed	No toxic effects observed

reported CNS toxicity. In contrast, in rats, no toxicity was identified in the four organs (Table 11.7).

11.3.6.4   SMA-838   SMA-838 caused toxicity in zebrafish CNS and liver. In rat, no toxicity was observed in any of the four organs. Dog studies identified toxicity in liver and kidney. No human clinical trial data were available (Table 11.8).

Table 11.8   Comparison of SMA-838 Toxicity in Zebrafish and Mammals

Toxicity	Zebrafish	Rat	Dog
Heart	No toxic effects observed	No toxic effects observed	No toxic effects observed
CNS	Cell death observed at 2 μM	No toxic effects observed	No toxic effects observed
Liver	Nonreversible degeneration at 1.25 μM Confirmed with cell death assay at 2 μM	No toxic effects observed	↑ ALT, AST, ALP, GGT observed at lethal (4 mg/kg; $C_{max} = 6\,\mu M$) and nonlethal (3 mg/kg; $C_{max} = 2\,\mu M$) doses Minimal to moderate necrosis
Kidney	No toxic effects observed	No toxic effects observed	↑ BUN, creatinine, phosphorus observed at lethal dose

11.3.6.5   Zebularine   After treatment with zebularine, similar to results in rat and dog preclinical studies, no toxicity was observed in zebrafish. No human clinical trial data were available.

11.4 CONCLUSIONS

Results for this small, targeted study to assess acute drug-induced organ toxicity showed good correlation between results in mammals and humans. Furthermore, toxicity induced by 17-DMAG, HDAC inhibitor X, and SMA838 was irreversible. Advantages of using zebrafish for toxicity testing include small amount of drug required, easy drug treatment, short treatment time, visual assessment of abnormalities in transparent animals without surgery, statistically significant number of animals per test, and low cost.

REFERENCE

WESTERFIELD M (1993). The Zebrafish Book: A Guide for the Laboratory Use of Zebrafish. University of Oregon Press, Eugene, OR.

Chapter 12

Locomotion and Behavioral Toxicity in Larval Zebrafish: Background, Methods, and Data*

Robert C. MacPhail[1], Deborah L. Hunter[2], Terra D. Irons[3], and Stephanie Padilla[2]

[1]*Toxicity Assessment Division, National Health and Environmental Effects Research Laboratory, Office of Research and Development, U.S. Environmental Protection Agency, Research Triangle Park, NC, USA*
[2]*Integrated Systems Toxicology Division, National Health and Environmental Effects Research Laboratory, Office of Research and Development, U.S. Environmental Protection Agency, Research Triangle Park, NC, USA*
[3]*Curriculum in Toxicology, University of North Carolina, Chapel Hill, NC, USA*

12.1 INTRODUCTION

This chapter focuses on evaluating the behavior of very young (larval) zebrafish (*Danio rerio*) and the effects of chemical exposures. The study of behavior in animals, including humans, has always been popular, even to the casual observer. In everyday life, we observe the behavior of all those we encounter, including pets. Our familiarity with behavior has often proven to be an impediment, however, to creating a scientific understanding of behavior (Kelleher and Morse, 1968) that is stripped of anecdotes and casual observations that characterize everyday descriptions.

---

* This manuscript has been reviewed by the National Health and Environmental Effects Research Laboratory and approved for publication. Approval does not signify that the contents reflect the views of the Agency, nor does mention of trade names or commercial products constitute endorsement or recommendation for use.

Zebrafish: Methods for Assessing Drug Safety and Toxicity, Edited by Patricia McGrath.
© 2012 John Wiley & Sons, Inc. Published 2012 by John Wiley & Sons, Inc.

12.2 BACKGROUND

12.2.1 History

The modern science of behavior can be traced to around the turn of the twentieth century, when researchers began objectively to study the behavior of animals under laboratory conditions. Impressions and narratives about behavior were replaced with numbers, experimental conditions were controlled, and variables were identified and manipulated leading to repeatable demonstrations of behavior. Automation of behavioral testing equipment contributed to scientific advances because, in a sense, it removed the experimenter from the experiment, and prevented any inadvertent influence on the results due to his or her expectations of the outcome. Automation and commercialization also increased the breadth of the science by permitting investigation of the same behavioral phenomena in multiple laboratories.

The behavior of any organism is unitary in name only. Several basic processes comprise the behavioral repertoire of all organisms, and countless studies have shown the essential conservation of these processes across phyla and species (e.g., Kelleher and Morse, 1968; Brembs et al., 2002; Greenspan, 2007). There are many schemes for categorizing behavior. The simplest scheme distinguishes naturally occurring and acquired behavior. Naturally occurring behavior emerges during development, is common to a species, and requires no explicit training for its occurrence. Reflexes and locomotion are good examples of naturally occurring behavior. Acquired behavior, on the other hand, refers to what is ordinarily called learned behavior because it emerges with training and experience. Acquired behavior comprises the vast majority of the repertoire of mammals and several other species, and is considered to result from cognitive processes originating in the nervous system. Even such a simple scheme is, however, somewhat misleading because it implies that learned behavior is somehow not a natural occurrence, when, in fact, all animals learn. It also ignores the fact that under appropriate conditions reflexes and locomotion can be learned (i.e., modified through experience).

12.2.2 Development and Toxicity

The commonality of behavioral processes among species is also seen during development, although the time frames differ widely. Infancy can last for hours to years, depending on the species. Regardless of the species, the development of behavior proceeds through well-defined stages involving the emergence of reflexes, locomotion, sensory competence, neuromuscular coordination and strength, and learning and memory. The nervous system also proceeds developmentally through well-defined stages including cellular proliferation, differentiation, migration, synapse formation, and myelination. Proper development of the nervous system and behavior ensures that an organism is able to adapt to its changing environment, and consequently grow and survive at least long enough to have the opportunity to reproduce. The almost clockwork regularities of behavioral and nervous system development, and their commonalities across species, remain one of the greatest marvels of biology and evolution.

Concerns over the adverse impact of chemical exposures on human development intensified following the tragedies caused by thalidomide, methylmercury, and ethanol in the mid-1900s. The developmental toxicity of lead was also provoking suspicion at this time, but it would take years for unequivocal proof (Needleman et al., 1979). Concerns over the adverse impact of chemicals on wildlife development intensified following the decline in American eagle and brown pelican populations due to widespread use of the insecticide DDT (Grier, 1982; Stickel et al., 1996). Partly as a result, testing guidelines were created by regulatory agencies for assessing the potential impact of toxicants on human development and on wildlife. For example, the U.S. Environmental Protection Agency (U.S. EPA) promulgated guidelines for evaluating the potential neurotoxicity for humans resulting from developmental exposures to commercial chemicals (U.S. EPA, 1985) that were subsequently modified and expanded to cover pesticides (U.S. EPA, 1991). The U.S. EPA also promulgated guidelines for evaluating the ecological impact of pesticides (U.S. EPA, 1998), which included specific tests for developmental toxicity. In addition, the Association for Testing Materials published guidelines for evaluating behavioral toxicity in fishes, amphibians, and macroinvertebrates (ASTM, 1994), and one guideline specifically describing behavioral toxicity testing in fishes (ASTM, 1995). These guidelines were, however, prepared separately, with little appreciation of the potential benefits of employing tests that could be used in both human health and ecological toxicity testing. It is significant that recent developments, especially in the use of fishes in toxicity testing, have begun to erode the barriers that have long divided these efforts.

12.3 LOCOMOTION

Of all the types of behavior that organisms display, locomotion is probably one of the most ancestral and biologically important. All nonsessile organisms must locomote in order to obtain food and shelter, mate, and avoid predators. Locomotion is also readily measured in the laboratory in a variety of species using a number of devices (Reiter and MacPhail, 1979, 1982; Ossenkopp et al., 1996). Given its availability and ease of recording, it is no wonder locomotion is a bedrock test of behavior that is used in toxicology as well as in pharmacology, neurobiology, genetics, and ethology. Tests of locomotor activity are now used routinely for screening chemicals for neurotoxicity, including developmental neurotoxicity. In this regard, the tests have undergone extensive evaluation for reliability and validation as an indicator of damage to the nervous system (MacPhail et al., 1989; Crofton et al., 1991; Moser et al., 1997).

This chapter focuses on the use of locomotor activity during larval development in zebrafish for screening environmental chemicals. A growing body of research has also introduced rapid screening methods for identifying developmental toxicants and prioritizing them for in-depth evaluation of their toxic potential. This area of research has seen broad support both in toxicology and in drug development in the pharmaceutical industry, but with different purposes. We have focused on developing screening methods for identifying and prioritizing the types of environmental chemicals of concern for the U.S. EPA in estimating risks to the human population.

12.4 ZEBRAFISH MODELS

Zebrafish have become a popular organism in ecological research (Magalhaes et al., 2007; Gunnarsson et al., 2008; Scholz et al., 2008; Seok et al., 2008), in addition to their increasing use as a model organism in toxicological, pharmacological, and biomedical research (reviewed in Hill et al., 2005; Rubinstein, 2006; Kari et al., 2007; Barros et al., 2008; Peterson et al., 2008; Scholz et al., 2008). There are many advantages of working with zebrafish, especially during development. Large stocks of fish can be maintained due to their small size and relatively simple husbandry requirements. Each mating of zebrafish can produce potentially hundreds of eggs, which can be raised in the small wells of microtiter plates. The embryos and larvae are also transparent, allowing detailed observation of their internal development. Development progresses rapidly, making the study of large numbers of chemicals practicable. In addition, direct exposure of the embryos eliminates the potential confounding influence of maternal toxicity that can occur with mammalian models. Finally, selective breeding and gene manipulation techniques have provided a broad array of strains for probing the mechanisms of organ system development and disease (Fetcho and Liu, 1998; Grunwald and Eisen, 2002; Lieschke and Currie, 2007).

The growing use of zebrafish in toxicology and pharmacology has highlighted the need for a firm understanding of their behavior. Behavioral studies with developing zebrafish may seem to pose greater difficulties than those encountered with adult fish, due to their small size and rapid development. It is these very properties, however, that make the larvae an attractive alternative model for assessing developmental toxicity in a screening context. For example, individual zebrafish can be monitored in the small wells of microtiter plates that are standard issue in biochemistry laboratories using miniaturized assays. Considerable research is now available on the effects of chemicals on zebrafish raised and tested in 96-well plates. Advances in studying zebrafish behavior have also benefited from automated testing procedures for studying locomotion, reflexes, sensory capacity, circadian rhythms, and learning and memory (see Fetcho and Liu, 1998; Orger et al., 2004; Lieschke and Currie, 2007; Levin and Cerutti, 2008; MacPhail et al., 2009 for references). Most of these tests rely on optical recording of behavior, followed by analysis using commercial data software programs.

Despite the rise in popularity of behavioral studies on zebrafish, a cautionary note is in order. Behavior occupies a unique niche in biology, including toxicology. Damage to internal organs can result in behavioral change, but so can the environmental conditions in which the behavior occurs. The main function of behavior is to allow an organism to adapt to the ever-changing conditions of its environment. A failure to understand and appreciate the *mutual* influence of the intrinsic (i.e., organ system) and the extrinsic (i.e., environmental) forces on behavior can lead to incomplete accounts and faulty interpretations of both well-being and toxicity (Morse, 1975; MacPhail, 1990). For this reason, we detail methods for assessing the locomotion of developing zebrafish larvae and the effects of chemical exposures.

12.5 ANALYZING LARVAL LOCOMOTION

12.5.1 Methods

In these studies, each embryo was placed in one of the 96 wells of a microtiter plate. Testing began at 6 days post fertilization (dpf) when the larvae were swimming yet still feeding off their yolk sacs. Larvae were maintained in an incubator (26°C with a 14:10 h light/dark cycle) until the morning of testing, at which time they were transferred to the (darkened) testing room (ambient temperature 26°C), and placed in a light-tight drawer for several hours. Early transfer to the testing room minimized any disturbance that might be associated with handling and movement of the plate. Just before testing, the plate was carefully placed on a platform (light box) that could provide both infrared and visible light (from light-emitting diodes) (Noldus Information Technology, Leesburg, VA). The light box provided infrared (800–950 nm, with a peak at 860 nm) or visible (430–700 nm) light as measured with a wideband spectroradiometer. A camera was placed above the platform to record simultaneously locomotion from each of the wells. A baffle was also lowered over the platform to prevent extraneous light from impinging on the plate and interfering with activity. Recordings of locomotion were stored on DVDs as MPEG-2 files, and later decoded using commercial software (EthoVision version 3.1), from which locomotion was calculated as distance moved per unit of time. The testing equipment allowed recording of locomotion in both visible light and infrared light, which is considered darkness because zebrafish do not see in the infrared region of the spectrum. As elaborated below, locomotor activity is remarkably sensitive to lighting conditions and light–dark transitions; this observation became a central feature in the paradigms for assessing chemical substances.

12.5.2 Variables Affecting Larval Locomotion

Preliminary experiments were conducted to determine a basic protocol for screening chemicals. These experiments are the primary focus of this chapter. They included determining whether well location had an effect on locomotion that might be due to differences in optical resolution of the camera or lighting intensity at the perimeter, or possibly due to differences in activity in larvae housed in the perimeter wells not being "surrounded" by conspecifics. An analysis of plates containing control larvae indicated rare instances when activity due to row or column was statistically significant (see MacPhail et al., 2009). These rare instances were considered biologically insignificant, but subsequent dosing studies arranged concentrations diagonally on the plate to minimize any potential contribution of well location.

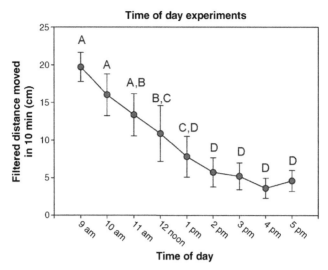

Figure 12.1   Effect of time of day on locomotor activity in larval zebrafish. Larvae ($n = 59$) were tested repeatedly for 10 min every hour in dark (infrared illumination). Values represent mean $\pm$ SEM distance moved in cm. Letters represent statistical results (one-way repeated-measures ANOVA followed by Fisher's PLSD test); different letters denote statistically significant differences in activity. Locomotor activity was highest early in the morning and then decreased to a stable low level by early afternoon.

We also determined the influence of the time of day when testing took place. This study recorded the activity of individual larva for 10 min every hour from mid-morning until late afternoon. Figure 12.1 shows the results. Activity was highest when tested in morning hours, and then decreased to a low and stable level in the afternoon. These results are consistent with earlier findings (MacPhail et al., 2009). It is intriguing to note that visual thresholds for zebrafish are also lowest in the afternoon hours (Li and Dowling, 1998); this observation may lead to a promising future research direction.

We next determined the effect of changes in lighting on locomotion. Testing always began in darkness (infrared light) to allow any disturbances to dissipate from transferring the plate to the recording platform. Switching from dark to light and back to dark had a substantial effect on activity. Activity was low in light, and then reached a high level shortly after the larvae were returned to darkness. These results are shown in Fig. 12.2. This pattern proved to be highly reproducible. It also seemed paradoxical at first, since zebrafish are diurnal organisms that are most active during daylight and quiescent at night (e.g., Cahill et al., 1998; Hurd et al., 1998). Studies have shown, however, that changes in lighting can completely override circadian activity rhythms in zebrafish. Figure 12.2 also highlights an important methodological feature in recording locomotion. The optical recording device can sample activity every 30 ms, and it is the change in location from one sampling period to the next that determined the activity of a larva. With such a high level of resolution, it is possible that many minor changes in the location could be recorded but would not reflect bona fide locomotion in the sense of movement of the whole organism in the

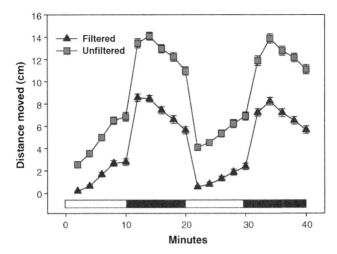

Figure 12.2   Effect of lighting conditions on activity. Larvae ($n = 164$) were tested under alternating 10 min periods of light and dark. The bar in the lower portion of the figure denotes lighting condition. Activity is presented as unfiltered and filtered data; filtering (see the text for details) eliminated minor movements that did not qualify as locomotion to a trained observer. Each point represents mean ± SEM distance moved in 2 min. Filtering the data reduced the magnitude of activity, but did not distort the pattern of activity over time or lighting condition.

well. For this reason, a sampling rate of 200 ms and a minimum distance of 0.135 cm were used to measure locomotion, which agreed with visual observation of the larvae. The data collected with this constraint can therefore be technically referred to as filtered data. Figure 12.2 shows a close correspondence between filtered and unfiltered data, although filtering results in lower activity as expected. While Fig. 12.2 indicates that similar results can be obtained regardless of filtering, it is likely that some chemical exposures could produce divergent results owing to minute displacements similar to those produced by tremor or shaking in studies on locomotor activity in rodents.

An obvious question concerned the importance of the length of the light and dark periods. The results of these experiments have been published (MacPhail et al., 2009). In brief, duration of the initial dark period (either 10 or 20 min) had no effect on either the level of activity in subsequent light or the increase in activity when the larvae were returned to dark. Duration of the light period did, however, affect subsequent dark activity, being higher following 15 min of light than following 5 min of light.

Given the substantial effect on activity of switching between light and dark, especially the rapid initial activity decrease in light, further studies investigated the influence of light intensity on locomotion. Accordingly, the apparatus was modified electronically to allow manipulation of light intensity over a broad range of values. Figure 12.3 shows the results of one experiment. The bright light level is the same as used in the experiments described above, while the lower intensity is noticeably dim to an observer. Under both lighting conditions, activity increased to an equivalent stable level, although that level was reached faster at the lower intensity. In addition, the

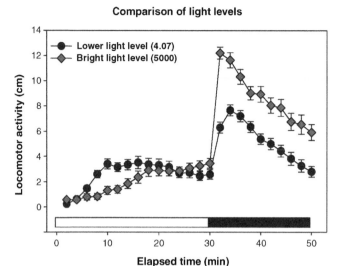

Figure 12.3   Influence of light level on locomotion. Larvae were tested for 30 min in one of the two light levels followed by 20 min in dark (denoted by the bar in the lower portion of the figure). Activity increased to a steady level in light, which was reached earlier under the lower light condition. Switching to dark produced a rapid rise in activity followed by a gradual decline. Dark activity was higher in larvae that had been exposed to the brighter light condition.

increase in activity on return to dark was greater in larvae that were exposed to brighter light. Interestingly, the subsequent decay in dark activity appeared equal under both lighting conditions.

The effect of light intensity on the level of activity during the return to dark resembled the effect produced by varying the duration of a constant intensity of light. These results suggest that it might be possible to investigate visual summation phenomena using the locomotion of larval zebrafish.

As a result of these preliminary studies, we arranged a testing protocol that included (1) an initial period of darkness, (2) a period of light, and (3) a return to darkness. The duration of these periods and the number of cycles (transitions) were left as variables that depended on the particular experiment. In addition, activity during the initial dark period was eliminated from further analysis, since it was considered a period of acclimation, during which the activity was not under the direct control of the variables we intended to manipulate.

12.6 CHEMICAL EFFECTS ON LARVAL LOCOMOTION

Recent experiments have determined the effects of acute drug administration on larval locomotion. The intent of these experiments was to administer compounds that are known to produce effects on the motor activity of laboratory rodents (mice, rats). We were particularly interested in determining whether similar qualitative effects of

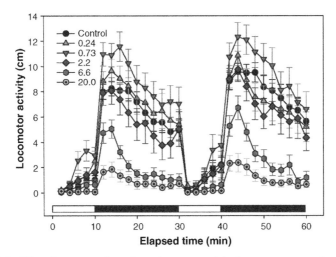

Figure 12.4   Effect of acute D-amphetamine on locomotor activity. Larvae were exposed to 10% Hanks' buffer solution (control, $n = 48$) or a concentration of drug (0.24–20 µM, $n = 22$–24/concentration) for 30 min before recording began. Effects of each drug dose are plotted as mean ± SEM distance moved in 2 min. Ten minutes of light was followed by 20 min of dark for two cycles (denoted by the bar in the lower portion of the figure). At the lower doses, D-amphetamine increased activity in dark; however, at the higher doses activity was decreased. Changes in activity due to drug were considerably muted during the light periods.

the drugs would be produced in the larvae. D-Amphetamine, for example, is a psychomotor stimulant that produces a characteristic biphasic effect on motor activity in both rats and mice; low doses increase activity while higher doses decrease it (Campbell et al., 1969; Porrino et al., 1984; Antoniou et al., 1998; Niculescu et al., 2005; Badanich et al., 2008).

In this study, rearing was carried out as previously described until the afternoon of testing. The larvae were exposed to either 10% Hanks' solution (control) or a range of D-amphetamine concentrations (0.24–20.0 µM), which were evenly distributed (diagonally) throughout the plate. The plate was then placed on the platform, and recording of activity began 30 min later. Testing included two cycles of 10 min of light and 20 min of dark. Figure 12.4 shows the effect of acute D-amphetamine exposure. The lowest dose produced no effect on activity, while the next higher dose increased activity. Further increases in dose produced decreases in activity. This biphasic drug effect was especially clear for dark activity; the pattern is the same for activity in light, although the magnitude is considerably less. These results are similar to those found in numerous studies on the effects of D-amphetamine on the activity of rats and mice. Moreover, the rapid change in activity with lighting conditions, at all doses, indicates that the drug primarily affected locomotion rather than sensory (visual) function. Additional work in our laboratory (Irons et al., 2010; MacPhail et al., 2009) determined the acute effects of cocaine and ethanol. These drugs also produced effects that resembled those reported for rats and mice. The effects of ethanol were especially intriguing in that locomotion was increased considerably at an intermediate dose in both light and dark periods, indicating an almost complete loss of visual control. Brief dips in activity at the beginning of the light periods, however, suggested some slight residual in visual competence (see Matsui et al., 2006, for ethanol effects on visual thresholds in larval zebrafish).

The results of these studies indicate the remarkable versatility of the behavioral testing paradigm. The effects of several variables on larval locomotion have been briefly summarized. There are more variables, for example, variations in lighting intensity and duration, that will further enrich our understanding of larval locomotion and its determinants.

The ultimate goal of these studies, to reiterate, was to establish a testing procedure for rapidly evaluating developmental exposures to neurotoxic chemicals. To this end, data are presented next for valproate, an anticonvulsant that is toxic to the developing nervous system in mammals, including humans (Bescoby-Chambers et al., 2001; Alsdorf and Wyszynski, 2005; Genton et al., 2006; Magalhaes et al., 2007; Yochum et al., 2008). Larvae were exposed to varying concentrations of valproate from day 1 through day 5 post fertilization. Following one day of washout, the larvae were tested under alternating 10 min periods of light and dark. The results of this study are shown in Fig. 12.5. Results are presented only for larvae that appeared structurally normal; all abnormal embryos were eliminated from the analyses. In particular, larvae receiving 100 µM valproate showed prominent signs of structural abnormalities and developmental delay, and were therefore not included. Figure 12.5 shows that

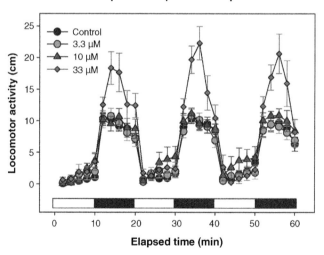

Figure 12.5   Effects of developmental exposure to valproate on locomotor activity. Larvae ($n = 13 - 28$) were exposed to Hanks' buffer solution (control) or a dose of valproate from 8 h post fertilization to 5 days post fertilization. Activity testing occurred on day 6 post fertilization. Each symbol represents mean ± SEM distance moved in 2 min. Lighting conditions alternated, with 10 min of light followed by 10 min of dark for three cycles (denoted by the bar in the lower portion of the figure). Valproate noticeably increased activity in larvae that had been exposed to 33 µM during development. Hyperactivity was, however, obtained only in the dark periods.

the two lowest concentrations of valproate were without effect, while the highest concentration (33 µM) produced robust increases in locomotion in the dark periods only. These results are similar to those reported for the effects of developmental exposure to valproate in mammals, where behavioral alterations were seen in the offspring of both humans and experimental animals (e.g., Vorhees, 1987; Moore et al., 2000). It is interesting to note that hyperactivity is reported for children of mothers taking anticonvulsants (including valproate) during gestation. These data are encouraging as they indicate that the larval zebrafish behavioral assay can detect abnormal behavioral effects at exposure levels below those that produce frank teratogenic effects.

12.7 CONCLUSIONS

12.7.1 Summary

The increasing use of larval zebrafish in toxicology, pharmacology, and biology emphasizes the need for a firm understanding of their behavior and the variables that influence it. Behavior is unique in that it represents the interface between the internal environment of an organism *and* the external environment in which the organism lives. From a practical view, therefore, care must be taken in research to identify and control the environmental variables that influence behavior. Stringent control over the testing environment will enhance reproducibility of results. In addition, systematic manipulation of environmental variables can lead to a fuller account of the effects on behavior of chemical treatments, whether environmental pollutants or drugs, and improve our understanding of the specificity of chemical effects on behavior. The behavioral preparation described in this chapter is versatile in allowing repeated manipulation of key variables such as lighting condition and duration, which can affect the response of larval zebrafish to chemical exposures. It is, however, reasonable to assume that we have only begun to exploit the preparation.

12.7.2 Future Avenues

Further studies are needed on the influence of light intensity on behavior, and its interaction with toxicant effects. The visual system of fish, including larvae, is highly developed and intimately related to behavior and survival. It is conceivable that visual thresholds may be obtained with suitable modification of the testing paradigm. Still other studies could investigate the habituation of behavior. Habituation is a primitive form of learning that appears to be virtually universal throughout the animal kingdom. Considerable information is currently available on the variables that influence habituation, including many drugs, and on its neurobiological substrates. It is likely that tests of habituation in toxicant-exposed zebrafish will increase in the coming years.

12.7.3 A Final Word on Behavioral Screening

Screening for developmental toxicity, or for any adverse health effect, necessarily requires a compromise between the efficiency of a test method and the completeness of the data it can provide. While there may be, perhaps, myriad avenues for future exploitation of the behavioral preparation described in this chapter, it is impossible to say at this time which will be most useful in identifying and prioritizing chemicals for adverse effects on the nervous system. It is likely, however, that the current behavioral preparation will be a valuable addition to screening studies, and, given its versatility, in further characterizing the effects of toxicants on the developing nervous system.

ACKNOWLEDGMENTS

The authors thank Beth Padnos, Brenda Proctor, and Dr. David Kurtz for maintenance and upkeep of the zebrafish colony. T.D. Irons is supported by the following NIH predoctoral traineeships: the NIGMS Initiative for Maximizing Student Diversity and the NIEHS National Research Service Award (T32 ES007126). The authors also thank Drs. Kevin Crofton and William Mundy for reviewing earlier versions of this manuscript.

REFERENCES

ALSDORF R and WYSZYNSKI DF (2005). Teratogenicity of sodium valproate. Expert Opin Drug Saf 4(2): 345–353.

ANTONIOU K, KAFETZOPOULOS E, PAPADOPOULOU-DAIFOTI Z, HYPHANTIS T, and MARSELOS M (1998). D-Amphetamine, cocaine and caffeine: a comparative study of acute effects on locomotor activity and behavioural patterns in rats. Neurosci Biobehav Rev 23(2): 189–196.

ASTM (1994). (E1604-64) Standard E1604-64: Standard Guide for Behavioral Testing in Aquatic Toxicology. ASTM International, West Conshohocken, PA.

ASTM (1995). (E1711-95) Standard E1711-95: Standard Guide for Measurement of Behavior During Fish Toxicity Tests. ASTM International, West Conshohocken, PA.

BADANICH KA, MALDONADO AM, and KIRSTEIN CL (2008). Early adolescents show enhanced acute cocaine-induced locomotor activity in comparison to late adolescent and adult rats. Dev Psychobiol 50(2): 127–133.

BARROS TP, ALDERTON WK, REYNOLDS HM, ROACH AG, and BERGHMANS S (2008). Zebrafish: an emerging technology for *in vivo* pharmacological assessment to identify potential safety liabilities in early drug discovery. Br J Pharmacol 154(7): 1400–1413.

BESCOBY-CHAMBERS N, FORSTER P, and BATES G (2001). Foetal valproate syndrome and autism: additional evidence of an association. Dev Med Child Neurol 43(3): 202–206.

BREMBS B, LORENZETTI FD, REYES FD, BAXTER DA, and BYRNE JH (2002). Operant reward learning in *Aplysia*: neuronal correlates and mechanisms. Science 296(5573): 1706–1709.

CAHILL GM, HURD MW, and BATCHELOR MM (1998). Circadian rhythmicity in the locomotor activity of larval zebrafish. Neuroreport 9(15): 3445–3449.

CAMPBELL BA, LYTLE LD, and FIBIGER HC (1969). Ontogeny of adrenergic arousal and cholinergic inhibitory mechanisms in the rat. Science 166(905): 635–637.

CROFTON KM, HOWARD JL, MOSER VC, GILL MW, REITER LW, TILSON HA, and MACPHAIL RC (1991). Interlaboratory comparison of motor activity experiments: implications for neurotoxicological assessments. Neurotoxicol Teratol 13(6): 599–609.

FETCHO JR and LIU KS (1998). Zebrafish as a model system for studying neuronal circuits and behavior. Ann NY Acad Sci 860: 333–345.

GENTON P, SEMAH F, and TRINKA E (2006). Valproic acid in epilepsy: pregnancy-related issues. Drug Saf 29(1): 1–21.

GREENSPAN RJ (2007). An Introduction to Nervous Systems. Cold Spring Harbor Press, Cold Spring Harbor, NY.

GRIER JW (1982). Ban of DDT and subsequent recovery of reproduction in Bald Eagles. Science 218: 1232–1234.

GRUNWALD DJ and EISEN JS (2002). Headwaters of the zebrafish: emergence of a new model vertebrate. Nat Rev Genet 3(9): 717–724.

GUNNARSSON L, JAUHIAINEN A, KRISTIANSSON E, NERMAN O, and LARSSON DG (2008). Evolutionary conservation of human drug targets in organisms used for environmental risk assessments. Environ Sci Technol 42(15): 5807–5813.

HILL AJ, TERAOKA H, HEIDEMAN W, and PETERSON RE (2005). Zebrafish as a model vertebrate for investigating chemical toxicity. Toxicol Sci 86(1): 6–19.

HURD MW, DEBRUYNE J, STRAUME M, and CAHILL GM (1998). Circadian rhythms of locomotor activity in zebrafish. Physiol Behav 65(3): 465–472.

IRONS TD, MACPHAIL RC, HUNTER DL, and PADILLA S (2010). Acute neuroactive drug exposures alter locomotor activity in larval zebrafish. Neurotoxicol Teratol 32(1): 84–90.

KARI G, RODECK U, and DICKER AP (2007). Zebrafish: an emerging model system for human disease and drug discovery. Clin Pharmacol Ther 82(1): 70–80.

KELLEHER RT and MORSE WH (1968). Determinants of the specificity of behavioral effects of drugs. Ergeb Physiol 60: 1–56.

LEVIN ED and CERUTTI DT (2008). Behavioral neuroscience of zebrafish. In: BUCCAFUSCO JJ (Ed.), Methods of Behavior Analysis in Neuroscience. CRC Press, New York, NY, pp. 291–308.

LI L and DOWLING JE (1998). Zebrafish visual sensitivity is regulated by a circadian clock. Visual Neurosci 15: 851–857.

LIESCHKE GJ and CURRIE PD (2007). Animal models of human disease: zebrafish swim into view. Nat Rev Genet 8(5): 353–367.

MACPHAIL RC (1990). Environmental modulation of neurobehavioral activity. In: RUSSELL RW, FLATTAU PE, and POPE AM (Eds.), Behavioral Measures of Neurotoxicity. National Academy Press, Washington, DC, pp. 347–358.

MACPHAIL RC, BROOKS J, HUNTER DL, PADNOS B, IRONS TD, and PADILLA S (2009). Locomotion in larval zebrafish: influence of time of day, lighting and ethanol. Neurotoxicology 30(1): 52–58.

MACPHAIL RC, PEELE DB, and CROFTON KM (1989). Motor activity and screening for neurotoxicity. J Am Coll Toxicol 8: 117–125.

MAGALHAES D DEP, DA CUNHA RA, DOS SANTOS JAA, BUSS DF, and BAPTISTA DF (2007). Behavioral response of zebrafish *Danio rerio* Hamilton 1822 to sublethal stress by sodium hypochlorite: ecotoxicological assay using an image analysis biomonitoring system. Ecotoxicology 16(5): 417–422.

MATSUI JI, EGANA AL, SPONHOLTZ TR, ADOLPH AR, and DOWLING JE (2006). Effects of ethanol on photoreceptors and visual function in developing zebrafish. Invest Ophthalmol Vis Sci 47(10): 4589–4597.

MOORE S, TURNPENNY P, QUINN A, GLOVER S, LLOYD D, MONTGOMERY T, and DEAN J (2000). A clinical study of 57 children with fetal anticonvulsant syndromes. J Med Genet 37: 489–497.

MORSE WH (1975). Schedule-controlled behaviors as determinants of drug response. Fed Proc 34(9): 1868–1869.

MOSER VC, BECKING GC, CUOMO V, FRANTIK E, KULIG BM, MACPHAIL RC, TILSON HA, WINNEKE G, BRIGHTWELL WS, DE SALVIA MA, GILL MW, HAGGERTY GC, HORNYCHOVA M, LAMMERS J, LARSEN JJ, MCDANIEL KL, NELSON BK, and OSTERGAARD G (1997). The IPCS Collaborative Study on Neurobehavioral Screening Methods. V. Results of chemical testing. Neurotoxicology 18(4): 969–1055.

NEEDLEMAN HL, GUNNOE C, LEVITON A, REED R, PERESIE H, MAHER C, and BARRETT P (1979). Deficits in psychologic and classroom performance of children with elevated dentine lead levels. N Engl J Med 300(13): 689–695.

NICULESCU M, EHRLICH ME, and UNTERWALD EM (2005). Age-specific behavioral responses to psychostimulants in mice. Pharmacol Biochem Behav 82(2): 280–288.

ORGER MB, GAHTAN E, MUTO A, PAGE-MCCAW P, SMEAR MC, and BAIER H (2004). Behavioral screening assays in zebrafish. Methods Cell Biol 77: 53–68.

OSSENKOPP K-P, KAVALEIRS M, and SANBERG PR (1996). Measuring movement and locomotion: perspective and overview. In: Ossenkopp K-P, Kavaliers M, and Sanberg PR (Eds.), Measuring Movement and Locomotion: From Invertebrates to Humans. R.G. Landes Co., Austin, TX, pp. 1–9.

PETERSON RT, NASS R, BOYD WA, FREEDMAN JH, DONG K, and NARAHASHI T (2008). Use of non-mammalian alternative models for neurotoxicological study. Neurotoxicology 29(3): 546–555.

PORRINO LJ, LUCIGNANI G, DOW-EDWARDS D, and SOKOLOFF L (1984). Correlation of dose-dependent effects of acute amphetamine administration on behavior and local cerebral metabolism in rats. Brain Res 307 (1–2): 311–320.

REITER LW and MACPHAIL RC (1979). Motor activity: a survey of methods with potential use in toxicity testing. Neurobehav Toxicol 1 (Suppl 1): 53–66.

REITER LW and MACPHAIL RC (1982). Factors influencing motor activity measurements in neurotoxicology. In: MITCHELL CL (Ed.), Nervous System Toxicology. Raven Press, New York, NY, pp. 45–65.

RUBINSTEIN AL (2006). Zebrafish assays for drug toxicity screening. Expert Opin Drug Metab Toxicol 2(2): 231–240.

SCHOLZ S, FISCHER S, GUNDEL U, KUSTER E, LUCKENBACH T, and VOELKER D (2008). The zebrafish embryo model in environmental risk assessment: applications beyond acute toxicity testing. Environ Sci Pollut Res Int 15(5): 394–404.

SEOK SH, BAEK MW, LEE HY, KIM DJ, NA YR, NOH KJ, PARK SH, LEE HK, LEE BH, and PARK JH (2008). *In vivo* alternative testing with zebrafish in ecotoxicology. J Vet Sci 9(4): 351–357.

STICKEL LF, STICKEL WH, and CHRISTENSEN R (1996). Residues of DDT in brain and bodies of birds that died on dosage and in survivors. Science 151: 1549–1551.

U.S. EPA (1985). Toxic Substances Control Act Test Guidelines, Neurotoxicity Screening Battery (799.9620). Federal Register (50 FR 39252).

U.S. EPA (1991). Pesticide Assessment Guidelines, Subdivision F. Hazard Evaluation: Human and Domestic Animals. Neurotoxicity. EPA 540/09-91-123, Washington, DC.

U.S. EPA (1998). Guidelines for ecological risk assessment. Fed Reg 69(93): 26846–26924.

VORHEES CV (1987). Behavioral teratogenicity of valproic acid: selective effects on behavior after prenatal exposure to rats. Psychopharmacology (Berl) 92(2): 173–179.

YOCHUM CL, DOWLING P, REUHL KR, WAGNER GC, and MING X (2008). VPA-induced apoptosis and behavioral deficits in neonatal mice. Brain Res 1203: 126–132.

Chapter 13

Zebrafish: A Predictive Model for Assessing Seizure Liability

Demian Park, Joshua Meidenbauer, Breanne Sparta, Wen Lin Seng, and Patricia McGrath

Phylonix, Cambridge, MA, USA

13.1 INTRODUCTION

Several marketed drugs have been associated with increased seizure risk and the U.S. Food and Drug Administration (FDA) has included this end point in its Safety Pharmacology Guidance for Industry (ICH, 2001). Using a quantitative video-based motion tracking system, zebrafish has been shown to be a predictive animal model for assessing compound-induced seizure-like movement (Baraban et al., 2005; Winter et al., 2008). In a recent study, using methods similar to those reported by others, we examined movement of 6 day post fertilization (dpf) zebrafish after treatment with nine seizure inducing compounds, pentylenetetrazole (PTZ), 4-aminopyridine (4-AP), picrotoxin, strychnine hemisulfate, methoxychlor, amoxapine, aminophylline hydrate, bicuculline methiodide, and enoxacin. After drug treatment, distance traveled at high speed (greater than 20 mm/s) in 1 min intervals (D) for 60 min was measured.

The longest distance traveled in a single 1 min interval (peak D) during the 60 min recording period was identified at various concentrations. Then, peak D for compound-treated animals was compared with D of carrier control at the same time point and Student's t-test was used to determine if compound effects were significant ($P < 0.05$). Using this criterion, eight of the nine seizure inducing compounds caused significant effects on zebrafish movement and the negative control compound did not induce high-speed movement. These and other results underscore the potential value of incorporating this zebrafish assay in preclinical safety studies to assess seizure liability.

Zebrafish: Methods for Assessing Drug Safety and Toxicity, Edited by Patricia McGrath.
© 2012 John Wiley & Sons, Inc. Published 2012 by John Wiley & Sons, Inc.

13.1.1 Seizures Are Complex Disorders

Seizures are complex brain disorders caused by abnormal neuronal excitability accompanied by either a variety of mild neurological symptoms or severe life-threatening convulsions. Seizures, which can be global (generalized) or focal (partial), are classified by anatomical origin. Because a large region of the brain is affected, generalized seizures are characterized by severe, life-threatening symptoms. Types of generalized seizures include (1) absence (petit mal)—a staring spell, (2) myoclonic—single or repetitive brief jerking muscle movement without loss of consciousness, (3) clonic—rhythmic jerking movement of arms and legs, sometimes on both sides of the body, (4) tonic—muscle stiffening with loss of consciousness and body rigidity, and (5) tonic–clonic (grand mal)—generalized convulsions in which the tonic phase is followed by the clonic phase. Depending on the area of brain affected and severity, partial seizures can include motor, sensory, autonomic, emotional, or cognitive abnormalities; symptoms of partial seizures include a blank stare, loss of consciousness, distortion of sensory perception (scents, music, or flashes of light), or uncontrolled muscle contractions (Luders et al., 1998; Tuxhorn and Kotagal, 2008).

There are many underlying causes for seizures including epilepsy, genetics, congenital and developmental conditions, tumors, head trauma, and infectious diseases, including HIV/AIDS. Numerous compound classes including antimicrobials, psychotropic agents, antidepressants, antipsychotics, anesthetics and antiarrhythmic agents, sedative-hypnotic agents, and, unfortunately, antiepileptic compounds themselves can induce seizures (Alldredge and Simon, 1991). Due to complexity, seizure diagnosis is typically done by (1) taking a detailed medical history describing seizure-related neurological symptoms and chronicling episodes, and (2) imaging and electrical assessment, by magnetic resonance imaging (MRI), single-photon emission computed tomography (SPECT), positron emission tomography (PET), magnetoencephalography (MEG), or electroencephalography (EEG). Imaging and electrical assessment are necessary for physiological and anatomical characterization.

13.1.2 Conventional Models for Assessing Seizures

Since seizures are systemic disorders involving several biological processes and organs, studies are best performed in physiologically intact animals. Diverse animal models ranging from fruit flies to nonhuman primates have been invaluable for understanding disease etiology and developing antiepileptic treatment strategies (Sarkisian, 2001a, 2001b; Cole et al., 2002; Martin and Pozo, 2006). Criteria necessary for validating seizure animal models include (1) EEG activity, that is, interictal and ictal spike–wave discharges, (2) etiology—genetic predisposition and injury, (3) developmental age, (4) brain-specific pathologies, focal lesions, and cortical dysplasia, (5) response to anticonvulsants, and (6) behavioral characteristics. Not surprisingly, since there are more than 100 different types of human seizures and epileptic disorders, no single animal model recapitulates the complete range of phenotypes observed in humans.

13.1.3 Zebrafish Seizure Liability Model

Several recent studies describe use of a zebrafish seizure model for drug screening (Baraban et al., 2005; Berghmans et al., 2007; Winter et al., 2008). In these studies, after treatment with PTZ, a potent proconvulsant, and other seizure inducing drugs, zebrafish movement was quantified using automated motion detection systems, including EthoVision (Noldus Information Technology) or VideoTrack (ViewPoint Life Sciences). Three distinct stages of seizure-like behavior were observed: (1) increased total movement, (2) whirlpool-like high-speed movement, and (3) loss of normal body orientation and immobilization, resembling a clonic seizure (Baraban et al., 2005; Berghmans et al., 2007; Winter et al., 2008). Both increased total movement and whirlpool-like high-speed movement were observed after treatment with PTZ concentrations between 2.5 and 15 mM and loss of normal body orientation and immobilization were observed after treatment with 15 mM (Baraban et al., 2005; Berghmans et al., 2007; Winter et al., 2008). Winter et al. (2008) reasoned that D at high speed (greater than 20 mm/s) in zebrafish could be used to predict compound-induced seizures in humans; using this parameter, nonspecific movement unrelated to compound-induced seizure-like activity was eliminated. Further confirming that high-speed movement in zebrafish was predictive of seizures in humans, after PTZ treatment, zebrafish EEG recordings revealed ictal spike–wave discharges, similar to those observed in epilepsy patients (Baraban et al., 2005).

Use of video-based motion detection for assessing zebrafish behavior has now been adapted for several additional end points, including neurons (Chapter 10), muscles (Chapter 18), and overall development (Chapter 12). These highly versatile systems will make it possible to validate use of zebrafish motility for a wide variety of toxicity, safety, and efficacy studies.

13.2 MATERIALS AND METHODS

13.2.1 Embryo Handling

Phylonix AB zebrafish were generated by natural pairwise mating in our aquaculture facility. Four to five pairs were set up for each mating; on average, 50–100 embryos per pair were generated. Embryos were maintained in embryo water (5 g of Instant Ocean Salt with 3 g of $CaSO_4$ in 25 L of distilled water) at 28°C for approximately 24 h before sorting for viability. The seizure assay was performed using 6dpf zebrafish. Because the early-stage embryo receives nourishment from an attached yolk sac, no additional maintenance was required.

13.2.2 Recording Zebrafish Movement Using a Motion Detector

To analyze zebrafish movement, we used a video motion tracking device (VideoTrack System, ViewPoint Life Sciences, Lyons, France). This instrument, adapted for zebrafish research, consists of a high-resolution black and white video camera (Phillips LT385) that can scan a video image in 40 ms, that is, at the rate of 25 video images per second, and a customized zebrafish behavior chamber that holds microplates (up to 96 wells) containing individual zebrafish. The VideoTrack System uses bright light to view zebrafish in microwells and to monitor movement continuously, and an infrared light to track zebrafish in a dark chamber. The video camera analog signal is digitized to 8 bits with luminosity values from 0 (black) to 255 (white).

From digitized video images, the VideoTrack System can analyze multiple movement parameters including time spent in peripheral versus central area of each well, number of movements per unit of time, number and angle of rotations of each movement, and D at low, medium, or high speed.

For these experiments, the motion tracker recorded activity from 20 wells in 24-well microplates, the maximum number of wells that can be imaged due to the format of the video camera (NTSC standard). The signal threshold setting for each well, which ranged from 20 to 140, was adjusted as follows. Briefly, after clicking on the full-scale image icon, 20 circles were oriented over each well and movement of single zebrafish was recorded. To reduce reflection or other shadows that could affect thresholding, area size analyzed in each well corresponded as closely as possible to the actual diameter of each well. After adjusting the size of each well, "draw areas" were deselected. Then, the first well to be analyzed was selected using the "tiles" button. To ensure that zebrafish were identified, we used the detection threshold window that displayed number of red pixels over the body of each zebrafish. If a zebrafish was identified in the well, the movement threshold window was used to examine each animal on the screen. If a small white dot was visible on the body of the zebrafish, then the instrument was ready to track movement. If no white dot was identified or a number of lines appeared in a well with no zebrafish movement, the size of the recording area was adjusted to exclude any dark areas.

13.2.3 Compounds and Treatment Conditions

To validate the zebrafish seizure assay, we tested nine compounds known to cause seizures in mammals and zebrafish, including PTZ, 4-AP, picrotoxin, strychnine hemisulfate, methoxychlor, amoxapine, aminophylline hydrate, bicuculline methiodide, and enoxacin (ATSDR, 1994; Winter et al., 2008), and lidocaine as a negative control compound. Fish water with 0.1% DMSO was used as carrier control. Compounds, mechanisms of action, and concentrations assessed are shown in Table 13.1. Stock drug solutions were made in 100% DMSO and stored at $-20°C$. On the day of the experiments, drug solution was diluted to a $2\times$ concentration in fish water (5 mM NaCl, 0.17 mM KCl, 0.33 mM CaCl$_2$, 0.33 mM MgSO$_4$, 10 mM HEPES, pH 7.5). For treatment with each compound concentration, a single 6dpf zebrafish was deposited in each well of 24-well plates containing 250 μL fish water. Then, immediately before image acquisition, we added 250 μL fish water containing $2\times$ drug solution. Final DMSO concentration was 0.1% and final fluid volume in each well was 500 μL. For this study, initial test concentrations were selected based on

Table 13.1 Compounds Used to Validate Zebrafish Seizure Assay

Compound	Mechanism of action	Concentrations (mM)
PTZ	GABA antagonist	1.25, 2.5, 5
4-AP	Norepinephrine reuptake inhibitor	0.05, 0.15, 0.3
Picrotoxin	Blocks potassium channel	0.01, 0.05, 0.1, 0.5, 1
Strychnine hemisulfate	Synthetic organochlorine	0.001, 0.005, 0.02. 0.05, 0.1, 0.5
Methoxychlor	Adenosine antagonist	0.001, 0.005, 0.025
Amoxapine	Glycine receptor antagonist	0.028, 0.056, 0.084
Aminophylline hydrate	GABA antagonist	1, 5, 10
Bicuculline methiodide	GABA antagonist	0.8, 2.5, 5
Enoxacin	GABA antagonist	0.1, 0.5, 1, 2, 5
Lidocaine	Blocks sodium channel	0.25, 1, 5

previous studies. If no effects were observed in initial tests, additional lower or higher concentrations were assessed. Compound concentrations are shown in Table 13.1.

13.2.4 Assessment of Drug Effects on Distance Traveled

The VideoTrack System separated movement of single zebrafish in each well into three speeds: low (less than 4 mm/s), medium (between 4 and 20 mm/s), and high (greater than 20 mm/s), and movements were color coded (Fig. 13.1). Black lines represent low-speed movement. Green lines (and green areas) indicate medium-speed movement. Red lines (and red areas) indicate high-speed movement. PTZ exposure induced high-speed movement (left panel, red areas).

0.1% DMSO   5 mM PTZ

Figure 13.1 Graphical representation of movement of single larval zebrafish (left, DMSO control; right, 5 mM PTZ) during 60 min. Recording commenced immediately after treatment with PTZ. Black lines represent low-speed movement (less than 4 mm/s). Green lines (areas) indicate medium-speed movement (between 4 and 20 mm/s). Red lines (areas) indicate high-speed movement (greater than 20 mm/s). PTZ exposure induced high-speed movement (right). (See the color version of this figure in Color Plates section.)

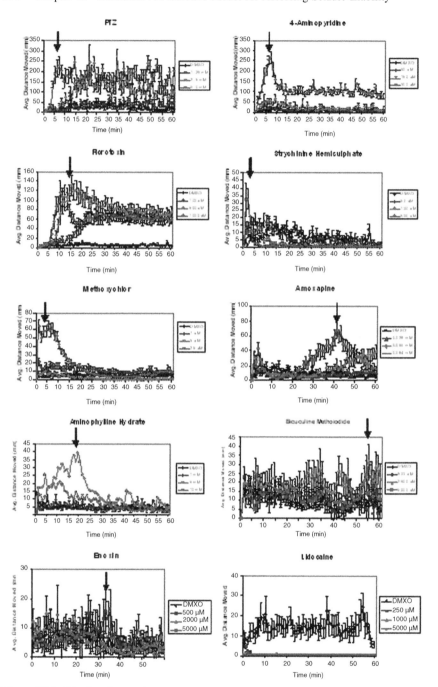

Figure 13.2   Kinetics of *D* traveled in 1 min intervals for 60 min. *D* was measured for 60 min for each compound concentration. Each point represents mean ± SE (*N* = 20 or 30). Black arrows show peak *D* used to determine percent increase in peak D compared to carrier control. (See the color version of this figure in Color Plates section.)

For the seizure assay, we analyzed high-speed movement (greater than 20 mm/s, Fig. 13.1, red lines) (Winter et al., 2008). Distance traveled was measured in 1 min intervals (D) for 60 min for each zebrafish. An Excel spreadsheet containing analyzed data was generated at the end of each experiment. Incorrect tracking data (e.g., detection of microwell wall instead of animals) were eliminated from the final data set prior to analysis. D ($N = 20$ or 30) was plotted against recording time (Fig. 13.2). From this data, we identified peak D for each concentration. We then compared peak D for compound-treated animals and D for carrier control zebrafish at the same time point. Compound effects were normalized to percent increase in peak D using formula (13.1):

$$\text{percent increase in peak } D = \left(\frac{D_{\text{compound}}}{D_{\text{control}}} - 1 \right) \times 100\%. \qquad (13.1)$$

Student's t-test was then used to determine if compound effects were significant ($P < 0.05$).

13.3 RESULTS

Using quantitative video-based movement tracking systems, zebrafish have been shown to be a predictive animal model for assessing compound-induced seizure-like locomotion (Baraban et al., 2005; Berghmans et al., 2007; Winter et al., 2008). For these studies, similar to previously reported methods (Winter et al., 2008), we used the VideoTrack System (ViewPoint Life Sciences) to measure distance traveled in 1 min intervals (D) for 60 min for nine seizure inducing compounds, including PTZ, 4-AP, picrotoxin, strychnine hemisulfate, methoxychlor, amoxapine, aminophylline hydrate, bicuculline methiodide, and enoxacin; lidocaine was used as a negative control compound (Table 13.1). The system measured D at three speeds: low (less than 4 mm/s), medium (between 4 and 20 mm/s), and high (greater than 20 mm/s). To assess seizure-like movement, we focused on D at high speed. We identified peak D during the 60 min recording period and the concentration used. Then, we compared peak D for compound-treated animals and D for carrier controls at the same time point. Student's t-test was used to determine if compound effects were significant ($P < 0.05$).

13.3.1 High-Speed Distance Traveled in 1 min Intervals

Using varying compound concentrations (Table 13.1), D was recorded for 60 min and mean D was calculated and plotted against time for each concentration ($N = 20$ or 30). Carrier control was included in each experiment. Figure 13.2 shows the kinetics of high-speed movement for compound-treated and carrier control animals for three compound concentrations.

13.3.2 Concentration Effect on Peak *D*

Using data shown in Fig. 13.2, we identified the longest D traveled (peak D) for each compound concentration (Table 13.2). Since the negative control compound,

Table 13.2 Concentration Effects on Peak D and Time Point at Peak D

Compound	Concentration (mM)	Peak D (mm)	Time point at peak D (min)
PTZ	1.25	66	30
	2.5	245	21
	5	226	7
4-AP	0.05	50	4
	0.15	63	9
	0.3	266	7
Picrotoxin	0.1	73	37
	0.5	131	15
	1	116	10
Strychnine hemisulfate	0.05	18	9
	0.1	13	7
	0.5	33	2
Methoxychlor	0.001	N.O.	N.O.
	0.005	N.O.	N.O.
	0.025	65	5
Amoxapine	0.028	N.O.	N.O.
	0.056	39	60
	0.084	67	41
Aminophylline hydrate	1	N.O.	N.O.
	5	39	19
	10	21	11
Bicuculline methiodide	0.8	20	34
	2.5	31	55
	5	25	53
Enoxacin	0.5	16	36
	2	16	35
	5	15	11
Lidocaine	0.25	N.O.	N.O.
	1	N.O.	N.O.
	5	N.O.	N.O.

N.O., no peak observed.

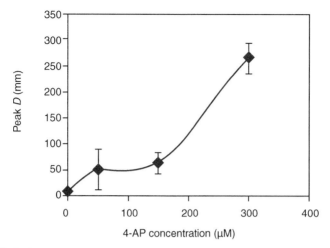

Figure 13.3 Peak D was concentration dependent for 4-AP. Using three 4-AP concentrations, peak D was concentration dependent. Each point represents mean $\pm$ SE ($N = 20$).

lidocaine, a sedative, did not induce movement, D was "0" and no peak D was observed.

Based on these results, for 4-AP, strychnine hemisulfate, methoxychlor, and amoxapine, increase in peak D was concentration dependent (Fig. 13.3, example, 4-AP). For PTZ, picrotoxin, aminophylline hydrate, and bicuculline methiodide, peak D plateaued at the second highest compound concentration.

Based on results shown in Table 13.2, we then compared peak D for compound-treated animals with D for control animals at the same time point, as described in Section 13.2. Percent increase in D compared to controls was 1776% (PTZ), 3707% (4-AP), 1777% (picrotoxin), 1531% (strychnine hemisulfate), 1521% (methoxychlor), 650% (amoxapine), 461% (aminophylline hydrate), 92% (bicuculline methiodide), 174% (enoxacin), and -82% (lidocaine). Statistical analysis using Student's t-test showed that PTZ, 4-AP, picrotoxin, strychnine hemisulfate, methoxychlor, amoxapine, aminophylline hydrate, and bicuculline methiodide induced significant effects ($P < 0.05$), whereas effect of enoxacin was insignificant ($P > 0.05$). Lidocaine, the negative control compound, did not induce movement and D was "0". Therefore, compared to movement of carrier control animals at the same time point, percent increase was negative, as shown in Fig. 13.4.

Based on percent increase in peak D, we ranked compound effects on zebrafish high-speed movement as 4-AP > picrotoxin = PTZ = strychnine hemisulfate = methoxychlor > amoxapine = aminophylline hydrate > bicuculline methiodide. Taking into consideration the error bars in Fig. 13.4, effects on high-speed movement for PTZ, picrotoxin, strychnine, and methoxychlor were considered the same, and effects of amoxapine and aminophylline hydrate on high-speed movement were considered the same. The effect of enoxacin was insignificant.

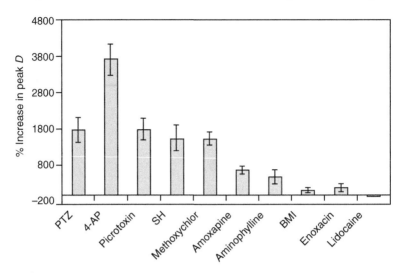

Figure 13.4   Comparison of percent increase in peak D after compound treatment. Y-axis is percent increase in peak D. Data presented as mean ± SE ($N = 20$ or 30).

13.4 CONCLUSIONS

In recent studies, PTZ, which induces seizures in rodents (Ferrendelli et al., 1989), induced seizure-like zebrafish movement as well as a characteristic seizure EEG profile in zebrafish brain (Baraban et al., 2005; Winter et al., 2008). Winter et al. (2008) assessed additional compounds known to cause seizures in mammals focusing on high-speed movement (greater than 20 mm/s). In this study, to further validate the zebrafish seizure assay, we assessed nine compounds known to cause seizures in mammals and zebrafish including PTZ, 4-AP, picrotoxin, strychnine hemisulfate, methoxychlor, amoxapine, aminophylline hydrate, bicuculline methiodide, and enoxacin. The assay correctly predicted results for eight of the nine seizure inducing compounds and lidocaine, a negative control compound. Therefore, overall predictivity was 90% and sensitivity was 89%.

In addition to distance traveled at high speed, we examined the kinetic patterns of high-speed seizure-like movement, including time to peak D, duration of peak D, and time to return to base level of high-speed movement, which varied by compound (Fig. 13.2). Differences may be due to rate of compound absorption or bioavailability, compound solubility or stability, experimental conditions, receptor binding affinity, or other factors. In subsequent studies, we are investigating if kinetic patterns can be used as fingerprints to categorize potential seizure inducing compounds by target or mechanism of action.

REFERENCES

ALLDREDGE BK and SIMON RP (1991). Drugs that can precipitate seizures. In: RESOR SR and KUTT H (Eds.), The Medical Treatment of Epilepsy. Informa Healthcare.

ATSDR (Agency for Toxic Substances and Disease Registry) (1994). Toxicological profile for methoxychlor. U.S. Department of Health and Human Services, Public Health Service, Atlanta, GA.

BARABAN SC, TAYLOR MR, CASTRO PA, and BAIER H (2005). Pentylenetetrazole induced changes in zebrafish behavior, neural activity and c-fos expression. Neuroscience 131(3): 759–768.

BERGHMANS S, HUNT J, ROACH A, and GOLDSMITH P (2007). Zebrafish offer the potential for a primary screen to identify a wide variety of potential anticonvulsants. Epilepsy Res 75(1): 18–28.

COLE AJ, KOH S, and ZHENG Y (2002). Are seizures harmful: what can we learn from animal models? Prog Brain Res 135: 13–23.

FERRENDELLI JA, HOLLAND KD, MCKEON AC, and COVEY DF (1989). Comparison of the anticonvulsant activities of ethosuximide, valproate, and a new anticonvulsant, thiobutyrolactone. Epilepsia 30(5): 617–622.

ICH (2001). S7A Safety Pharmacology Studies for Human Pharmaceuticals. International Conference on Harmonisation of Technical Requirements for Registration of Pharmaceuticals for Human Use (ICH), U.S. Department of Health and Human Services, Food and Drug Administration, Center for Drug Evaluation and Research (CDER), Center for Biologics Evaluation and Research (CBER).

LUDERS H, ACHARYA J, BAUMGARTNER C, BENBADIS S, BLEASEL A, BURGESS R, DINNER DS, EBNER A, FOLDVARY N, GELLER E, HAMER H, HOLTHAUSEN H, KOTAGAL P, MORRIS H, MEENCKE HJ, NOACHTAR S, ROSENOW F, SAKAMOTO A, STEINHOFF BJ, TUXHORN I, and WYLLIE E (1998). Semiological seizure classification. Epilepsia 39(9): 1006–1013.

MARTIN E and POZO M (2006). Animal models for the development of new neuropharmacological therapeutics in the status epilepticus. Curr Neuropharmacol 4(1): 33–40.

SARKISIAN MR (2001a). Animal models for human seizure and epileptic activity. Reply. Epilepsy Behav 2(5): 506–507.

SARKISIAN MR (2001b). Overview of the current animal models for human seizure and epileptic disorders. Epilepsy Behav 2(3): 201–216.

TUXHORN I and KOTAGAL P (2008). Classification. Semin Neurol 28(3): 277–288.

WINTER MJ, REDFERN WS, HAYFIELD AJ, OWEN SF, VALENTIN JP, and HUTCHINSON TH (2008). Validation of a larval zebrafish locomotor assay for assessing the seizure liability of early-stage development drugs. J Pharmacol Toxicol Methods 57(3): 176–187.

Chapter 14

Zebrafish: A New *In Vivo* Model for Identifying P-Glycoprotein Efflux Modulators

Demian Park, Maryann Haldi, and Wen Lin Seng
Phylonix, Cambridge, MA, USA

14.1 INTRODUCTION

In order to address the rapidly increasing incidence of central nervous system (CNS) diseases, new approaches for delivering and retaining drugs in the brain are needed. P-glycoprotein (Pgp), one of the most important members of the ABC transporter family, has been shown to inhibit absorption and retention of more than 50% of frequently prescribed drugs (Bauer et al., 2005). Development of Pgp inhibitors is one strategy for increasing drug accumulation in the brain. Although drug transporters are well described in mammals (Jeong et al., 2001), blood brain barrier (BBB) formation and Pgp expression and function are not well characterized in zebrafish. Here we describe development of a novel *in vivo* zebrafish assay to rapidly identify Pgp efflux inhibitors. By injecting a fluorescent Pgp substrate directly into the transparent zebrafish brain, we determined that Pgp is fully functioning by 7 days post fertilization (dpf). We then monitored Pgp efflux kinetics by capturing images of fluorescent dye efflux from the brain at varying time points. Using this approach, potential drug candidates can be cotreated with Pgp substrate to visualize retention in the brain. Next, to identify potential Pgp efflux inhibitors, we developed a quantitative functional assay that relies on morphometric image analysis. To validate this bioassay, we confirmed that six positive control drugs, verapamil, phenytoin, loperamide, cyclosporine, RU486, and quinidine, inhibited Pgp efflux in zebrafish brain and one negative control drug, caffeine, did not cause significant inhibitory effects. These results were similar to results in mammalian models and 100% correct prediction is considered "excellent" by the Interagency Coordinating Committee on the Validation of Alternative Methods (ICCVAM). A compelling advantage of using zebrafish for this application is that animals are transparent throughout development and level of fluorescence, which correlates with drug efflux in the brain region, can be quantitated *in vivo*.

14.1.1 Brain ABC Transporters Play a Role in Drug Resistance

ABC efflux transporters, present in several tissue compartments, including the gastrointestinal tract and brain microcapillary endothelial cells, have been shown to inhibit drug absorption and retention. Seven different brain ABC transporters, including ABCB1, ABCC1, ABCC2, ABCC3, ABCC4, ABCC5, and ABCG2 (Dean et al., 1997), are known to play a role in drug resistance; ABCB1/MDR1/Pgp, ABCC1/MRP1, and ABCG2 are the most well characterized. Several ABC transporters have been identified at the BBB and the blood cerebrospinal fluid barrier (BCSFB) (Loscher and Potschka, 2008), where they are found in the plasma membranes on both the luminal and abluminal sides of microcapillary endothelial cells (de Vries et al., 2001). ABCB1/MDR1, also known as Pgp, is the most widely studied efflux transporter; it is an important component of active efflux transport (AET) and serves as a gatekeeper to actively efflux small molecules from the brain to the blood (Tsuji et al., 2006; Stewart et al., 2001; Schinkel, 1999). Three factors make Pgp a key factor regulating drug entry into the brain: location, potency, and broad specificity (Begley, 2004). Localization of Pgp on brain microvessels is consistent with its role as an efflux modulator. Potent ATP-driven pumping prevents drug accumulation in the brain. Broad Pgp specificity ensures that it can pump out a large number of drugs from different classes (Bauer et al., 2005) and Pgp inhibition has been shown to lead to drug retention. Underscoring the important role Pgp plays in drug efflux, a Pgp knockout mouse exhibited a hundred-fold increase in drug absorption and a significant decrease in drug elimination from the CNS (Schinkel et al., 2002; van Asperen et al., 1998).

14.1.2 Zebrafish: A Transparent Model for Investigating Drug Effects on ABC Transporters

The BBB, present in all vertebrates, has been confirmed in zebrafish (Cserr and Bundgaard, 1984; Jeong et al., 2001). In previous studies (Fig. 10.6), using ZO-1 antibody staining and Evans blue dye injection into the circulation, at 3dpf, presence of tight junctions was confirmed and blockage of dye diffusion from the vasculature to brain tissue was observed (McGrath et al., 2004). Using a monoclonal antibody, Pgp expression was detected in zebrafish liver at 5dpf (Sadler et al., 2010). In addition, a phylogenetic study identified two genes encoding Pgp, *ABCB1a* and *ABCB1b* (Annilo et al., 2006), and Pgp expression has also been detected in adult zebrafish

(Bresolin et al., 2005). Furthermore, in fish, xenobiotic efflux pumps, including Pgp, have been identified in brain microcapillaries, which are sensitive to different glycoprotein inhibitors and substrates, and pharmacological response is similar to mammals (Miller et al., 2005). Taken together, these data support use of zebrafish as an animal model for assessing drug effects on Pgp inhibition.

14.2 MATERIALS AND METHODS

14.2.1 Embryo Handling

Zebrafish were generated by natural pairwise mating in our aquaculture facility as described by Westerfield (1999). Zebrafish were maintained in embryo water (5 g of Instant Ocean Salt with 3 g of $CaSO_4$ in 25 L of distilled water) at 28°C for approximately 24 h before sorting for viability. Because early-stage zebrafish receive nourishment from an attached yolk sac, no additional maintenance was required.

14.2.2 Reagents

Unless otherwise described, reagents were purchased from Sigma-Aldrich Co. (St. Louis, MO).

14.2.3 Circulation Dye Injection

Zebrafish were placed on an injection stage containing fish water and immobilized by 0.32 mM tricaine (ethyl 3-aminobenzoate methanesulfonate). A microinjection needle was fabricated from a 1 mm (outer diameter) glass capillary (WPI, Sarasota, FL) using a P-30 vertical pipette puller (Sutter Instrument, Novato, CA). Using an M-330 micromanipulator (WPI), the tip of the injection needle was inserted into the common cardinal vein (CCV) (Isogai et al., 1994). A solution of 10 mM rhodamine 123 was injected into CCV under a stereo dissecting microscope (Stemi, 2000, Zeiss, Thornwood, NY). Following injection, zebrafish were transferred to a Petri dish containing fish water and incubated at 28°C for 2 h. Zebrafish were then positioned laterally on a depression glass slide containing 1% methylcellulose for image acquisition.

14.2.4 Fluorescence Microscopy

Fluorescence microscopy was performed using a Zeiss M2Bio fluorescence microscope (Zeiss), equipped with a rhodamine cube (excitation: 540 nm, emission: 605 nm), and a chilled CCD camera (AxioCam MRm, Zeiss). Fluorescence images were acquired at 25× magnification and processed using AxioVision software Rel 4.0 (Zeiss). ImageJ software (NIH, Bethesda, MD) was used for image analysis.

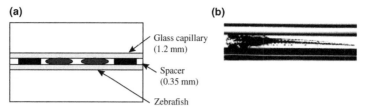

Figure 14.1   Specially designed slide facilitated dye injection into zebrafish brain.
(a) Cartoon illustrating slide design for retaining zebrafish in dorsal orientation to facilitate brain injection and image analysis. Three rows can be included on each slide and five zebrafish can be maintained in each row. (b) Image of 4dpf zebrafish, dorsal orientation.

14.2.5 Specialized Zebrafish Slide for Brain Injection and Imaging

Since maintaining individual zebrafish in a dorsal orientation to facilitate brain injection and image capture using a conventional depression slide was difficult, we designed a specialized zebrafish slide that increased throughput. As shown in Fig. 14.1a, rows of glass capillaries, separated by 0.35 mm spacers, were attached to slides. Zebrafish were placed between capillaries in the dorsal position, five zebrafish per row, and at least three rows can be placed on each slide. Using this specialized slide, the number of zebrafish injected and imaged in 1 h increased from ~6 to 30.

14.2.6 Brain Dye Injection

Zebrafish were placed on slides with 0.32 mM tricaine mixed with appropriate solution for each condition to immobilize animals. Using a stereo dissecting microscope (Stemi 2000, Zeiss) for visualization, an injection needle containing fluorescent rho-HRP (rhodamine-labeled horseradish peroxidase) Pgp substrate (50 mM) (21st Century Biochemicals, Marlboro, MA) was inserted into the midline of the optic tecta bordering the cerebellum, and 30 nL of substrate was delivered using a pressure-controlled microinjector (PV-830 Pneumatic PicoPump, WPI) (Meng et al., 2004) (Fig. 14.2).

14.2.7 Quantitative Morphometric Analysis of Brain Images

Dorsal view images of brain ROI (region of interest) (25×) were captured using the same exposure time and fluorescence gain. Fluorescence was quantified using ImageJ software. A constant threshold was applied to fluorescent brain images of each zebrafish. This threshold was automatically set by the software and adjusted manually for artifacts, when necessary. Total fluorescence from threshold-processed images

(a)

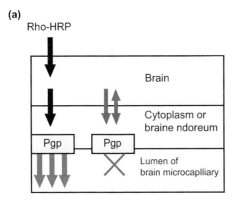

(b)

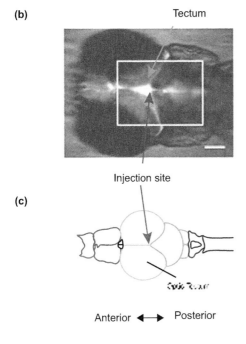

Figure 14.2 Rho-HRP efflux from the brain. (a) A schematic diagram of rho-HRP efflux. During Pgp efflux (blue arrows), rho-HRP was transported from the brain into the cytoplasm of the brain endothelium (black arrows) where it was pumped into the lumen of brain microcapillaries. After treatment with Pgp inhibitors (red X), rho-HRP accumulates in the brain and the cytoplasm of the brain endothelium (red arrows). (b) Dorsal view of brain of 7dpf zebrafish injected with fluorescent rho-HRP (25×). Region of interest: midbrain and hindbrain region; white box. White scale bar is 100 μm. (c) Cartoon of zebrafish brain in dorsal position (Wullimann and Reichert, 1996). Injection site is marked by red arrows. (See the color version of this figure in Color Plates section.)

was quantified using the automated measurement function. Total fluorescent signal = area of ROI × average fluorescence intensity in ROI. For each animal, fluorescence intensity at (a) 15 min post-dye injection (T0) and (b) 1, 2, and 4 h post-dye injection (T1, T2, and T4) was compared. Percent retention was then calculated for each animal at each time point using formula (14.1):

$$\text{percent retention} = \frac{\text{FL}(Tn)}{\text{FL}(T0)} \times 100\%. \tag{14.1}$$

14.2.8 Drug Treatment

For drug treatment, 7dpf zebrafish were deposited into 6-well plates containing 3 mL of sterile fish water per well. Drugs were dissolved in DMSO and diluted by fish water to 1, 10, and 100 µM for testing; final DMSO concentration was 0.1%. Animals were treated with drugs 1 h before injecting rho-HRP. To ensure zebrafish remained in either control or drug-treated condition during dye injection and image capture, zebrafish were placed on slides with 0.32 mM tricaine mixed with either fish water (control condition) or drug solution (drug-treated condition) to immobilize animals, and image acquisition began 15 min after rho-HRP injection, T0. Zebrafish were maintained in drug solution throughout image acquisition at T0, T1, T2, and T4. 0.1% DMSO was used as vehicle control. Mean % retention at the optimal time point for 10 zebrafish in each condition was then determined and used to assess drug effects.

14.2.9 Statistics

ANOVA was used to determine if drug effects were significant ($P < 0.05$), followed by Dunnett's test (pairwise comparison) to identify which concentrations that caused significant effects.

14.3 RESULTS

The overall aim of this research was to develop a zebrafish bioassay to identify potential Pgp efflux inhibitors. Pgp has been shown to be the most important efflux transporter, affecting drug absorption, distribution, metabolism, and excretion (ADME). Design of drugs that are not Pgp substrates is one strategy for increasing drug retention in the brain (Mahar Doan et al., 2003). Alternatively, a Pgp efflux inhibitor can be coadministered with drug to enhance retention in the brain (Kemper et al., 2001).

14.3.1 Identification of Optimal Zebrafish Stage for Assessing Pgp Efflux

To optimize assay conditions, we first determined the stage at which Pgp efflux is fully functioning. We injected rhodamine 123, a fluorescent Pgp substrate shown to cross BBB in rats and cows (Wang et al., 2005; Fontaine et al., 1996; Rose et al., 2005), into zebrafish circulation at 3, 4, 5, 6, and 7dpf stages. Lateral images of zebrafish head were captured 2 h post injection (hpi). Fluorescence was detected in zebrafish brain (yellow boxed region) at 3, 4, and 5dpf (Fig. 14.3, yellow lines), indicating that dye crossed BBB and was retained in the brain region. However, no fluorescence was observed in 6 and 7dpf zebrafish brains (blue arrows), confirming that Pgp efflux system is fully functioning by 7dpf.

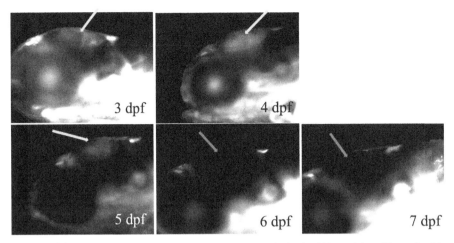

Figure 14.3   Determine optimal stage to assess Pgp efflux. Rhodamine 123 was injected into zebrafish circulation at 3, 4, 5, 6, and 7dpf. (b) Fluorescence was detected in the brain of 3, 4, and 5dpf zebrafish (yellow arrows) at 2 hpi, indicating that dye crossed BBB and was retained in the brain region. However, fluorescence signal was absent in the brain region of 6 and 7dpf zebrafish (blue arrows), indicating dye efflux. White scale bar is 300 μm. (See the color version of this figure in Color Plates section.)

14.3.2 Assessment of Pgp Efflux

After establishing that Pgp was fully functioning by 7dpf, we next assessed Pgp efflux kinetics. Since rhodamine 123 injected into the circulation was not retained in brain tissue at 7dpf stage (Fig. 14.4), we attempted direct brain injection. However, due to variable clearance of rhodamine 123 from the brain, we used an alternative fluorescent

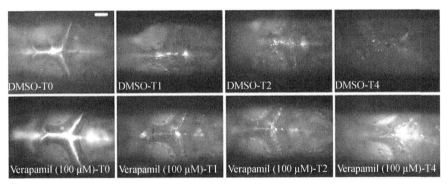

Figure 14.4   Kinetics of rho-HRP efflux. Sevendpf zebrafish were soaked in either 0.1% DMSO or verapamil for 1 h before injecting rho-HRP into the brain. Dorsal view ROI images were captured at 15 min post injection (designated as T0) and at 1, 2, and 4 hpi, designated as T1, T2, and T4, respectively (25×). By T4, minimal fluorescence signal was detected in 0.1% DMSO control animals. In contrast, fluorescence remained high in verapamil-treated animals, indicating decreased Pgp-mediated rho-HRP efflux. Anterior to the left; white scale bar is 100 μm.

Pgp substrate, rho-HRP. After injecting rho-HRP into the midline of the optic tecta bordering the cerebellum, we assessed Pgp efflux by measuring active efflux of rho-HRP from the brain tissue into the lumen of brain microcapillaries. Since rho-HRP was injected directly into brain tissue, the concentration was higher than in brain microcapillaries, creating a concentration gradient that facilitated active rho-HRP efflux from brain tissue to the lumen of microcapillaries (Fig. 14.2a). Moreover, since rho-HRP does not cross the BBB (Jeong et al., 2001), after it is effluxed, it cannot diffuse back into brain tissue. To monitor variability in the brain microinjection procedure, initially we injected rho-HRP into various sites in the brain and examined level of fluorescence in ROI, which included the midbrain and hindbrain. In these studies, we determined that the optimum injection site was the midbrain–hindbrain junction (Fig. 14.2b, red arrow).

Since rho-HRP efflux kinetics can vary depending on treatment conditions, in order to select the optimal time point for drug assessment, we next treated 7dpf animals with DMSO or verapamil, a known Pgp inhibitor, and compared efflux kinetics (Choo et al., 2000; Potschka and Loscher, 2002; Hung et al., 1996). As shown in Fig. 14.4, dorsal view images of ROI were captured at T0, T1, T2, and T4. Compared to signal at T0, level of fluorescence decreased at T1 and T2 for both control and verapamil-treated zebrafish. At T4, low fluorescence intensity was detected in 0.1% DMSO control animals; however, fluorescence intensity remained strong in verapamil-treated animals, indicating inhibition of Pgp efflux.

14.3.3 Optimization of Quantitative Morphometric Analysis

Although inhibition of rho-HRP efflux could be assessed visually, quantitative analysis was expected to increase accuracy. Using quantitative morphometric analysis, fluorescence image was quantified from threshold-processed brain images using the automated measurement function of ImageJ software, as described in Section 14.2. We quantified total fluorescence in ROI for DMSO control and verapamil-treated zebrafish (Fig. 14.4). Images for 10 animals from each group were analyzed and mean and standard error (SE) of the mean were calculated and compared (Table 14.1).

Table 14.1   Quantitation of Rho-HRP Efflux in Control and Verapamil-Treated Zebrafish

Time point	Percentage of dye retention	
	0.1% DMSO	Verapamil (100 μM)
T0	100	100
T1	48 ± 6	61 ± 7
T2	29 ± 5	44 ± 6
T4	39 ± 6	46 ± 10

Data expressed as mean $\pm$ SE ($n = 10$).

Using this quantitative data, efflux kinetic curves were plotted (data not shown). Based on these kinetic curves, at T2, the difference in percent dye retention between DMSO control and verapamil-treated zebrafish was significant (nonoverlapping error bars). After T2, percent dye retained in the brain plateaued (the higher percent retention observed at T4 compared with T2 may be due to drug toxicity). Percent dye retained in verapamil-treated zebrafish was higher than in DMSO control zebrafish. Based on these results and taking into consideration (a) higher experimental variability at T2 and (b) the potential for drug toxicity at T4, we decided to use T0 and T3 to quantify rho-HRP efflux for all subsequent studies.

14.3.3.1 Validation of Assay Conditions Using an Additional Pgp Inhibitor
To further validate optimum conditions for drug screening, we assessed an additional Pgp inhibitor, phenytoin. As shown in Fig. 14.5, left panels, using the same gain and exposure time we acquired fluorescence images of ROI (25×) in each animal for each treatment condition. Next, we applied a constant threshold to brain images, which were pseudocolored red (Fig. 14.5, right panels). As previously described, to eliminate artifacts, the threshold was set automatically by the software and adjusted manually, when necessary. We then calculated percent fluorescence retention at each time point. As shown in Fig. 14.5, percent

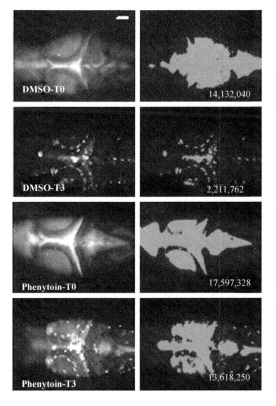

Figure 14.5 Morphometric analysis of fluorescent images. Fluorescent images of ROI in individual zebrafish were captured at 25×. T0 and T3 images of DMSO control and 100 μM phenytoin-treated zebrafish were acquired, as shown in left panels. The same threshold was applied to images that were then pseudocolored red (right panels). Total fluorescence intensity (area of ROI × average fluorescence intensity) in the pseudocolored ROI was quantitated using ImageJ software. Dye retention was 15.8% (2,211,762/14,132,040 × 100%) for DMSO-treated animals and 77.4% (13,618,250/17,597,328 × 100%) for phenytoin-treated animals. Anterior to the left; white scale bar is 100 μm. (See the color version of this figure in Color Plates section.)

fluorescence retention at T3 was 15.8% for DMSO control zebrafish (2,211,762/ 14,132,040 × 100%) and 77.4% for phenytoin-treated zebrafish (13,618,250/ 17,597,328 × 100%).

14.3.4   Assessment of Drug Effects on Pgp Efflux

Using optimum conditions, we next quantitated effects of six drugs shown to inhibit Pgp in mammals, verapamil, phenytoin, loperamide, RU486, quinidine, and cyclosporine, and one negative drug, caffeine; 0.1% DMSO was used as vehicle control. Three concentrations, 1, 10, and 100 μM, were used for each drug. Drug treatment and image capture and analysis were performed as described. Representative T3 images of ROI in brain in control and drug-treated animals are shown in Fig. 14.6. After treatment with all six positive drugs, fluorescence intensity was higher than DMSO control, indicating Pgp efflux inhibition. At T3, caffeine-treated zebrafish

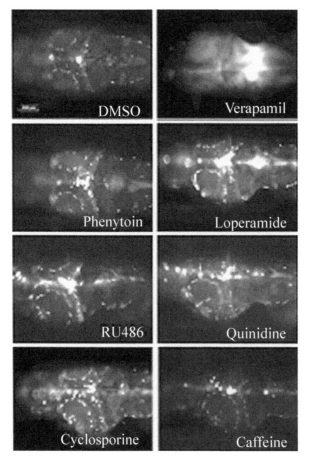

Figure 14.6 Drug effects on Pgp efflux. Pgp efflux was assessed by quantitating rho-HRP clearance from the brain. Seven dpf zebrafish were incubated with varying drug concentrations for 1 h before injecting rho-HRP into the brain. Representative dorsal view images of ROI in brain at T3 are shown. At T3, all six positive compounds, verapamil, phenytoin, loperamide, RU486, quinidine, and cyclosporine, increased fluorescence intensity compared to DMSO control, indicating that these drugs inhibited Pgp efflux. At T3, DMSO- and caffeine-treated zebrafish exhibited low fluorescence intensity, indicating no Pgp efflux inhibition. Anterior to the left. White scale bar is 100 μm.

Table 14.2 Drug Effects on Rho-HRP Retention in the Brain

Compound	Optimal concentration[a] (µM)	Drug effect (%)	P-value
Verapamil	75	57	<0.0001
Phenytoin	100	49	0.0003
Loperamide	1	41	0.0065
RU486	10	17.5	0.0389
Quinidine	100	11	<0.0001
Cyclosporine	10	30	0.0005
Caffeine	12.5	16	0.1007

[a] Optimal concentration was the concentration that induced the highest effect.

exhibited low fluorescence similar to the level in control DMSO-treated zebrafish, indicating that this negative compound did not inhibit Pgp efflux.

14.3.4.1 Quantitation of Drug Effects Using Morphometric Image Analysis

To quantitate drug effects on Pgp efflux, we then performed image-based morphometric analysis in control and drug-treated animals. DMSO control exhibited $44 \pm 5.3\%$ (mean $\pm$ SE) dye retention ($N = 19$). Due to difficulty in controlling injection site, volume, and depth, experimental variability was high. Therefore, in order to compare results, we normalized data by assigning percent retention in DMSO control in each experiment as baseline, and calculated drug effect using formula (14.2):

$$\text{drug effect} (\%) = \text{percent retention} (\text{drug}) - \text{percent retention} (\text{DMSO}). \quad (14.2)$$

Drug effects are summarized in Table 14.2: verapamil (75 µM), phenytoin (100 µM), loperamide (10 µM), and cyclosporine (10 µM) caused statistically significant inhibitory effects; 57% ($P < 0.0001$), 50% ($P < 0.0001$), 49% ($P = 0.0003$), 41% ($P = 0.0065$), and 30% ($P = 0.0005$), respectively. RU486 (10 µM) and quinidine (100 µM) exhibited moderate inhibitory effects; 17.5% ($P = 0.0389$) and 11% ($P < 0.0001$), respectively. In comparison, although caffeine, a negative control, exhibited 16% greater retention than the DMSO control, the difference was not statistically significant ($P = 0.1007$), indicating that it is not a Pgp efflux inhibitor.

14.3.4.2 Comparison of Drug Effects in Zebrafish with Results in Cells and Mammals

We then compared results in zebrafish with results in cells/tissues and mammals (Table 14.3). We confirmed that six positive drugs, verapamil, phenytoin, loperamide, RU486, quinidine, and cyclosporine, inhibited Pgp efflux in zebrafish brain. Negative control, caffeine, did not induce significant inhibitory effects. Therefore, the zebrafish brain efflux assay correctly predicted results for all seven compounds (100%) (Table 14.3).

Table 14.3 Comparison of Pgp Efflux Inhibitors in Zebrafish, Cells/Tissues, and Mammals

Compound	Function in mammals	Pgp efflux in zebrafish (*in vivo*)	Pgp efflux in cells/tissues (*in vitro*)	Pgp efflux in mammals (*in vivo*)	Correct prediction
Verapamil	Calcium channel blocker	Inhibition	Inhibition[a]	Inhibition[b]	Yes
Phenytoin	Antiepileptic agent	Inhibition	Inhibition[c]	Inhibition[d]	Yes
Loperamide	Immunosuppressant	Inhibition	Inhibition[e]	NA	Yes
RU486	Progesterone antagonist	Inhibition	Inhibition[f]	NA	Yes
Quinidine	Antiarrhythmic agent	Inhibition	Inhibition[g]	Inhibition[h]	Yes
Cyclosporine	Immunosuppressant	Inhibition	Inhibition[i]	Inhibition[j]	Yes
Caffeine	Neural stimulator	None	None	NA	Yes

NA, not available.

[a] Muller et al. (1994), Rose et al. (2005), Yumoto et al. (1999), Choo et al. (2000), Mahar Doan et al. (2003), Xia et al. (2007), and Hung et al. (1996).
[b] Choo et al. (2000), Potschka and Loscher (2002), and Hung et al. (1996).
[c] Weiss et al. (2003) and Maines et al. (2005).
[d] Hung et al. (1996).
[e] Mahar Doan et al. (2003) and Wandel et al. (2002).
[f] Gruol et al. (1994), Fardel et al. (1996), Li et al. (2004), Xia et al. (2007).
[g] Yumoto et al. (1999), Choo et al. (2000), Rautio et al. (2006), and Xia et al. (2007).
[h] Choo et al. (2000).
[i] Rose et al. (2005), Yumoto et al. (1999), Choo et al. (2000), Hamilton et al. (2001), Maines et al. (2005), Rautio et al. (2006), and Xia et al. (2007).
[j] Wang et al. (2005), Choo et al. (2000), Kemper et al. (2003), (2004), and Hung et al. (1996).

14.4 CONCLUSIONS

In order to address the rapidly increasing incidence of CNS diseases, new approaches for drug delivery are a critical unmet need. Modulation of Pgp efflux is a significant factor affecting drug retention. Here we report development of a novel *in vivo* zebrafish model to assess drug effects on Pgp efflux in the brain. Although compared to other *in vivo* models (Schinkel et al., 1994, 1996; Choo et al., 2000, 2006; Potschka and Loscher, 2002; Kemper et al., 2003, 2004; Hung et al., 1996), the zebrafish Pgp efflux assay offers several experimental advantages, we are assessing additional compounds to validate this model for drug screening and we are investigating approaches for automating animal processing.

ACKNOWLEDGMENT

This research was supported by a grant from the National Institutes of Health: 1R43MH082456.

REFERENCES

ANNILO T, CHEN ZQ, SHULENIN S, COSTANTINO J, THOMAS L, LOU H, STEFANOV S, and DEAN M (2006). Evolution of the vertebrate ABC gene family: analysis of gene birth and death. Genomics 88(1): 1–11.

BAUER B, HARTZ AM, FRICKER G, and MILLER DS (2005). Modulation of p-glycoprotein transport function at the blood-brain barrier. Exp Biol Med (Maywood) 230(2): 118–127.

BEGLEY DJ (2004). ABC transporters and the blood-brain barrier. Curr Pharm Des 10(12): 1295–1312.

BRESOLIN T, DE FREITAS REBELO M, and CELSO DIAS BAINY A (2005). Expression of PXR, CYP3A and MDR1 genes in liver of zebrafish. Comp Biochem Physiol C Toxicol Pharmacol 140(3–4): 403–407.

CHOO EF, KURNIK D, MUSZKAT M, OHKUBO T, SHAY SD, HIGGINBOTHAM JN, GLAESER H, KIM RB, WOOD AJ, and WILKINSON GR (2006). Differential in vivo sensitivity to inhibition of P-glycoprotein located in lymphocytes, testes, and the blood-brain barrier. J Pharmacol Exp Ther 317(3): 1012–1018.

CHOO EF, LEAKE B, WANDEL C, IMAMURA H, WOOD AJ, WILKINSON GR, and KIM RB (2000). Pharmacological inhibition of P-glycoprotein transport enhances the distribution of HIV-1 protease inhibitors into brain and testes. Drug Metab Dispos 28(6): 655–660.

CSERR HF and BUNDGAARD M (1984). Blood-brain interfaces in vertebrates: a comparative approach. Am J Physiol 246 (3 Pt 2): R277–R288.

DE VRIES HE, KUIPER J, DE BOER AG, VAN BERKEL TJ, and BREIMER DD (1997). The blood-brain barrier in neuroinflammatory diseases. Pharmacol Rev 49(2): 143–155.

DEAN M, RZHETSKY A, and ALLIKMETS R (2001). The human ATP-binding cassette (ABC) transporter superfamily. Genome Res 11(7): 1156–1166.

FARDEL O, COURTOIS A, DRENOU B, LAMY T, LECUREUR V, LE PRISE PY, and FAUCHET R (1996). Inhibition of P-glycoprotein activity in human leukemic cells by mifepristone. Anticancer Drugs 7(6): 671–677.

FONTAINE M, ELMQUIST WF, and MILLER DW (1996). Use of rhodamine 123 to examine the functional activity of P-glycoprotein in primary cultured brain microvessel endothelial cell monolayers. Life Sci 59(18): 1521–1531.

GRUOL DJ, ZEE MC, TROTTER J, and BOURGEOIS S (1994). Reversal of multidrug resistance by RU 486. Cancer Res 54(12): 3088–3091.

HAMILTON KO, TOPP E, MAKAGIANSAR I, SIAHAAN T, YAZDANIAN M, and AUDUS KL (2001). Multidrug resistance-associated protein-1 functional activity in Calu-3 cells. J Pharmacol Exp Ther 298(3): 1199–1205.

HUNG C-C, CHEN C-C, LIN C-J, and LIOU H-H (2008). Functional evaluation of polymorphisms in the human ABCB1 gene and the impact on clinical responses of antiepileptic drugs. Pharmacogenetics & Genomics 18: 390–402.

ISOGAI S, HORIGUCHI M, and WEINSTEIN BM (2001). The vascular anatomy of the developing zebrafish: an atlas of embryonic and early larval development. Dev Biol 230(2): 278–301.

JEONG JY, KWON HB, AHN JC, KANG D, KWON SH, PARK JA, and KIM KW (2008). Functional and developmental analysis of the blood-brain barrier in zebrafish. Brain Res Bull 75(5): 619–628.

KEMPER EM, VAN ZANDBERGEN AE, CLEYPOOL C, MOS HA, BOOGERD W, BEIJNEN JH, and VAN TELLINGEN O (2003). Increased penetration of paclitaxel into the brain by inhibition of P-Glycoprotein. Clin Cancer Res 9(7): 2849–2855.

KEMPER EM, VERHEIJ M, BOOGERD W, BEIJNEN JH, and VAN TELLINGEN O (2004). Improved penetration of docetaxel into the brain by co-administration of inhibitors of P-glycoprotein. Eur J Cancer 40(8): 1269–1274.

LI DQ, WANG ZB, BAI J, ZHAO J, WANG Y, HU K, and DU YH (2004). Reversal of multidrug resistance in drug-resistant human gastric cancer cell line SGC7901/VCR by antiprogestin drug mifepristone. World J Gastroenterol 10(12): 1722–1725.

LOSCHER W and POTSCHKA H (2005). Blood-brain barrier active efflux transporters: ATP-binding cassette gene family. NeuroRx 2(1): 86–98.

MAHAR DOAN KM, HUMPHREYS JE, WEBSTER LO, WRING SA, SHAMPINE LJ, SERABJIT-SINGH CJ, ADKISON KK, and POLLI JW (2002). Passive permeability and P-glycoprotein-mediated efflux differentiate central nervous system (CNS) and non-CNS marketed drugs. J Pharmacol Exp Ther 303(3): 1029–1037.

MAINES LW, ANTONETTI DA, WOLPERT EB, and SMITH CD (2005). Evaluation of the role of P-glycoprotein in the uptake of paroxetine, clozapine, phenytoin and carbamazapine by bovine retinal endothelial cells. Neuropharmacology 49(5): 610–617.

MCGRATH P, PARNG C, and SERBEDZIJA G (2010). Methods of screening agents using teleosts, US Patent 7,767,880, issued August 2010.

MENG A, JESSEN JR, and LIN S (1999). Transgenesis. Methods Cell Biol 60: 133–148.

MILLER DS, GRAEFF C, DROULLE L, FRICKER S, and FRICKER G (2002). Xenobiotic efflux pumps in isolated fish brain capillaries. Am J Physiol Regul Integr Comp Physiol 282(1): R191–198.

MULLER C, BAILLY JD, GOUBIN F, LAREDO J, JAFFREZOU JP, BORDIER C, and LAURENT G (1994). Verapamil decreases P-glycoprotein expression in multidrug-resistant human leukemic cell lines. Int J Cancer 56(5): 749–754.

POTSCHKA H and LOSCHER W (2001). In vivo evidence for P-glycoprotein-mediated transport of phenytoin at the blood-brain barrier of rats. Epilepsia 42: 1366–1368.

RAUTIO J, HUMPHREYS JE, WEBSTER LO, BALAKRISHNAN A, KEOGH JP, KUNTA JR, SERABJIT-SINGH CJ, and POLLI JW (2006). In vitro p-glycoprotein inhibition assays for assessment of clinical drug interaction potential of new drug candidates: a recommendation for probe substrates. Drug Metab Dispos 34(5): 786–792.

ROSE JM, PECKHAM SL, SCISM JL, and AUDUS KL (1998). Evaluation of the role of P-glycoprotein in ivermectin uptake by primary cultures of bovine brain microvessel endothelial cells. Neurochem Res 23(2): 203–209.

SADLER KC, AMSTERDAM A, SOROKA C, BOYER J, and HOPKINS N (2005). A genetic screen in zebrafish identifies the mutants vps18, nf2 and foie gras as models of liver disease. Development 132(15): 3561–3572.

SCHINKEL AH (1999). P-Glycoprotein, a gatekeeper in the blood-brain barrier. Adv Drug Deliv Rev 36(2–3): 179–194.

SCHINKEL AH, SMIT JJ, VAN TELLINGEN O, BEIJNEN JH, WAGENAAR E, VAN DEEMTER L, MOL CA, VAN DER VALK MA, ROBANUS-MAANDAG EC, TE RIELE HP, et al. (1994). Disruption of the mouse mdr1a P-glycoprotein gene leads to a deficiency in the blood-brain barrier and to increased sensitivity to drugs. Cell 77(4): 491–502.

SCHINKEL AH, WAGENAAR E, MOL CA, and VAN DEEMTER L (1996). P-glycoprotein in the blood-brain barrier of mice influences the brain penetration and pharmacological activity of many drugs. J Clin Invest 97(11): 2517–2524.

STEWART PA, BELIVEAU R, and ROGERS KA (1996). Cellular localization of P-glycoprotein in brain versus gonadal capillaries. J Histochem Cytochem 44(7): 679–685.

TSUJI A, TERASAKI T, TAKABATAKE Y, TENDA Y, TAMAI I, YAMASHIMA T, MORITANI S, TSURUO T, and YAMASHITA J (1992). P-glycoprotein as the drug efflux pump in primary cultured bovine brain capillary endothelial cells. Life Sci 51(18): 1427–1437.

VAN ASPEREN J, SCHINKEL AH, BEIJNEN JH, NOOIJEN WJ, BORST P, and VAN TELLINGEN O (1996). Altered pharmacokinetics of vinblastine in Mdr1a P-glycoprotein-deficient Mice. J Natl Cancer Inst 88(14): 994–999.

WANDEL C, KIM R, WOOD M, and WOOD A (2002). Interaction of morphine, fentanyl, sufentanil, alfentanil, and loperamide with the efflux drug transporter P-glycoprotein. Anesthesiology 96(4): 913–920.

WANG Q, YANG H, MILLER DW, and ELMQUIST WF (1995). Effect of the p-glycoprotein inhibitor, cyclosporin A, on the distribution of rhodamine-123 to the brain: an in vivo microdialysis study in freely moving rats. Biochem Biophys Res Commun 211(3): 719–726.

WEISS J, KERPEN CJ, LINDENMAIER H, DORMANN SM, and HAEFELI WE (2003). Interaction of antiepileptic drugs with human P-glycoprotein in vitro. J Pharmacol Exp Ther 307(1): 262–267.

WESTERFIELD M (1993). The Zebrafish Book: a guide for the laboratory use of zebrafish. The University of Oregon Press.

WULLIMANN MF and REICHERT H (1996). Neuroanatomy of the zebrafish brain. Birkhauser, Basel, 144 pp.

XIA CQ, MILTON MN, and GAN LS (2007). Evaluation of drug-transporter interactions using in vitro and in vivo models. Curr Drug Metab 8(4): 341–363.

YUMOTO R, MURAKAMI T, NAKAMOTO Y, HASEGAWA R, NAGAI J, and TAKANO M (1999). Transport of rhodamine 123, a P-glycoprotein substrate, across rat intestine and Caco-2 cell monolayers in the presence of cytochrome P-450 3A-related compounds. J Pharmacol Exp Ther 289(1): 149–155.

Chapter 15

Assessment of Effects on Visual Function in Larval Zebrafish

Wendy Alderton

CB1 Bio Ltd, Cambridge, UK

15.1 INTRODUCTION

Numerous marketed drugs cause adverse ocular events by affecting the function of the retina or visual pathways or by causing overt retinal toxicity. Medications that have been reported to cause adverse ocular effects in the clinic include bisphosphonates; antiepileptic drugs such as topiramate and vigabatrin; antituberculosis treatments such as ethambutol and isoniazid; isotretinoin and other retinoids; and amiodarone (Santaella and Fraunfelder, 2007). The relatively small mass of the eye and its rich blood supply make it susceptible to adverse effects on visual function for drugs that cross the blood–retinal barrier. While effects on the eye are reversible if detected early, if undetected, toxic effects may progress to serious and irreversible ocular damage (Chiou, 1999). With a few exceptions, the mechanisms of retinal toxicity are poorly understood. The incidence and impact of retinal toxicity in a typical pharmaceutical development portfolio has been reported as 6.8% of the failures between 1993 and 2006 were due to retinal toxicity. Although this is a small but measurable incidence, ocular toxicity does not lend itself to "risk management," so the impact on a candidate drug project is potentially serious. Currently, ocular safety is assessed at a late stage in the preclinical development of a compound. Conventional studies, such as the measurement of the electroretinogram (ERG) in dogs or rodents, are technically difficult, labor intensive, and often poorly predictive. Consequently, the lack of a predictive, convenient method of assessing visual function preclinically has been recognized by the industry.

The organization of the genome and the genetic pathways controlling signal transduction and development are highly conserved between zebrafish and man. The

Zebrafish: Methods for Assessing Drug Safety and Toxicity, Edited by Patricia McGrath.
© 2012 John Wiley & Sons, Inc. Published 2012 by John Wiley & Sons, Inc.

zebrafish larva is amenable to medium-to-high-throughput *in vivo* screening due the following attributes:

- Relative ease of maintaining large stocks of animals and their high fecundity.
- Larvae can live in only 200 μL of fluid; therefore, screening can be undertaken in multiwell plates and only milligrams of compound are needed.
- Zebrafish are DMSO tolerant and can readily absorb compounds from the water in which they swim.
- Rapid embryonic development *ex utero*, which facilitates experimental manipulation and allows the direct observation of organ function *in vivo*.

These properties have established the zebrafish as an *in vivo* model system that is relevant to studies of human diseases (Zon and Peterson, 2005; Lieschke and Currie, 2007) and for the assessment of safety liabilities (Barros et al., 2008). Thus, *in vivo* analysis of the effects of compounds, for example, on visual function, can be undertaken at much earlier stages in the drug development process (hit to lead or lead optimization) with lower compound requirement and at a higher throughput in zebrafish larvae than is usually possible with rodents.

15.2 DEVELOPMENT OF VISUAL SYSTEM IN ZEBRAFISH

The zebrafish retina is structurally very similar to the human retina and therefore the zebrafish visual system has been evaluated for both the modeling of eye diseases (Goldsmith, 2001) and the assessment of the effects of drugs on visual function. Zebrafish have a cone dense retina and thus, like humans, have rich color vision, providing a potential advantage over testing compounds for effects on visual function in nocturnal rodents, which have rod-dominant retinas. Figure 15.1 shows the cell layout in an embryonic zebrafish retina (a) and a comparison of sections through an adult human (left) and embryonic zebrafish (right) retina (b). The relative positions of the various cell types are the same in both human and zebrafish retinas.

Visual system development is very rapid in zebrafish embryos and is imperative for orientation and to enable predator avoidance and feeding behavior. The initial event of eye development is the evagination of the optic lobes from the diencephalon at around 10 h post fertilization (hpf). The optic primordium appears at about 12hpf (Schmitt and Dowling, 1994) and the first ganglion cells appear in the ventronasal retina at 30hpf (Schmitt and Dowling, 1994; Burrill and Easter, 1995). Cone outer segments appear at 60hpf in a restricted area of ventral retina (Branchek and Bremiller, 1984) and then gradually elongate. Signal transmission from photoreceptors to second-order neurons starts around 3.5 days post fertilization (dpf) and is fully functional at 5dpf (Biehlmaier et al., 2003). The earliest quantifiable visual behavior is the visual startle response, whereby larvae respond to a sudden decrease in illumination with a rapid body movement. This behavioral response starts at around 68hpf, just at the time when outer segments of photoreceptors and synaptic ribbons have

(a)

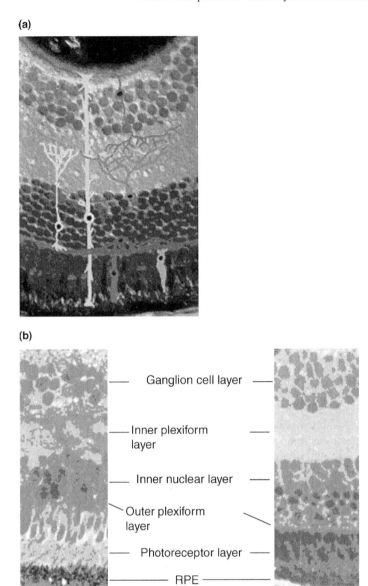

(b)

Ganglion cell layer

Inner plexiform layer

Inner nuclear layer

Outer plexiform layer

Photoreceptor layer

RPE

Figure 15.1   (a) The cell layout in the retina is highly stereotypical. This plastic section through an embryonic zebrafish retina has been overlaid for illustration. The lens is at the top and retinal pigment epithelium at the bottom. Red: ganglion cell; light blue: Mueller cell; green: amacrine cell; dark blue: horizontal cell; yellow: bipolar cell; pink: rod photoreceptor; orange: cone photoreceptor. (b) Plastic sections through an adult human (left) and embryonic zebrafish (right) retina. The relative positions of the various cell types are the same in both. (Reprinted from Goldsmith and Harris (2003) with permission from Elsevier.) (See the color version of this figure in Color Plates section.)

formed in the retina. By 5dpf, the visual system is well developed according to electrophysiological, morphological, and behavioral criteria. However, it should be noted that while cone receptors in the zebrafish retina develop and become fully functional relatively early, the rod receptors do not become adult-like until 30dpf (Bilotta et al., 2001).

15.3 METHODS FOR ASSESSING VISUAL FUNCTION IN LARVAL ZEBRAFISH

To assess visual function in zebrafish, behavioral assays have been developed that take advantage of inherent visual reflexes in the fish, such as the optokinetic response (OKR) and optomotor response (OMR). These assays have been successfully employed in a number of screens to identify mutant zebrafish with defects in the visual system (reviewed in Neuhauss, 2003). These assays have more recently also been applied to the evaluation of the effect of compounds on visual function (Berghmans et al., 2008; Richards et al., 2008). The OMR and OKR methods and validation are described in Sections 15.3.1 and 15.3.2, respectively. Larvae used for both the OMR and OKR assays are reared in the dark since rearing in light can cause abnormal effects on the development of the zebrafish retina (Saszik et al., 2002). Circadian rhythms can affect visual sensitivity (Li and Dowling, 1998) that can affect assay performance. Therefore, the OMR and OKR assays should be run at a similar time of day on each occasion. The pH, conductivity, and dissolved oxygen of the media are monitored during the 3–8dpf exposure of larvae to compound prior to OMR evaluation. The OECD Guideline 212 for testing of chemicals: fish, short-term toxicity test on embryo and sac fry stages (www.oecd.org and Chapter 4) recommends that pH should remain within the range of ±0.5 pH units in an assay. The Guideline also recommends that dissolved oxygen should be between 60% and 100% saturation throughout a study. There is little literature on the conductivity requirements for larval zebrafish but adults are tolerant to conductivity ranging from 400 µS to more than 1000 µS. These conditions are monitored to eliminate the possibility that effects on visual function could be attributable to the changes in pH, conductivity, or dissolved oxygen rather than directly to the compound.

ERG measurement of retinal processing and histology of the larval zebrafish retina have been applied as secondary, mechanistic studies. The methodology for ERG is described in Section 15.3.3.

Visual background adaptation (VBA) is the aggregation and dispersal of pigment granules in the larval zebrafish skin in response to ambient light levels and is a camouflage mechanism. VBA is controlled in response to signals from the retina via the hypothalamus by the pituitary secreting hormones to aggregate or disperse the melanosomes (reviewed in Balm and Groneveld, 1998). Blind zebrafish cannot sense the light level and therefore appear darker than control larvae and this has been used to identify blind mutant zebrafish (Neuhauss et al., 1999). However, while VBA provides a high-throughput screening method for compound treatments, it was found to be unsatisfactory for screening of compound for effects on visual function, since it gave poor correlation both with the results in OMR and OKR assays and with the expected effects from clinical studies (Richards, personal communication). This is likely to be due to the compounds having effects on pigmentation via toxicity mechanisms not related to visual function.

15.3.1 Optomotor Response Assay

The OMR is the locomotor behavior of an animal induced in response to a repetitive pattern. This response can be elicited in zebrafish by moving horizontal stripes below long transparent chambers in which the larvae swim (Orger et al., 2000; Krauss and Neumeyer, 2003; Maaswinkel and Li, 2003; Roeser and Baier, 2003). The larvae swim to maintain a constant position relative to a stripe and therefore visually normal fish accumulate at one end of the channel. The number of larvae in the final 25%, or "pass area," of the channel can then be quantified and the proportion of fish in the pass area will be reduced in groups with defective visual function. At 6–7dpf, larvae will respond to this stimulus >90% of the time (Orger et al., 2004). The OMR technique has been used to investigate motion perception in larval zebrafish. It was shown that inputs to motion vision derive predominantly from L (red) and M (green) cones, but not from short-wavelength (UV and S (blue)) cones, as is also the case in higher vertebrates (Orger and Baier, 2005). The OMR assay has a higher throughput for evaluating the effect of compounds on zebrafish vision than the OKR assay described in Section 15.3.2. However, compounds that affect the locomotor ability of the zebrafish would be false positives in the OMR assay, since hypomotility will decrease the number of larvae scored in the pass area of the chamber. Therefore, the OMR assay may be used as a primary screen with positive compounds being further evaluated for defects in visual function by secondary screening in OKR and for hypomotility in a locomotor assay. The OMR assay can also be adapted for adult zebrafish although the throughput of such an assay is very low (Fleisch and Neuhauss, 2006). To avoid schooling behavior, single adult zebrafish are placed inside a round testing chamber surrounded by a rotating drum containing alternating black and white stripes and the fish then swims in the direction of the stripes.

15.3.1.1 OMR Method The maximum tolerated concentration (MTC) for each compound is first determined in larval zebrafish to identify toxic concentrations such that testing for OMR is not compromised by general toxicity. This approach has recently been reviewed by Hutchinson et al. (2009). The MTC is determined as the concentration at which no lethality was observed above that seen in control (untreated) siblings. Death is assessed by the absence of a heartbeat. In addition, the following tests are used to assess toxicity at nonlethal concentrations: (i) locomotor activity and startle response (gross score by eye in response to tapping the dish), (ii) swim position; loss of dorsoventral balance, and (iii) any morphological abnormalities, the most common of which are pericardial edema, bent body, and failure to inflate swim bladder.

Larvae are exposed to the test compound in 1% (v/v) DMSO final solvent concentration in swimming medium in 24-well plates for 5 days (from 3 to 8dpf)

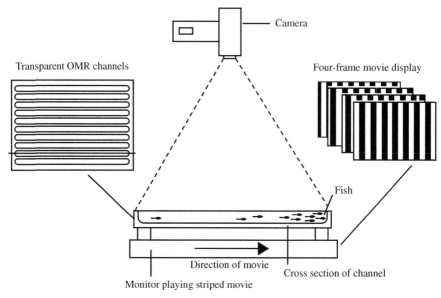

Figure 15.2   Schematic of the OMR apparatus, viewed from above and in cross section. Larvae at 8dpf swim in transparent channels on a flat screen monitor playing a movie with a high-contrast black and white grating (bandwidth: 2 cm; drift speed: 1 cycle/s). The movie is played for 45 s in the forward and reverse directions three times and the position of each larva is recorded by digital photography. (Reprinted from Richards et al. (2008) with permission from Elsevier.)

at $28.5 \pm 0.5°C$ with compound replenishment at 6dpf. The larvae are then transferred to a transparent, multi-lane acrylic block containing embryo medium, with 10 larvae per lane and with two groups of vehicle controls in each assay. The experimental apparatus for assessing OMR is shown in a schematic representation in Fig. 15.2. OMR is assessed by subjecting the larvae to a movie played with a high-contrast black and white grating on a horizontal screen underneath the transparent channels that contained the larvae. The larvae are shown the movie such that the stripes pass in the forward and reverse directions three times for 45 s each. The position of each larva in relation to the end of the channel is then recorded. Larvae with normal vision swim in the direction of the moving grating toward the end of the channel. At the end of each run, the number of larvae in the "pass area," which represents the furthest 25% of the channel, is calculated and the mean percentage in the pass area for all six runs obtained. If either of the vehicle controls scored <60% overall, then the assay is considered invalid (Richards et al., 2008).

15.3.1.2   OMR Validation Studies   Richards et al. (2008) assessed compounds that cause adverse effects on visual function by a variety of different mechanisms in a zebrafish OMR assay. This study reported the results of the blinded testing of 27 compounds, and revealed a good concordance between the effects of

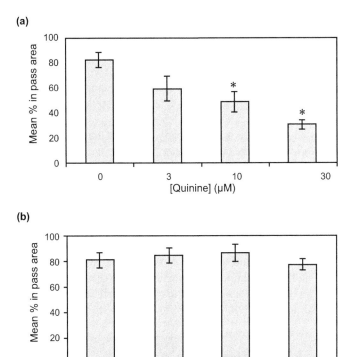

Figure 15.3   Assessment of OMR in zebrafish larvae at 8dpf after 5 days treatment with (a) quinine, rated as positive, and (b) allopurinol, rated as negative. The mean and standard error are shown for six runs. Each run used a group of 10 larvae. Statistical significance was calculated using a Student's t-test, comparing control to treated groups. $^*P < 0.05$. (Reprinted from Richards et al. (2008) with permission from Elsevier.)

compounds in larval zebrafish with the data available from other *in vivo* and *in vitro* models or the clinic. Thirteen out of 19 positive compounds produced the expected effect while 6 of the 8 negative compounds were correctly predicted. Representative data obtained for quinine (positive) and allopurinol (negative) are shown in Fig. 15.3. This study gave an overall predictivity of 70% for the OMR assay with a sensitivity of 68% and a specificity of 75%. Two false positives were observed, bisoprolol and spironolactone. Visual inspection of the larvae treated with these two compounds had suggested that a motility defect might be present at the highest concentrations tested. This was confirmed by testing larvae treated with bisoprolol and spironolactone in a locomotor activity assay and OKR assay as described in Section 15.3.2.2. There were two false negatives, sodium iodate and indomethacin, detected in the study. False negatives could arise where compounds are poorly absorbed by the larvae and tissue exposure levels are too low to have an effect. This is particularly likely to be the case where an MTC could not be determined. Determination of concentrations within the larvae using liquid chromatography–mass spectrometry would allow for the detection of false negatives arising from lack of compound penetration. This would also provide the effective concentrations in the larvae allowing a comparison with the effects of compounds in other model systems and with plasma concentrations in humans. In addition, a comparison of four standard compounds in the OMR assay in WIK and TL zebrafish wild-type strains revealed no difference in sensitivity between the two strains.

In a second study, the OMR assay correctly predicted the effect on visual function in mammalian or clinical studies of seven compounds of the nine tested (Berghmans et al., 2008). In agreement with previously observed mammalian preclinical or clinical effects, chloroquine, chlorpromazine, diazepam, nicotine, ouabain, and phenytoin showed inhibition of OMR. Atropine and lithium were the two compounds tested that did not affect zebrafish larval OMR but are known to affect vision in humans by affecting binocular vision or accommodation. However, zebrafish have no binocular vision and a spherical lens with an accommodation process unlike that of man (Vihtelic et al., 1999) and this may explain the false negative data for these compounds. In addition, lithium is known to cause nystagmus and decreased accommodation only after prolonged treatment in the clinic, and a 5-day exposure in larval zebrafish may not be long enough to replicate this effect.

15.3.2 Optokinetic Response Assay

The optokinetic response, which maintains optimal visual acuity and measures the ability of larval zebrafish to track movements in their environment, is observed as early as 3dpf and is present in 98% of larvae at 5dpf (Brockerhoff et al., 1995, 1997; Neuhauss et al., 1999). The OKR assay is carried out with larvae immobilized in methylcellulose inside a drum on which black and white stripes are rotated, which elicits the optokinetic nystagmus (Brockerhoff et al., 1995, 1997; Easter and Nicola, 1996; Roeser and Baier, 2003; Brockerhoff, 2006; Huang and Neuhauss, 2008). A schematic diagram of the OKR apparatus is shown in Fig. 15.4. Nystagmus is a stereotyped behavior in which a series of smooth ocular pursuits track the movement of the stripe, followed by a rapid saccade as the eyes flick onto the next stripe as the first leaves the visual field. Eye movements are instantly reversed when the direction of the rotating stripes is reversed. The movement of the large, pigmented zebrafish eye can be easily followed under a microscope and the number of saccades counted. Since larval zebrafish absorb oxygen through the skin, they can be maintained in the methylcellulose for several hours without adverse effects. Larvae with defective visual function are expected to show a reduced number of saccades compared to control (untreated) larvae. Changing the stripe width, angular velocity, and contrast also allows quantification of visual acuity, contrast sensitivity, and light adaptation in zebrafish larvae. A study by Rinner et al. (Rinner et al., 2005) revealed that optokinetic response in zebrafish larvae, as in cats and humans, is a function of angular velocity rather than temporal frequency. The OKR assay can also be adapted to test for color vision as demonstrated by the isolation of a red light-insensitive mutant

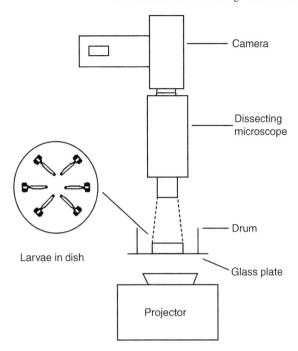

Camera

Dissecting microscope

Figure 15.4 A schematic of the OKR apparatus. A projector plays a movie of black and white grating onto the inside of the drum, which surrounds a dish containing the 8dpf zebrafish larvae immobilized in methylcellulose (inset). A camera records movies of the larvae for subsequent analysis of eye position and quantification of the number of saccades for the eye of each larva.

Drum

Larvae in dish

Glass plate

Projector

zebrafish (partial optokinetic nystagmus (poa)) using gratings illuminated with long-wavelength red instead of white light (Brockerhoff et al., 1997).

15.3.2.1 OKR Method The MTC for the compound is first determined as described in Section 15.3.1.1. The zebrafish larvae are treated with compound in 1% (v/v) DMSO final solvent concentration in swimming medium for 5 days from 3 to 8dpf with compound refreshment at 6dpf before assessment by OKR. The OKR assay is undertaken by immobilizing groups of up to 10 zebrafish 8dpf larvae in 4% (w/v) methylcellulose in a 2.5 cm plastic Petri dish with the dorsal side up, tail toward the center, and head facing outward, taking care not to introduce air bubbles that would impede vision. The larvae are allowed to habituate for 10 min prior to assessment. The methylcellulose does not contain test compound but the OKR assessment is completed within 15 min of the larvae being removed from the compound solution. A white drum is then placed around the Petri dish and larvae are video-recorded from above while a movie of vertical black and white gratings is played by projection from below in a clockwise and then counterclockwise direction inside the drum. The stripes are run for 1 min in each direction at a speed of 78 cycles/min, 0.05 cycles/degree. Eye movement of each individual larva is then analyzed by counting the number of saccades. If the vehicle-treated larvae control scored less than 20 saccades, the experiment is considered invalid (Richards et al., 2008).

15.3.2.2 OKR Validation Studies The OKR assay can be used to detect the adverse effects of compounds on visual function. However, the OKR assay requires each fish to be mounted individually, dorsal side up, in methylcellulose and is therefore lower throughput than the OMR assay. The OKR assay has been used as a secondary assay to analyze specific effects on visual function as a follow-up to primary screening in the OMR assay, which may be indicative of effects on either visual function or locomotor activity. Richards et al. (2008) employed the OKR assay in such a manner to test two compounds, bisoprolol and spironolactone, which had been identified as false positives in the OMR assay. The data obtained for bisoprolol are shown in Fig. 15.5. No significant effect of the compound on OKR was observed (b) but significant inhibition of locomotor activity at concentrations at which inhibition of OMR was observed (c). Therefore, in this case it was demonstrated that bisoprolol was a false positive in the OMR assay due to detrimental effects on locomotor activity in the OMR assay rather than specific effects on visual function.

In a further study on the effects of compounds on larval zebrafish visual function, marketed sedatives and muscle relaxants were assessed by OKR (Liu, unpublished data, 2007). The assay correctly predicted the effects of 8 out of the 10 drugs. The sedatives trazodone, diphenhydramine, and haloperidol all cause blurred vision as a known, clinical adverse effect and caused inhibition of OKR at 3–10 μM. The muscle relaxants tizanidine and methocarbamol, both of which cause adverse visual effects in the clinic, showed inhibition of OKR in zebrafish. Chlorzoxazone and dantrolene have no reported adverse visual effects in the clinic and did not inhibit OKR. Diazepam was found to be a false negative in this study as it has been reported to cause blurred vision in the clinic, although it did exhibit the expected sedative effect on larval zebrafish locomotor function.

15.3.3 Electroretinography

Measurement of the electroretinograms of larval zebrafish retinas has been used to localize defects in the visual pathways of mutant larval zebrafish (Brockerhoff et al., 1995; Li and Dowling, 1997; Neuhauss et al., 1999). The ERG measures light-evoked sum potentials at the corneal surface of the retina. The most prominent features of the ERG are the short, negative a-wave, which originates from the photoreceptor, followed by the extended, positive b-wave, which reflects the activity of bipolar neurons (Neuhauss et al., 1999). Adaptation of the larval ERG apparatus has also been made for the measurement of ERG in adult zebrafish (Makhankov et al., 2004). The effects of a few compounds on retinal processing or development have been assessed. The compound APB (DL-2-amino-4-phosphonobutyric acid) has been found to affect retinal processing of many vertebrate species by suppressing the b-wave of the ERG, as is also the case for zebrafish (Saszik et al., 2002; Ren and Li, 2004). Also, zebrafish exposed to ethanol from 2 to 5dpf showed a reverse reduction in a- and b-waves and OKR response at concentrations that did not cause obvious morphological changes to the retina (Matsui et al., 2006). However, due to the low-throughput nature of the assay, ERG has not been used extensively to study the effects of compounds on retinal processing.

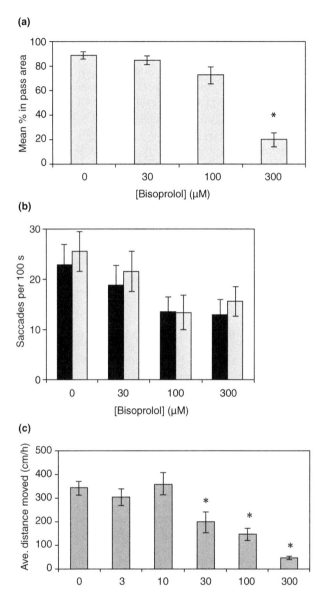

Figure 15.5 Assessment of visual function and locomotor activity of zebrafish larvae at 8dpf after 5 days of treatment withbisoprolol in: (a) OMR assay: The mean and standard errors are shown for 6 runs. Each run used a group of 10 larvaetreated with compound or DMSO control.* $P<0.005$ in a Student's t test comparing control with treated groups. Treatment with 300 μM bisoprolol significantly inhibited OMR; (b) OKR assay: Data represents the mean andstandard error of groups of 12 larvae assessed for the number of saccades in 1 min in the ■ clockwise and ▨ counterclockwise directions. No group was significantly different from control in one-way ANOVA and Dunnett's multiple comparison tests; (c) Locomotor activity: Data represents the mean and standard error of groups of 10 larvae assessed by livevideo-tracking of movement in 96-well plates for 45 minutes. *$P<0.01$ in one-way ANOVA and Dunnett's multiple comparison tests. Treatment with 30-300 μM bisoprolol significantly inhibited locomotor activity. (Reprinted from Richards et al. (2008) with permission from Elsevier)

15.4 CONCLUSIONS

Several well-established methods are available for the safety testing of compounds for adverse effects on visual function in larval zebrafish, including OMR, OKR and ERG measurement. These methodologies, along with other larval zebrafish safety assays, are gaining acceptance in nonclinical safety evaluation, as a means of "frontloading" *in vivo* studies, because the assays are amenable to medium/high-throughput screening and chronic dosing (Valentin and Hammond, 2008). Preliminary published observations are encouraging with predictivity of the OMR assay being reported as 70%. However, the visual function assays require further validation using a wide range of compounds to ensure the reliability, robustness, reproducibility, and predictive value of the OMR and OKR assays. Further automation and miniaturization of the OMR and OKR assays would improve throughput and the determination of the amount of compound in the zebrafish larvae by mass spectrometry would help to eliminate false negatives and transform these assays from a qualitative to quantitative prediction of effects.

REFERENCES

BALM PH and GRONEVELD D (1998). The melanin-concentrating hormone system in fish. Ann NY Acad Sci 839: 205–209.

BARROS TP, ALDERTON WK, REYNOLDS HM, ROACH AG, and BERGHMANS S (2008). Zebrafish: an emerging technology for *in vivo* pharmacological assessment to identify potential safety liabilities in early drug discovery. Br J Pharmacol 154(7): 1400–1413.

BERGHMANS S, BUTLER P, GOLDSMITH P, WALDRON G, GARDNER I, GOLDER Z, RICHARDS FM, KIMBER G, ROACH A, ALDERTON W, and FLEMING A (2008). Zebrafish based assays for the assessment of cardiac, visual and gut function: potential safety screens for early drug discovery. J Pharmacol Toxicol Methods 58(1): 59–68.

BIEHLMAIER O, NEUHAUSS SC, and KOHLER K (2003). Synaptic plasticity and functionality at the cone terminal of the developing zebrafish retina. J Neurobiol 56(3): 222–236.

BILOTTA J, SASZIK S, and SUTHERLAND SE (2001). Rod contributions to the electroretinogram of the dark-adapted developing zebrafish. Dev Dyn 222(4): 564–570.

BRANCHEK T and BREMILLER R (1984). The development of photoreceptors in the zebrafish, Brachydanio rerio. I. Structure. J Comp Neurol 224(1): 107–115.

BROCKERHOFF SE (2006). Measuring the optokinetic response of zebrafish larvae. Nat Protoc 1(5): 2448–2451.

BROCKERHOFF SE, HURLEY JB, JANSSEN-BIENHOLD U, NEUHAUSS SC, DRIEVER W, and DOWLING JE (1995). A behavioral screen for isolating zebrafish mutants with visual system defects. Proc Natl Acad Sci USA 92(23): 10545–10549.

BROCKERHOFF SE, HURLEY JB, NIEMI GA, and DOWLING JE (1997). A new form of inherited red-blindness identified in zebrafish. J Neurosci 17(11): 4236–4242.

BURRILL JD and EASTER SS Jr. (1995). The first retinal axons and their microenvironment in zebrafish: cryptic pioneers and the pretract. J Neurosci 15(4): 2935–2947.

CHIOU G (1999). Ocular Toxicity, 2nd edition. Taylor & Francis, Philadelphia, pp. 43–86, 225–354.

EASTER SS Jr. and NICOLA GN (1996). The development of vision in the zebrafish (*Danio rerio*). Dev Biol 180(2): 646–663.

FLEISCH VC and NEUHAUSS SC (2006). Visual behavior in zebrafish. Zebrafish 3(2): 191–201.

GOLDSMITH P (2001). Modelling eye diseases in zebrafish. Neuroreport 12(13): A73–A77.

GOLDSMITH P and HARRIS WA (2003). The zebrafish as a tool for understanding the biology of visual disorders. Semin Cell Dev Biol 14(1): 11–18.

HUANG YY and NEUHAUSS SC (2008). The optokinetic response in zebrafish and its applications. Front Biosci 13: 1899–1916.

HUTCHINSON TH, BOGI C, WINTER MJ, and OWENS JW (2009). Benefits of the maximum tolerated dose (MTD) and maximum tolerated concentration (MTC) concept in aquatic toxicology. Aquat Toxicol 91(3): 197–202.

KRAUSS A and NEUMEYER C (2003). Wavelength dependence of the optomotor response in zebrafish (*Danio rerio*). Vision Res 43(11): 1273–1282.

LI L and DOWLING JE (1997). A dominant form of inherited retinal degeneration caused by a non-photoreceptor cell-specific mutation. Proc Natl Acad Sci USA 94(21): 11645–11650.

LI L and DOWLING JE (1998). Zebrafish visual sensitivity is regulated by a circadian clock. Vis Neurosci 15(5): 851–857.

LIESCHKE GJ and CURRIE PD (2007). Animal models of human disease: zebrafish swim into view. Nat Rev Genet 8(5): 353–367.

MAASWINKEL H and LI L (2003). Spatio-temporal frequency characteristics of the optomotor response in zebrafish. Vision Res 43(1): 21–30.

MAKHANKOV YV, RINNER O, and NEUHAUSS SC (2004). An inexpensive device for non-invasive electroretinography in small aquatic vertebrates. J Neurosci Methods 135(1-2): 205–210.

MATSUI JI, EGANA AL, SPONHOLTZ TR, ADOLPH AR, and DOWLING JE (2006). Effects of ethanol on photoreceptors and visual function in developing zebrafish. Invest Ophthalmol Vis Sci 47(10): 4589–4597.

NEUHAUSS SC (2003). Behavioral genetic approaches to visual system development and function in zebrafish. J Neurobiol 54(1): 148–160.

NEUHAUSS SC, BIEHLMAIER O, SEELIGER MW, DAS T, KOHLER K, HARRIS WA, and BAIER H (1999). Genetic disorders of vision revealed by a behavioral screen of 400 essential loci in zebrafish. J Neurosci 19(19): 8603–8615.

ORGER MB and BAIER H (2005). Channeling of red and green cone inputs to the zebrafish optomotor response. Vis Neurosci 22(3): 275–281.

ORGER MB, GAHTAN E, MUTO A, PAGE-MCCAW P, SMEAR MC, and BAIER H (2004). Behavioral screening assays in zebrafish. Methods Cell Biol 77: 53–68.

ORGER MB, SMEAR MC, ANSTIS SM, and BAIER H (2000). Perception of Fourier and non-Fourier motion by larval zebrafish. Nat Neurosci 3(11): 1128–1133.

REN JQ and LI L (2004). Rod and cone signaling transmission in the retina of zebrafish: an ERG study. Int J Neurosci 114(2): 259–270.

RICHARDS FM, ALDERTON WK, KIMBER GM, LIU Z, STRANG I, REDFERN WS, VALENTIN JP, WINTER MJ, and HUTCHINSON TH (2008). Validation of the use of zebrafish larvae in visual safety assessment. J Pharmacol Toxicol Methods 58(1): 50–58.

RINNER O, RICK JM, and NEUHAUSS SC (2005). Contrast sensitivity, spatial and temporal tuning of the larval zebrafish optokinetic response. Invest Ophthalmol Vis Sci 46(1): 137–142.

ROESER T and BAIER H (2003). Visuomotor behaviors in larval zebrafish after GFP-guided laser ablation of the optic tectum. J Neurosci 23(9): 3726–3734.

SANTAELLA RM and FRAUNFELDER FW (2007). Ocular adverse effects associated with systemic medications: recognition and management. Drugs 67(1): 75–93.

SASZIK S, ALEXANDER A, LAWRENCE T, and BILOTTA J (2002). APB differentially affects the cone contributions to the zebrafish ERG. Vis Neurosci 19(4): 521–529.

SCHMITT EA and DOWLING JE (1994). Early eye morphogenesis in the zebrafish, *Brachydanio rerio*. J Comp Neurol 344(4): 532–542.

VALENTIN JP and HAMMOND T (2008). Safety and secondary pharmacology: successes, threats, challenges and opportunities. J Pharmacol Toxicol Methods 58(2): 77–87.

VIHTELIC TS, DORO CJ, and HYDE DR (1999). Cloning and characterization of six zebrafish photoreceptor opsin cDNAs and immunolocalization of their corresponding proteins. Vis Neurosci 16(3): 571–585.

ZON LI and PETERSON RT (2005). *In vivo* drug discovery in the zebrafish. Nat Rev Drug Discov 4(1): 35–44.

Chapter 16

Development of a Hypoxia-Induced Zebrafish Choroidal Neovascularization Model

Wen Lin Seng, Yingxin Lin, Susie Tang, and Lisa Zhong
Phylonix, Cambridge, MA, USA

16.1 INTRODUCTION

In this research, using $CoCl_2$, a hypoxia mimetic, we describe development of a zebrafish choroidal neovascularization (CNV) model for drug screening. We demonstrate that abnormal CNV can be observed 4 days after adding $CoCl_2$ directly to the fish water, facilitating rapid screening of compound libraries. To eliminate nonspecific staining from angiogenesis present in other sites, we describe methods for removing intact eyes from whole zebrafish using collagenase enzyme treatment. This method facilitates both visual assessment of CNV, which occurs in the back of the eye, and quantitative assessment of compound effects on isolated eyes using a comparatively high-throughput microplate format. This convenient model will facilitate drug discovery and assist in elucidating the pathogenesis of choroidal neovascularization.

16.1.1 Hypoxia and Ocular Neovascularization

Oxygen homeostasis plays a crucial role in supporting normal development and physiology in all animals. Pathological conditions can develop if the balance between oxidative phosphorylation and oxidative stress is disrupted. In eye diseases, hypoxia has been shown to be a major mechanism of abnormal neovascularization (NV)

Zebrafish: Methods for Assessing Drug Safety and Toxicity, Edited by Patricia McGrath.
© 2012 John Wiley & Sons, Inc. Published 2012 by John Wiley & Sons, Inc.

(Semenza, 2000; Schwesinger et al., 2001; Smith, 2002; Shih et al., 2003). Diabetic retinopathy (DR) and age-related macular degeneration (AMD) are the two leading causes of adult blindness in developed countries and retinopathy of prematurity (ROP) is a major cause of vision loss in premature infants. DR occurs when diabetes damages small blood vessels in the retina. As this disease progresses and enters the proliferative stage, newly formed blood vessels grow along the retina and in the clear, gel-like vitreous body that fills the interior of the eyes. These blood vessels can bleed, cloud vision, and destroy the retina. Macular degeneration (MD) targets the central area of the retina macula. AMD is usually categorized as either "wet" or "dry." In the more common "dry" form, fat deposits damage macular cells. In the "wet" form, abnormally proliferating blood vessels beneath the retina in the choroid rupture and leak fluid that damages light-sensitive cells in the macula region. This disease can progress rapidly, eventually leading to vision loss.

16.1.2 Hypoxia and $CoCl_2$

HIF-1α is a significant hypoxia transcription factor. Under hypoxic conditions, ubiquitination decreases dramatically, causing HIF-1α accumulation. HIF-1α dimerizes with HIF-1β, binds to the hypoxia response element located in the promoter region of the corresponding genes, and upregulates gene expression. Several genes, including glucose/energy metabolism genes and vascular development/remodeling genes, such as *VEGF*, which promotes progression of AMD and DR, are known to respond to HIF transcription factor. Increased *VEGF* has been observed in both human retinopathy and MD (Rasmussen et al., 2001). Mounting evidence suggests that genes involved in vascular development/remodeling, including *IGF*, also play pivotal roles in pathological angiogenesis (Bouck, 2002). Oxygen level, which correlates with iron level and heme protein synthesis, can be modulated by iron chelators, and Co^{2+}, produced by $CoCl_2$, has been found to be the most potent iron chelator (Rafii et al., 2000). In human cells, $CoCl_2$ has been shown to induce *HIF-1α* expression and inhibit HIF-1α ubiquitination (Jiang et al., 1997; Yuan et al., 2003). $CoCl_2$ has also been shown to increase activity in the HIF-1α transcriptional domain and upregulate *VEGF* (Van Lieshout et al., 2003). $CoCl_2$ mimics the hypoxia mechanism and upregulates downstream genes that are reactive to HIF-1.

16.1.3 Current Neovascularization Models

Although HIF and VEGF have been shown to be involved in abnormal ocular neovascularization (Schwesinger et al., 2001; Smith, 2002; Shih et al., 2003), drug development has been hindered by the lack of suitable animal models. A ROP model has been generated by initially exposing animals to a high oxygen atmosphere, simulating hypoxic conditions, followed by return to normal levels (Shih et al., 2003). Current *in vivo* CNV rodent models are highly manual and rely on lengthy surgical procedures performed by well-trained personnel, limiting our understanding of the underlying mechanisms of CNV and inhibiting drug discovery.

16.1.4 Advantages of Zebrafish Ocular Neovascularization Model

Development of the ocular vasculature in zebrafish is indistinguishable from development in other vertebrates (Fouquet et al., 1997). Two days post fertilization (dpf), both the choroid vasculature located at the back of the eye and the hyaloid vasculature surrounding the lens are present. By 5dpf, the choroidal vascular plexus (CVP) is well formed and hyaloid vessels appear to have regressed (Isogai et al., 2001). Since choroidal vessels can supply nutrients to the entire retina, zebrafish do not require retinal vessels. A number of genes involved in vertebrate angiogenesis, including *VEGF*, *Flk-1*, *Ang-1*, *Ang-2*, *Tie-1*, and *Tie-2*, have been identified and shown to perform the same functions in zebrafish and mammals (Fouquet et al., 1997; Liang et al., 1998; Lyons et al., 1998). Because early-stage zebrafish are completely transparent and they develop in 96-well microplates in 100 µL media with no feeding, assessing effects of chemicals on all aspects of vascular formation is straightforward.

Recent research showed that early-stage zebrafish (<24 h post fertilization (hpf)) can survive in anoxic or hypoxic atmospheres for 24 h (Padilla and Roth, 2001; Ton et al., 2003). Although maintenance in a low oxygen atmosphere for 24 h increased *HIF-1α* and other oxygen homeostasis-related gene expression, these conditions did not increase expression of genes involved in vascular development and remodeling, including *VEGF* (Christopher Ton, personal communication). An alternative method to induce hypoxia response that can lead to changes in vascular development in zebrafish was required. To generate a zebrafish CNV model, we then assessed use of $CoCl_2$ as a hypoxia induction agent. Using this method, we demonstrated that compound treatment upregulated *HIF 1α* and *VEGF* expression and induced abnormal angiogenesis in CVP.

16.2 MATERIALS AND METHODS

16.2.1 Zebrafish Breeding

Zebrafish were generated by natural pairwise mating according to *The Zebrafish Book* (Westerfield, 1993) and maintained in embryo water (5 g of Instant Ocean Salt in 25 L of distilled water) at 28°C. Since embryos receive nourishment from an attached yolk sac for the first 5–6 days of development, no additional maintenance was required. For ELISA experiments, Phylonix AB zebrafish embryos were used. For imaging experiments, Phylonix albino zebrafish were used.

16.2.2 Chemicals and Reagents

All chemicals and reagents were purchased from Sigma (St. Louis, MO), except Celebrex was purchased from Sequoia Research Product Ltd (Oxford, UK) and SU5416 was purchased from Calbiochem, a division of EMD Biosciences Inc. (San Diego, CA). Phy-V antibody, which labels activated endothelial cells

(Seng et al., 2004), was generated by Phylonix (Cambridge, MA). Rhodamine-conjugated secondary antibody was purchased from Jackson ImmunoResearch Laboratories, Inc. (West Grove, PA). Horseradish peroxidase (HRP) suppressor, HRP-conjugated secondary antibody, and Ps-atto were purchased from Beckman Coulter/Lumigen, Inc. (Southfield, MI).

16.2.3 CoCl$_2$ Treatment to Induce Hypoxia

CoCl$_2$ (Sigma, St. Louis, MO) was dissolved in fish water and serially diluted to the desired concentration and used to treat 24hpf zebrafish continuously for 4 days. On day 5, treated zebrafish were processed to perform the following procedures.

16.2.4 Whole Mount Immunofluorescence Staining and Fluorescence Microscopy

To visualize vessel pattern in zebrafish eyes, whole mount immunofluorescence staining was performed. Zebrafish were fixed by Dent's fixative (DMSO/methanol = 1/4) and processed for whole mount immunochemical staining following standard procedures (Westerfield, 1993). Phy-V antibody was used as the primary antibody. Rhodamine-conjugated secondary antibody was then used to stain zebrafish. Animals were examined using a Zeiss M2Bio fluorescence microscope (Zeiss, Thornwood, NY) equipped with a red rhodamine cube, and a chilled CCD camera (AxioCam MRm, Zeiss). Images were analyzed with AxioVision software Rel 4.0 (Zeiss) and Photoshop 7.0 (Adobe, San Jose, CA).

16.2.5 Eye Extraction

Whole mount Phy-V-stained zebrafish were incubated with collagenase (150 unit/mL) at 37°C for 40 min in an Eppendorf tube to loosen eyes from sockets without damaging vessels. An Eppendorf tube containing treated zebrafish was then gently shaken a few times and loosened, intact eyes were then dissociated from whole zebrafish. The CVP, located in the back of the eyes, was then accessible for assessing drug effects.

16.2.6 Immunohistochemistry and Microscopy

Immunohistochemistry was used to validate the CNV model and to assess drug effects. Briefly, whole mount immunostaining was performed as described above; however, instead of rhodamine-conjugated secondary antibody, HRP-conjugated secondary antibody and DAB (both from Jackson ImmunoResearch Laboratories, Inc.) were used to stain zebrafish. After staining, zebrafish were embedded in JB-4 (Polysciences, Warrington, PA), a transparent medium, following manufacturer's instructions. Whole mount immunostained zebrafish were sectioned sagitally (5 μm thick). Slides were then counterstained by nuclear fast red (Sigma, St. Louis, MO). Since JB-4 ordinarily inhibits penetration of large molecules, such as antibodies, instead of conventional immunohistochemistry, we stained and then sectioned zebrafish. Sections were examined using a Zeiss Axiostar compound microscope (Zeiss).

16.2.7 RT-PCR

To determine if *HIF-1α* was upregulated after $CoCl_2$ treatment, total RNA was isolated from control and treated zebrafish and reverse transcribed with MMLV reverse transcriptase (GIBCO/Invitrogen, Carlsbad, CA) primed with oligo dT and subjected to PCR using zebrafish-specific primers. Primers included *HIF* (left: 5′-GAC GTG GAA GGT TCT TCA CTG-3′; right: 5′-TCA AGA GGT CAT CTG GCT CAT-3′) and VGEF (left: 5′-GTA AAG GCT GCC CAC ATA CC-3′; right: 5′-GCT TTG ACT TCT GCC TTT GG-3′). β-actin was used as an internal control. Semiquantitative PCR was performed using Advantage 2 Taq Polymerase (BD Biosciences, Palo Alto, CA) in a PTC-100 thermocycler (Bio-Rad, Hercules, CA) using the following cycling parameters: 94°C for 2 min followed by 30 cycles of 94°C/1 min, 59°C/30 s, and 72°C/1 min, followed by 72°C for 10 min. PCR products were visualized following electrophoresis in a 2.0% agarose gel and stained with ethidium bromide.

16.2.8 Whole Mount *In Situ* Hybridization

To determine the site of *HIF-1α* upregulation, using the manufacturer's protocol (Roche Diagnostics, Inc., Indianapolis, IN) and *HIF-1α* cDNA as template, a RNA probe was synthesized and digoxigenin-labeled for whole mount *in situ* hybridization. Zebrafish were fixed with 4% paraformaldehyde in PBS and rehydrated with PBST. RNA probe was hybridized at 65°C in hybridization solution (50% formamide, $5 \times$ SSC, 0.1% Tween 20, 0.05 mg/mL heparin, 0.5 mg/mL tRNA, 10 mM sodium citrate buffer, pH 6.0). Alkaline phosphatase-conjugated anti-digoxigenin antibody (Santa Cruz Biotechnology, Inc., Santa Cruz, CA) was used for detection. For staining, zebrafish were equilibrated in NTMT buffer (0.1 M Tris–HCl, pH 9.5, 50 mM MgCl, 0.1 M NaCl, 0.1% Tween 20) at room temperature. Then, 4.5 μL of 75 mg/mL NBT (nitro blue tetrazolium) and 3.5 μL of 50 mg/mL BCIP (5-bromo-4-chloro-3-indolyl phosphate) per mL were added to the staining solution. The staining reaction was stopped by washing zebrafish with PBST. Zebrafish were then examined on a Zeiss Stemi 2000C stereomicroscope (Zeiss).

16.2.9 Drug Treatment

Celebrex, genistein, and SU5416 (Calbiochem/EMD, Gibbstown, NJ) are known to cause antiangiogenic effects in mice (Fong et al., 1999; Leahy et al., 2002). We next assessed effects in the zebrafish CNV model. Thirty 1dpf zebrafish were incubated simultaneously with 0.1 mg/mL $CoCl_2$ and test drug continuously in 6 mL of fish water for 4 day. Drugs were dissolved and serially diluted in DMSO before adding to microwells; final DMSO concentration was 1%. After drug treatment, zebrafish were washed with fish water and then processed for either whole mount immunostaining or ELISA processing.

16.2.10 CNV ELISA

Whole mount immunostaining and eye extraction were performed as described above; however, for the CNV ELISA, Phy-V directly conjugated with HRP (Phy-V-HRP) was used as the antibody. Isolated intact eyes were distributed into 96-well micro/plates. To generate chemiluminescence signal, 150 μL of PS-atto (Beckman Coulter/Lumigen, Inc.) was added to wells and signal was measured immediately using a Synergy HT microplate reader (Bio-Tek Instruments, Inc., Winooski, VT). A linear relationship was observed between number of eyes/well and chemiluminescence signal 4 eyes/well was found to be optimal for ELISA processing and drug assessment.

16.2.11 Statistics

Two-way ANOVA was performed for all drug-treated zebrafish. Drug effect was considered significant if $P < 0.05$ (95% confidence). All calculations were performed using Excel (Microsoft Corporation, Seattle, WA).

16.3 RESULTS

In this research, we optimized conditions for generating a zebrafish CNV model for drug screening by treatment with $CoCl_2$. Next, we developed a method for dissociating intact eyes from whole zebrafish. Then, we optimized a CNV ELISA for drug screening. Since reducing oxygen level failed to upregulate *VEGF* expression in zebrafish (Ton et al., 2003), we decided that it was unlikely that a CNV model could be developed using this approach. $CoCl_2$, an hypoxia mimetic, is known to upregulate *VEGF* expression (Van Lieshout et al., 2003) and we hypothesized that a CNV phenotype can be induced in zebrafish by adding $CoCl_2$ directly to fish water.

To develop a reproducible zebrafish CNV model, we initially determined optimum conditions for $CoCl_2$ treatment including compound concentration and embryo stage. Recent research showed that anoxic and hypoxic atmosphere treatment using zebrafish older than 1dpf stage resulted in rapid death (Padilla and Roth, 2001). In addition, in early experiments, we observed that $CoCl_2$ treatment at 2dpf caused instantaneous death. We therefore focused on finding the optimal $CoCl_2$ treatment conditions using 1dpf zebrafish.

16.3.1 Optimization of CoCl₂ Treatment for Generating a Hypoxia-Induced Zebrafish CNV Model

We treated 1dpf zebrafish with $CoCl_2$ concentrations ranging from 0.01 to 25 mg/mL. Using whole mount immunostaining with Phy-V antibody, which is specific for activated endothelial cells (ECs) (Seng et al., 2004), we observed that $63.6 \pm 11.1\%$ treated zebrafish exhibited specific CNV phenotypes. Phy-V antibody staining was observed throughout the vasculature including vessels, eyes, CVP, trunk/tail region, and intersegmental vessels (ISVs) (Serbedzija et al., 2006). Although the ISV pattern was normal (data not shown), a network of new vessels exhibiting strong staining in the CVP indicated abnormal angiogenesis. In some zebrafish, edema was observed in the region surrounding abnormal vessel growth, similar to the phenotype present in AMD. Overall morphology of zebrafish treated with 0.1 mg/mL $CoCl_2$ was similar to untreated zebrafish (data not shown), indicating that the pro-angiogenic effect in the CVP was specific and there was no obvious toxicity. Although a higher $CoCl_2$ concentration (1 mg/mL) did not increase CNV severity or frequency, it did increase overall toxicity (data not shown). Therefore, treatment of 1dpf zebrafish with 0.1 mg/mL $CoCl_2$ continuously for 4 days was selected as the optimal condition for generating a hypoxia-induced zebrafish CNV model.

16.3.2 Removal of Intact Eyes from Whole Zebrafish

For optimal assessment of abnormal angiogenesis in CVP, which is located in the back of the eyes, this region must be assessable. Initially, we used forceps to extract eyes from whole animals; however, this procedure was tedious and often resulted in damage. We then tried treating fixed, whole mount Phy-V-stained zebrafish with collagenase (100 units/mL, at 37°C for 40 min) in an Eppendorf tube (McGrath and Seng, 2009). This procedure loosened eyes from their sockets without damaging vessels. Eppendorf tubes containing treated zebrafish were gently shaken a few times and loosened, intact eyes were then dissociated from whole zebrafish (Fig. 16.3); the CVP, located in the back of the eyes, was then accessible (Fig. 16.1). Compared to untreated zebrafish eyes (Fig. 16.1a), ECs were highly activated (overall bright fluorescence and red arrows) in vascularized CVP (yellow arrows) $CoCl_2$-treated zebrafish eyes (Fig. 16.1b and c). Abnormal activated EC aggregates (green box in Fig. 16.1c) were visible in CVP of some $CoCl_2$-treated zebrafish (Fig. 16.1d). These observations confirm that $CoCl_2$ induced abnormal angiogenesis in the CVP.

16.3.3 Immunohistochemistry of CoCl₂-Treated Zebrafish Eyes

To further characterize the zebrafish CNV model, we performed immunohistochemistry. Untreated 5dpf zebrafish eyes exhibited the characteristic five-layer retinal structure (Fig. 16.2a, left panel, ND5) comprised of densely packed cells in the ganglion cell layer (GCL), wide inner plexiform layer (IPL, long black bar), and

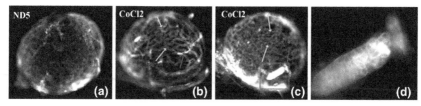

Figure 16.1 Full view of intact CVP in the back of isolated eyes. Untreated (ND5, a) and $CoCl_2$-treated (b and c) 5dpf zebrafish were whole mount immunostained with Phy-V-Alexa 568. After extensive washing, zebrafish were then incubated with 50 U/mL collagenase at 37°C for 40 min and eyes were dissociated from whole animals by gentle shaking. Isolated eyes were then placed on depression slides with the CVP side facing up (8×). Yellow arrows indicate individual vessels in the CVP of $CoCl_2$-treated eyes. Red arrows mark intense fluorescence staining, indicating highly activated ECs in $CoCl_2$-treated zebrafish eyes. Green box (c) indicates a large, abnormal, highly activated, EC aggregate. (d) High-magnification image (100×) of area defined by the green box in (c) clearly shows an abnormal EC cluster. (See the color version of this figure in Color Plates section.)

differentiated outer segment cells in the photoreceptor layer (PC, red arrow). Although $CoCl_2$-treated 5dpf eyes (Fig. 16.2a, right panel, $CoCl_2$D5) exhibited the distinctive five-layer retinal structure, abnormalities included few cells in ganglion layer (green arrow), narrow inner plexiform layer (short black bar), large vacuoles in inner nuclear layer (INL, yellow arrows), and oval-shaped, less differentiated outer segment cells in the photoreceptor layer (red arrow). Since 3 and 4dpf zebrafish do not exhibit a distinct five-layer retinal structure, presence of this structure indicated that

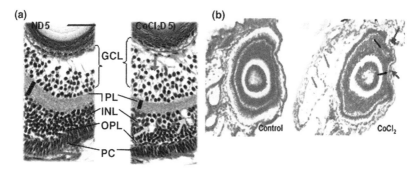

Figure 16.2 Histological assessment of $CoCl_2$-treated zebrafish eyes. One dpf zebrafish were treated for 4 days with 0.1 mg/mL $CoCl_2$. (a) Histology showed that untreated 5dpf eyes (ND5) exhibited a normal five-layer retinal structure comprised of densely packed cells in the ganglion cell layer, wide inner plexiform layer (long black bar), and differentiated outer segment cells in the photoreceptor layer (red arrow). $CoCl_2$-treated 5dpf eyes ($CoCl_2$D5) exhibited the distinctive five-layer retinal structure. However, few cells in the ganglion layer (green arrow), short inner plexiform layer (short black bar), large vacuoles in inner nuclear layer (yellow arrows), and oval-shaped, undifferentiated outer segment cells in the photoreceptor layer (red arrow) were observed, indicating retinal degeneration. (b) In severe cases, edema (green arrows), degenerated photoreceptors (blue arrows), and abnormal vessels (black arrows) were observed ($CoCl_2$). Scale bar in (a) = 50 μm. (See the color version of this figure in Color Plates section.)

CoCl$_2$-treated zebrafish eyes were at the 5dpf stage and that CoCl$_2$ treatment did not arrest zebrafish embryogenesis. Although under normal conditions, there are no vessels in the zebrafish retina, immunohistochemical staining using Phy-V mAb demonstrated that CoCl$_2$-treated zebrafish exhibited abnormal vessels in the retina (Fig. 16.2b, black arrows). Edema (Fig. 16.2b, green arrows), a phenotype present in human CNV, and degenerated photoreceptors (blue arrows) were also observed. These results show that CoCl$_2$ treatment induced abnormal vessels that penetrated deep into the retinal structure.

16.3.4   Confirmation of *HIF-1α* and *VEGF* Upregulation by RT-PCR and *In Situ* Hybridization

Since CoCl$_2$ is known to upregulate *HIF-1α*, which can affect downstream *VEGF* gene expression and induce abnormal angiogenesis, we assessed *HIF-1α* and *VEGF* gene expression at varying time points after CoCl$_2$ treatment. RT-PCR results showed that *HIF-1α* was upregulated 4, 6, and 48 h after CoCl$_2$ treatment (hpt). *VEGF* upregulation was observed at 48 and 72 hpt, which, as expected, was later than *HIF-1α* upregulation. At 96 hpt, level of *HIF-1α* and *VEGF* gene expression for untreated and CoCl$_2$-treated eyes was the same (data not shown), possibly because expression of both genes had plateaued. *In situ* hybridization confirmed *HIF-1α* upregulation in CoCl$_2$-treated zebrafish eyes, which was particularly prominent in the ganglion cell layer (data not shown). Using whole mount TUNEL staining, apoptosis was also detected in CoCl$_2$-treated zebrafish retinas (data not shown).

16.3.5   Inhibitory Effect of Antiangiogenic Compounds on New Vessel Formation in Zebrafish

Next, we assessed effects of three compounds on CoCl$_2$-induced CNV: Celebrex, a COX-2 inhibitor, genistein, a natural isoflavone kinase inhibitor, and SU5416, a VEGF receptor inhibitor. Celebrex and genistein have been shown to exhibit both anti-inflammatory and antiangiogenic properties (Wei et al., 2004; Wang et al., 2005). In contrast, SU5416 is antiangiogenic but not anti-inflammatory (Shaheen et al., 1999). To assess compound effects, we explored different drug treatment regimens: pre-treatment, co-treatment, and post treatment. We determined that co-treatment with CoCl$_2$ and antiangiogenic compounds induced optimal effects. Using visual assessment, co-treatment with CoCl$_2$, Celebrex, and genistein inhibited a higher level of CoCl$_2$-induced angiogenesis than SU5416. Furthermore, Celebrex and genistein also inhibited edema associated with abnormal angiogenesis, whereas SU5416 did not (data not shown).

16.3.6   Development of a Quantitative Eye-Specific ELISA

Taking advantage of (a) specificity of Phy-V mAb and (b) ability to dissociate intact eyes, we developed a quantitative eye ELISA to assess drug effects on angiogenesis.

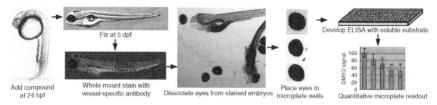

Figure 16.3   Zebrafish eye dissociation. Zebrafish were treated with CoCl₂ from 1dpf stage to 5dpf. Control and compound-treated 5dpf zebrafish were fixed and whole mount stained with Phy-V. Intact eyes were removed by gentle shaking after treatment with collagenase (150 U/mL, at 37°C for 40 min). Dissociated eyes were then distributed into 96-well microplates to perform the CNV ELISA.

Since individual zebrafish eyes are considerably smaller than whole animals, a sensitive method was required to assess vessels and we used a chemiluminescence HRP substrate to develop the CNV ELISA. An overview of the zebrafish CNV ELISA is shown in Fig. 16.3.

We initially determined the optimal number of eyes/well for the ELISA by establishing a linear relationship between chemiluminescence signal and number of eyes per well (data not shown). Four eyes/well generated signal in the middle of the linear portion of the curve, which permitted detection of both increase and decrease in signal within the range. Next, using the CNV ELISA, we quantitated effects of co-treatment of CoCl₂ and Celebrex, genistein, or SU5416. In order to compare results from different assays performed on different days, drug effects were normalized to percent inhibition of eye angiogenesis. Relative luminescent units (RLUs) for Phy-V-HRP-stained zebrafish eyes were used to calculate drug effects. Background signal from 5dpf zebrafish eyes with no antibody incubation was assessed and mean value was subtracted from each RLU measurement. Mean RLU of CoCl₂ + DMSO control was used as 100% control. Drug effect was calculated using formula (16.1):

$$\text{percent inhibition} = \left(1 - \frac{\text{RLU (drug-treated)}}{\text{RLU (control)}}\right) \times 100\%. \qquad (16.1)$$

A typical dose response was observed for Celebrex and SU5416, whereas genistein inhibited CNV only at a low concentration, 0.0001 μM. ELISA results agreed with visual results. Based on dose–response curves (Fig. 16.4), optimum drug concentrations were 20 μM (Celebrex), 0.0001 μM (genistein), and 2.5 μM (SU5416). These results demonstrated that the ELISA accurately quantitated angiogenesis level in drug-treated eyes.

16.3.7 Confirm Drug Effects by Immunohistochemistry

Since CVP in normal zebrafish forms from 2 to 5dpf, it is possible that inhibitory drug effects quantitated by ELISA were due to arrested eye development. To address this

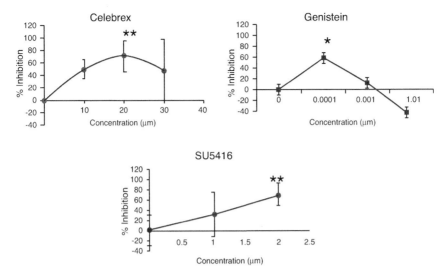

Figure 16.4 Drug effects on CNV in $CoCl_2$-treated zebrafish. One dpf zebrafish were co-treated with 0.1 mg/mL $CoCl_2$ and varying drug concentrations for 4 days. At 5 dpf, zebrafish were processed for the CNV ELISA, as described in Section 16.2. Each point represents mean ± SE ($n = 5$–18). * indicates $p < 0.1$ and ** indicates $p < 0.05$ by ANOVA.

concern, using immunohistochemistry, we examined zebrafish eye structure after Celebrex treatment. Normal 5 dpf zebrafish exhibit a distinctive five-layer retinal structure with no vessels. In contrast, 0.1 mg/mL $CoCl_2$-treated 5 dpf zebrafish exhibited disrupted inner nuclear layer, indicating degenerated retinal structure, and abnormal vessel structure. Five dpf zebrafish co-treated with $CoCl_2$ and DMSO showed degenerated retinal structure, newly formed vessels invading the retina from the choroidal layer, and a degenerated photoreceptor layer, indicating that DMSO alone had no effect on $CoCl_2$-induced CNV. Five dpf zebrafish co-treated with $CoCl_2$ and Celebrex exhibited a normal five-layer retina with no vessels present (data not shown). These results clearly demonstrate that Celebrex treatment inhibited $CoCl_2$-induced CNV phenotypes.

16.4 DISCUSSION

16.4.1 CNV Zebrafish Model Is Specific for Retinal and Choroidal Neoangiogenesis

Using a low $CoCl_2$ concentration (0.1 mg/mL), we generated a chemical induced zebrafish CNV model. As evidenced by overall morphology (data not shown), $CoCl_2$ treatment did not inhibit zebrafish development. Although a stimulatory effect on angiogenesis was present throughout zebrafish, the most striking effect of $CoCl_2$ treatment was the presence of a network of abnormal small vessels in the CVP, often accompanied by edema in the surrounding tissue, a phenotype present in AMD in humans. In normal zebrafish, there are no vessels present in the retina. In the CNV model, immunohistochemistry demonstrated that angiogenic vessels were present in the retina, indicating that neovascularization is a pathological phenotype. Furthermore, the ISV pattern was normal, indicating specificity of $CoCl_2$-induced neovascularization in CVP and absence of toxicity.

16.4.2 Isolation of Eyes from Whole Zebrafish Using Collagenase Treatment Facilitates CNV Assessment and Quantitation of Drug Effects

Since the CVP is located in the back of the eyes, examination of abnormal angiogenesis is difficult in intact, whole mount Phy-V-stained zebrafish. In order to optimize assessment of abnormal eye angiogenesis, we dissociated intact eyes from whole zebrafish. Initially, we used forceps to extract eyes from whole animal; however, this procedure was tedious and often caused eye damage. To loosen eyes, we then tried collagenase treatment on whole mount immunostained zebrafish followed by gentle shaking. This method permitted examination of abnormal neovascularization specifically in the CVP region in the back of the eye with no background signal from other sites in whole zebrafish.

16.4.3 Hypoxia Is a Major Mechanism of Abnormal Angiogenesis in the Zebrafish CNV Model

We hypothesized that hypoxia is the major mechanism of $CoCl_2$-induced CNV in zebrafish. To support this hypothesis, using RT-PCR and *in situ* hybridization, we demonstrated that *HIF-1α* and *VEGF* expression was upregulated in $CoCl_2$-treated zebrafish eyes.

16.4.4 Inflammation Plays an Important Role in the Zebrafish CNV Model

We tested three antiangiogenic compounds, Celebrex, genistein, and SU5416, in the $CoCl_2$-induced CNV model. Not only did Celebrex and genistein inhibit abnormal angiogenesis in CVP, but they also inhibited the edematous inflammatory response. SU5416, a VEGF receptor inhibitor, is antiangiogenic in mice (Shaheen et al., 1999). However, in clinical trials, this compound was shown to induce serious toxic side effects (Kindler et al., 2001). As expected, co-treatment with $CoCl_2$ and either Celebrex or genistein prevented CNV formation. In contrast, SU5416 did not. These results imply that, in addition to angiogenesis, inflammation plays a crucial role in CNV formation. Assessment of additional compounds involved in both angiogenesis and inflammation will further support this hypothesis and may lead to identification of macular degeneration drug candidates.

ACKNOWLEDGMENTS

This research was supported by grants from the National Institutes of Health: 2R44EY015335 and 2R44EY016254.

REFERENCES

Bouck N (2002). PEDF: anti-angiogenic guardian of ocular function. Trends Mol Med 8(7): 330–334.

Fong GH, Zhang L, Bryce DM, and Peng J (1999). Increased hemangioblast commitment, not vascular disorganization, is the primary defect in flt-1 knock-out mice. Development 126(13): 3015–3025.

Fouquet B, Weinstein BM, Serluca FC, and Fishman MC (1997). Vessel patterning in the embryo of the zebrafish: guidance by notochord. Dev Biol 183(1): 37–48.

Isogai S, Horiguchi M, and Weinstein BM (2001). The vascular anatomy of the developing zebrafish: an atlas of embryonic and early larval development. Dev Biol 230(2): 278–301.

Jiang BH, Zheng JZ, Leung SW, Roe R, and Semenza GL (1997). Transactivation and inhibitory domains of hypoxia-inducible factor 1alpha. Modulation of transcriptional activity by oxygen tension. J Biol Chem 272(31): 19253–19260.

Kindler H, VOgelzang N, Chien K, Stadler W, Karczmar G, Heimann R, and Vokes E (2001). SU5416 in malignant mesothelioma: a University of Chicago Phase II Consortium Study. 2001 ASCO Annual Meeting.

Leahy KM, Ornberg RL, Wang Y, Zweifel BS, Koki AT, and Masferrer JL (2002). Cyclooxygenase-2 inhibition by celecoxib reduces proliferation and induces apoptosis in angiogenic endothelial cells *in vivo*. Cancer Res 62(3): 625–631.

Liang D, Xu X, Chin AJ, Balasubramaniyan NV, Teo MA, Lam TJ, Weinberg ES, and Ge R (1998). Cloning and characterization of vascular endothelial growth factor (VEGF) from zebrafish, *Danio rerio*. Biochim Biophys Acta 1397(1): 14–20.

Lyons MS, Bell B, Stainier D, and Peters KG (1998). Isolation of the zebrafish homologues for the tie-1 and tie-2 endothelium-specific receptor tyrosine kinases. Dev Dyn 212(1): 133–140.

McGrath P and Seng W (2009). Methods of screening an agent for an activity in an isolated eye of a teleost. U.S. Patent Application 2009/0010849.

Padilla PA and Roth MB (2001). Oxygen deprivation causes suspended animation in the zebrafish embryo. Proc Natl Acad Sci USA 98(13): 7331–7335.

Rafii B, Coutinho C, Otulakowski G, and O'Brodovich H (2000). Oxygen induction of epithelial Na$^+$ transport requires heme proteins. Am J Physiol Lung Cell Mol Physiol 278(2): L399–L406.

Rasmussen H, Chu KW, Campochiaro P, Gehlbach PL, Haller JA, Handa JT, Nguyen QD, and Sung JU (2001). Clinical protocol. An open-label, phase I, single administration, dose-escalation study of ADGVPEDF.11D (ADPEDF) in neovascular age-related macular degeneration (AMD). Hum Gene Ther 12(16): 2029–2032.

Schwesinger C, Yee C, Rohan RM, Joussen AM, Fernandez A, Meyer TN, Poulaki V, Ma JJ, Redmond TM, Liu S, Adamis AP, and D'Amato RJ (2001). Intrachoroidal neovascularization in transgenic mice overexpressing vascular endothelial growth factor in the retinal pigment epithelium. Am J Pathol 158(3): 1161–1172.

Semenza GL (2000). HIF-1 and human disease: one highly involved factor. Genes Dev 14(16): 1983–1991.

Seng WL, Eng K, Lee J, and McGrath P (2004). Use of a monoclonal antibody specific for activated endothelial cells to quantitate angiogenesis *in vivo* in zebrafish after drug treatment. Angiogenesis 7(3): 243–253.

Serbedzija G, Semino C, and Frost D (2006). Methods of screening agents for activity using teleosts and screening for angiogenesis in zebrafish. U.S. Patent 7,041,276.

Shaheen RM, Davis DW, Liu W, Zebrowski BK, Wilson MR, Bucana CD, McConkey DJ, McMahon G, and Ellis LM (1999). Antiangiogenic therapy targeting the tyrosine kinase receptor for vascular endothelial growth factor receptor inhibits the growth of colon cancer liver metastasis and induces tumor and endothelial cell apoptosis. Cancer Res 59(21): 5412–5416.

SHIH SC, JU M, LIU N, and SMITH LE (2003). Selective stimulation of VEGFR-1 prevents oxygen-induced retinal vascular degeneration in retinopathy of prematurity. J Clin Invest 112(1): 50–57.

SMITH LE (2002). Pathogenesis of retinopathy of prematurity. Acta Paediatr Suppl 91(437): 26–28.

TON C, STAMATIOU D, and LIEW CC (2003). Gene expression profile of zebrafish exposed to hypoxia during development. Physiol Genomics 13(2): 97–106.

Van LIESHOUT T, STANISZ J, ESPIRITU V, RICHARDSON M, and SINGH G (2003). A hypoxic response induced in MatLyLu cells by cobalt chloride results in an enhanced angiogenic response by the chick chorioallantoic membrane. Int J Oncol 23(3): 745–750.

WANG B, ZOU Y, LI H, YAN H, PAN JS, and YUAN ZL (2005). Genistein inhibited retinal neovascularization and expression of vascular endothelial growth factor and hypoxia inducible factor 1alpha in a mouse model of oxygen-induced retinopathy. J Ocul Pharmacol Ther 21(2): 107–113.

WEI D, WANG L, HE Y, XIONG H, ABBRUZZESE J, and XIE K (2004). Celecoxib inhibits vascular endothelial growth factor expression in and reduces angiogenesis and metastasis of human pancreatic cancer via suppression of Sp1 transcription factor activity. Cancer Res 64(6): 2030–2038.

WESTERFIELD M (1993). The Zebrafish Book: A Guide for the Laboratory Use of Zebrafish. University of Oregon Press, Eugene, OR.

YUAN Y, HILLIARD G, FERGUSON T, and MILLHORN DE (2003). Cobalt inhibits the interaction between hypoxia-inducible factor-alpha and von Hippel-Lindau protein by direct binding to hypoxia-inducible factor-alpha. J Biol Chem 278(18): 15911–15916.

Chapter 17

Zebrafish Xenotransplant Cancer Model for Drug Screening

Chunqi Li, Liqing Luo, and Patricia McGrath

Phylonix, Cambridge, MA, USA

17.1 INTRODUCTION

Although immunodeficient mice have played an important role in cancer research, new cost-effective xenotransplant (Xt) animal models are needed to improve pre-clinical drug discovery and screening. In this research, we describe methods for developing a zebrafish xenotransplant cancer model and a whole animal ELISA for drug screening. For model development, we microinjected human colon cancer cells into 3 days post fertilization (dpf) zebrafish and assessed proliferation, migration, and mass formation. Next, using a human cell-specific antibody that exhibited limited cross-reactivity with zebrafish tissues, we developed a quantitative microplate-based whole zebrafish ELISA. Xenotransplant zebrafish were then treated with four FDA-approved cancer drugs (5-fluorouracil (5-FU), oxaliplatin, camptothecin, and leucovorin) and one drug combination (5-fluorouracil + leucovorin) and results for four of the five drugs and the drug combination were similar to results in humans.

17.2 BACKGROUND AND SIGNIFICANCE

17.2.1 Conventional Mouse Xenograft Models

Quantitating cancer cells *in vivo* to assess drug efficacy is a major challenge. Although mice remain the model system of choice for cell transplantation (Kelland, 2004), there are several inherent disadvantages associated with mouse

Zebrafish: Methods for Assessing Drug Safety and Toxicity, Edited by Patricia McGrath.
© 2012 John Wiley & Sons, Inc. Published 2012 by John Wiley & Sons, Inc.

xenograft models. After cell injection, presence of a tumor, defined as a subcutaneous lump, is assessed. Tumor growth is then quantitated by measuring tumor size after surgical removal (Yang et al., 1997). Because tumor development takes weeks and animal survival time varies, the experimental period can span several months. In order to avoid cell rejection, human tumor cells are typically transplanted into nude or SCID immunosuppressed mice (Yang et al., 1997; Greiner et al., 1998; Katsanis et al., 1998), which are expensive to maintain, less hardy than normal mice, and more susceptible to infection and drug toxicity. An additional disadvantage of mouse cancer models is that a large number of cells ($\sim 1 \times 10^6$) are required to generate a tumor. Although this is not a concern using cell lines, it is often difficult to procure a sufficient number of viable cells from primary tumors. Moreover, it is extremely difficult to generate mouse xenograft metastasis models (van Weerden and Romijn, 2000).

17.2.2 Zebrafish: A Predictive Cancer Model

In the past decade, zebrafish has emerged as an important model organism for biomedical research. Although the primary focus of zebrafish studies has been on developmental biology, most human diseases, including cancers, have now been modeled in this organism. Since 2000, more than 100 manuscripts describing zebrafish cancer models have been published. Research strategies have included mammalian cancer cell xenotransplant, forward genetic screens for assessing proliferation or genomic instability, reverse genetic target-selected mutagenesis to inactivate tumor suppressor genes, and generation of transgenic animals expressing human oncogenes. Zebrafish have been shown to exhibit comparable signaling pathways, assessed by microarrays, and similar morphology, assessed by histology, as mammalian cancer models. However, compared to rodents, zebrafish-specific tumor incidence is low and tumors develop at later stages. Despite these disadvantages, zebrafish has created its own niche in cancer research, complementing conventional animal models with unique experimental advantages.

17.2.3 Advantages of Zebrafish Xenotransplant Models

The technique for cell transplantation in zebrafish is well established. Genetic mosaics are usually generated by transplanting cells from a donor embryo labeled with a lineage-specific marker into unlabeled blastula stage host embryos (Ho and Kane, 1990). These studies have been crucial for addressing fundamental questions in developmental biology including (a) when do cells commit to a certain lineage? and (b) which cell interactions are involved in establishing this commitment? Lineage markers or tracer dyes such as fluorescent dextrans or lipophilic carbocyanine tracers, including CM-DiI, can be used to distinguish donor from host-derived cells and to visualize fate of transplanted cells in transparent zebrafish. In these studies, xenotransplant cells are usually homografts (cells from the same species).

The zebrafish thymus begins forming ∼65 h post fertilization (hpf) (Willett et al., 1997) and T-cell receptor (TCR) alpha gene expression has been used to follow thymus development and T-cell distribution. Gene expression is detectable at 4dpf in the thymus but not until 9dpf outside this organ (Danilova et al., 2004). Since the zebrafish xenotransplant cancer model terminates on 6dpf (3 days post xeno-transplant (dpx)), there is a rejection free period for drug screening (McGrath and Serbedzija, 2004). In contrast, since the mouse thymus develops *in utero*, it is difficult to introduce human cells prior to immune competence.

Another advantage of the zebrafish model is that significantly fewer cells (∼100–1000) are required for transplant than rodent models (1,000,000). Further-more, several hundred zebrafish can be microinjected in a single experiment permitting rapid assessment of multiple drug concentrations and drug combinations. Although patient biopsy samples usually contain few cells, in theory, these cells can be propagated in zebrafish. Histopathology can also be performed on zebrafish tissues and xenotransplant cells can be discriminated in host organs.

Recently, we and several other groups combined the advantages of cultured human cancer cells with the advantages of the transparent zebrafish to generate xenotransplant cancer models. In one study (Lee et al., 2005), fluorescently labeled human metastatic melanoma cells transplanted into zebrafish blastula stage embryos survived, proliferated, migrated, and remained viable through the adult stage. However, these cells did not from masses or metastasize. Another recent study showed that aggressive human melanoma cells transplanted into 3hpf zebrafish induced a secondary axis or an abnormal head induced by Nodal signaling (Lee and Herlyn, 2006; Topczewska et al., 2006). In contrast to these studies in which cancer development was not observed, in related research, we observed that xenotransplant human melanoma, colorectal, and pancreatic cancer cells induced tumor-like cell masses in zebrafish (Haldi et al., 2006). In one elegant study, researchers combined i.p. injection of fluorescently labeled human breast cancer cells into 30dpf zebrafish and three-dimensional modeling to elucidate interaction of human cells and zebrafish vessels and subsequent tissue invasion. In these studies, vascular endothelial growth factor (VEGF) expression induced openings in vessel walls through which RhoC expressing cancer cells entered the blood system (Stoletov et al., 2007). In this research, transplanted cancer cells were not quantified either visually or by image-based morphometric analysis. Development of a quantitative whole animal ELISA format for drug screening addresses a significant void.

17.3 MATERIALS AND METHODS

17.3.1 Zebrafish Handling

Zebrafish were generated by natural pairwise mating in our aquaculture facility, as described by Westerfield (1993). Four to five pairs were set up for each mating; on average, 50–100 zebrafish per mating were generated. Prior to xenotransplant at 2dpf, zebrafish were maintained in fish water at 28°C (5 g of Instant Ocean Salt in 25 L of distilled water). During microinjection, zebrafish were anesthetized with MESAB (0.5 mM 3-aminobenzoic acid ethyl ester, 2 mM Na_2HPO), and placed in fish water on their sides on a ramp of 1% agarose. Cancer cells were then injected into the yolk sac. Zebrafish recovered in fresh fish water at 28°C for 1 h and were transferred to 35°C. Although zebrafish are usually maintained at 28°C, after 2dpf stage, they have been shown to develop normally at 35°C (Detrich et al., 1999). Since this 2°C decrease from the standard temperature for human cell culture did not inhibit SW620 colorectal cancer cell proliferation or mass formation in our previous studies (Haldi et al., 2006), we used 35°C for subsequent experiments. Because zebrafish receive nourishment from an attached yolk sac, no additional feeding is required for 1 week post fertilization.

17.3.2 Cells Lines and Culture Conditions

Human colorectal cancer cell lines Colo320 and SW620 and human lymphoblastoid TK6 cells (Table 17.1) were purchased from the American Type Culture Collection (ATCC, Manassas, VA). Cell culture reagents were obtained from Invitrogen (Carlsbad, CA), and fetal bovine serum was obtained from Hyclone (Logan, UT). Colo320 and TK6 cells were cultured in RPMI 1640 medium and SW620 cells in Leibovitz's L-15 medium supplemented with 10% heated-inactivated calf serum, 100 units/mL penicillin, 100 µg/mL streptomycin, and 2 mM l-glutamine. Cell lines were subcultured and maintained in tissue culture flasks at 37°C in a humidified, 5% CO_2 atmosphere.

17.3.3 Cell Labeling to Assess Proliferation

Colo320 and SW620 cells were fluorescently labeled by incubating with 10 µg/mL CM-DiI (Molecular Probes, Eugene, OR) containing 0.5% DMSO and adjusted to a density of 100×10^6 cells/mL (100 cells/nL) in HBSS. Prior to injection, cells were in a single cell suspension comprised of uniformly fluorescently labeled cells that routinely exhibit ~95% viability, assessed by trypan blue exclusion. Labeled cells were injected within 2 h. Since CM-DiI tracking dye was transferred from mother to

Table 17.1 Cell Lines for Zebrafish Xt Colorectal Cancer Model Development

Cell line	Tissue or origin	Cancer cells	Tumorigenic in mice
Colo320	Human colorectal adenocarcinoma	Yes	Yes
SW620	Lymph node metastasis of human colorectal adenocarcinoma	Yes	Yes
TK6 cells	Human lymphoblasts isolated from the spleen of a hereditary spherocytosis patient	No	No

daughter cells, after several doublings, fluorescently labeled single cells were clearly visible on the cancer cell masses (Haldi et al., 2006).

17.3.4 Cell Transplantation

The transplantation protocol used was similar to the protocol described for homograft transplantation in zebrafish (Ho and Kane, 1990). CM-DiI-labeled cells were loaded into a pulled glass micropipette (VWR) that was drawn on an electrode puller and trimmed to form a needle with ~15 μm inner diameter and ~18 μm outer diameter. The microneedle was attached to an air-driven Cell Tram (Eppendorf) and the tip of the needle was then inserted into the yolk of 2 or 3dpf zebrafish. Using positive pressure, pulse time was controlled to deliver ~1500 cells in 15 nL volume. Number of injected cells was standardized by fixing cell density and injection volume. After a 1 h recovery period at 28°C, injected zebrafish were examined under a fluorescence microscope for the presence of xenotransplant cells located specifically in the yolk sac and then transferred to 35°C.

17.3.5 Antibody Sources

Sources for antibodies were as follows: HLA, mouse anti-human, and HLA-DR, mouse anti-human monoclonal antibodies (Invitrogen, Carlsbad, CA), survivin, rabbit anti-human monoclonal antibody, XIAP, rabbit anti-human monoclonal antibody, and Bax, rabbit anti-human polyclonal antibody (Cell Signaling Technology, Danvers, MA), and Keap1, rabbit anti-human polyclonal antibody (Santa Cruz Technology, Santa Cruz, CA).

17.3.6 Whole Mount Xenotransplant Zebrafish Immunostaining

Whole mount xenotransplant zebrafish immunostaining was performed to confirm antibody specificity to xenotransplant human colon cancer cells and to assess zebrafish cross-reactivity. Xt zebrafish were processed using Dent's fixative and whole mount immunochemical staining was performed using human cell-specific antibodies, followed by staining with Alexa Fluor 488-conjugated secondary antibody (Invitrogen). Zebrafish were deposited in methylcellulose and laterally oriented for image acquisition using a Zeiss Stemi SV11 Apo upright fluorescence microscope equipped with a Zeiss AxioCam (Zeiss, Thornwood, NY). CM-DiI-labeled cells were visualized in the rhodamine channel and human antibody-stained zebrafish were visualized in the FITC channel.

17.3.7 Quantitative Whole Xt Zebrafish ELISA

Xt zebrafish were fixed in Dent's fixative (DMSO/methanol = 1/4) and processed using a whole mount immunochemical staining protocol (Westerfield, 1993);

however, horseradish peroxidase (HRP)-conjugated secondary antibody (Bio-Rad, Hercules, CA) replaced Alexa Fluor 488-conjugated antibody. Peroxidase suppressor was used to inhibit endogenous HRP and Triton-100 treatment was used to increase antibody permeability. Xt zebrafish without primary antibody staining served as a control. After extensive washing, zebrafish were distributed into the wells of transparent, black, flat-bottom microplates (Corning Life Sciences, Lowell, MA) and PS-atto (Beckman Coulter/Lumigen Inc., Southfield, MI) was used as the enzyme substrate that was quantified in Xt cells by measuring chemiluminescence intensity of the end product using a Bio-Tek ELX-800 microplate reader (Bio-Tek, Winooski, VT).

17.3.8 Establish Assay Robustness and Reproducibility

In order to avoid skewing results in large-scale drug screening, it is essential to initially establish robustness and reproducibility of signal (*S*) generated in the biological process in the absence of test compounds, followed by compound screening to establish assay and screen quality. To optimize assay quality, the end signal must be consistent within and between plates and control compounds must exhibit expected activity. A general layout for assessing plate reproducibility is illustrated in Fig. 17.1, where H represents the highest signal, M represents medium signal, and L represents the lowest signal. H, M, and L were assessed in each position on the plate. This study was performed twice per day on two different days. We used 1% DMSO-treated Xt zebrafish as H because chemiluminescence signal was highest, 1000 μM 5-FU-treated Xt zebrafish as M because signal was in the middle of the range, and the Xt zebrafish without primary antibody incubation as L because signal was lowest.

Based on results of this study for assessing plate reproducibility, we determined mean, SE, and CV (%) for each signal (H, M, and L) on each plate. We determined signal-to-noise ratio (background measurement, S/N); S/N ratio ≥2.3 is considered acceptable. We then calculated the Z-factor following formula 17.1. An assay with

H	M	L	H	M	L	H	M	L	H	M	L
H	M	L	H	M	L	H	M	L	H	M	L
H	M	L	H	M	L	H	M	L	H	M	L
H	M	L	H	M	L	H	M	L	H	M	L
H	M	L	H	M	L	H	M	L	H	M	L
H	M	L	H	M	L	H	M	L	H	M	L
H	M	L	H	M	L	H	M	L	H	M	L
H	M	L	H	M	L	H	M	L	H	M	L

Figure 17.1 Microplate layout for assessing reproducibility. H, maximum signal sample; M, medium signal sample; L, minimum signal sample. One percent DMSO-treated Xt zebrafish, H; 1000 μM 5-FU-treated Xt zebrafish, M; Xt zebrafish without primary antibody incubation, L.

Z-factor ≥ 0.5 is considered robust (Zhang et al., 1999).

$$Z = \frac{(\text{mean } S(\text{H}) - 3\text{SD}(\text{H})) - (\text{mean } S(\text{L}) + 3\text{SD}(\text{L}))}{\text{mean } S(\text{H}) - \text{mean } S(\text{L})}. \tag{17.1}$$

17.3.9 Cancer Drugs

5-Fluorouracil, oxaliplatin, camptothecin, and leucovorin were purchased from Sigma-Aldrich Co. (St. Louis, MO). These small molecules were delivered directly in fish water.

17.3.10 LC$_{10}$ Determination

Three dpf zebrafish were distributed into six-well plates in 3 mL fish water, 30 zebrafish/well. Cancer drugs were added to fish water and zebrafish were exposed by semistatic immersion until 6dpf. Dead zebrafish were counted daily and removed. In initial tests, five drug concentrations (0.1, 1, 10, 100, and 250 μM) were assessed. If an LC$_{10}$ could not be determined, additional lower (down to 0.01 μM) or higher concentrations (up to 1000 μM) were tested. To estimate LC$_{10}$, lethality curves were generated using JMP 7.0 statistical software (SAS Institute, Cary, NC).

17.3.11 Drug Treatment and Assessment of Drug Efficacy in Xt Zebrafish

Three concentrations (LC$_{10}$, (1/2)LC$_{10}$, and (1/4)LC$_{10}$) were tested for each drug. If no lethality was observed at the highest test concentration (1000 μM), to assess efficacy, in addition to 1000 μM we assessed two additional concentrations: 500 and 250 μM. Drugs were delivered in DMSO to a final solvent concentration of no more than 1%, which we and others have shown does not affect development. One percent DMSO-treated Xt zebrafish were used as vehicle control and 50 Xt zebrafish were used for each condition. Test drug was added to fish water immediately after xenotransplant at 3dpf stage, and incubated at 35°C for 3 days; zebrafish were then fixed for Xt ELISA processing. For all experiments, signal from drug-treated Xt zebrafish was compared to signal from vehicle control. Data are presented as mean ± SE and drug effect was normalized to percent inhibition and calculated using the following formula:

$$\text{drug effect (percent inhibition)} = \left(1 - \frac{\text{signal(drug-treated)}}{\text{signal(DMSO control)}}\right) \times 100\%. \tag{17.2}$$

17.3.12 Statistics

Since multiple concentrations were tested, ANOVA was used to analyze data. If a significant effect ($P < 0.05$) was observed, Dunnett's test was then performed to determine which concentrations exhibited significant effects.

17.3.13 Vertebrate Animal Care and Safety

The Office of Laboratory Animal Welfare (OLAW), National Institutes of Health (NIH), has approved our Animal Welfare Assurance effective through February 2012. We euthanize zebrafish of all ages by overexposure to tricaine methanesulfonate. These procedures are consistent with the American Veterinary Medical Association's (AVMA) Panel on Euthanasia.

17.4 RESULTS

In order to develop a quantitative zebrafish xenotransplant colon cancer model for drug screening, we initially showed that two colon cancer cell lines, Colo320 and SW620, proliferated in zebrafish. Next, we theorized that a human cell-specific antibody that labels normal and cancer cells but has limited cross-reactivity with zebrafish could be used to develop a whole animal ELISA for drug screening (Herlyn and Koprowski, 1988; Seng et al., 2004; Li et al., 2008, 2009; McGrath and Li, 2008; Haldi et al., 2009). Using whole mount staining, we identified two human cell-specific antibodies, survivin and XIAP (Cell Signaling, Danvers, MA), for ELISA development. Specificity was validated by (1) limited cross-reactivity with zebrafish tissue, confirmed by whole mount immunostaining, and (2) a linear relationship between signal and number of Xt zebrafish per well. This whole animal ELISA quantitated signal from antigens constitutively expressed on xenotransplant human cells, followed by reaction with a HRP-conjugated secondary antibody. Chemiluminescent signal was then quantitated using a conventional microplate reader. Chemiluminescent light units produced in the reaction correlated with number of Xt cells. Next, we used the Xt zebrafish ELISA to assess effects of four FDA-approved colorectal cancer drugs, 5-FU, oxaliplatin, camptothecin, and leucovorin, and one drug combination, 5-FU + leucovorin. Inhibition of Xt cell proliferation by cancer drugs resulted in decreased signal. The overall prediction success rate was 80% (4/5) and 60% (3/5) for Colo320 and SW6320 cells, respectively, indicating good correlation with results in humans. This novel approach is rapid, sensitive, quantitative, and amenable to automation for drug screening. Because cancer patients are frequently treated with multiple drugs, combination drug treatment by direct addition to fish water is a significant advantage compared to drug treatment in mammalian models.

17.4.1 Human Colon Cancer Cell Line Xt in Zebrafish

To generate a zebrafish xenotransplant colon cancer model, we initially injected 2 and 3dpf zebrafish with 1500 CM-DiI-labeled human colon cancer Colo320 and SW620 cells into the yolk sac. In pilot experiments, we observed that Colo320, a colon cancer cell line derived from a moderately undifferentiated colon adenocarcinoma, metastasized and proliferated in zebrafish and exhibited an infiltrative growth pattern (Fig. 17.2). SW620 cell line, derived from a grade 3–4 colon adenocarcinoma that had metastasized to lymph nodes, was tumorigenic in nude mice. In zebrafish, Xt SW620

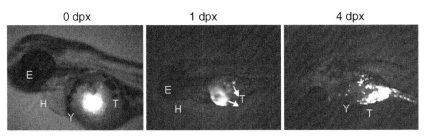

Figure 17.2 Human Colo320 colon cancer cells after xenotransplant in 2dpf zebrafish. One thousand five hundred CM-DiI-labeled Colo320 cells in 25 nL HBSS were injected into the zebrafish yolk sac and cancer cell proliferation and migration were tracked. Images are from the same Xt zebrafish at 0, 1, and 4 dpx. E = eye, H = heart, Y = yolk sac, and T = tumor.

cells migrated and formed masses. Since Colo320 and SW620 cells showed different cancer characteristics, they represented good cell lines for model development. Human lymphoblastoid TK6 cells were used as a noncancer control cell line.

Although we observed that both 2 and 3dpf zebrafish were suitable for xeno-transplant, 3dpf zebrafish tolerated a greater number of cells. To optimize model development, after xenotransplant, we tracked CM-DiI-labeled cancer cell progression in live zebrafish for 7 days using fluorescence microscopy. Although Xt zebrafish generally survived up to 3 dpx (>85% survival), death increased markedly at 4 dpx, likely due to increased cancer cell burden and nutrient depletion. In contrast, human TK6 lymphoblast control cells did not survive in zebrafish and by 5 dpx, few fluorescently labeled TK6 cells were visible (data not shown). These studies showed that early-stage zebrafish tolerate thousands of human cancer cells with no rejection response, supporting use as an *in vivo* animal model for drug screening. Because Colo320 cells have a short doubling time (~38 h) in Xt zebrafish, we selected this cell line for drug screening.

In a previous study, we determined that zebrafish injected with 50 human melanoma cells can be maintained at 35°C (Haldi et al., 2006) supporting normal development of the host as well as proliferation of human cancer cells. In these studies, we also established the yolk sac as the optimum injection site. For the first 7dpf, the yolk sac is the sole source of nutrition for the rapidly developing zebrafish. After this stage, zebrafish can ingest food.

17.4.2 Development of Xt ELISA

To quantitate drug effects in Xt zebrafish, we initially visually counted cancer cells labeled with CM-DiI tracking dye. However, in pilot experiments, we observed that similar to mammalian models, after xenotransplant, some colon cancer cells formed masses, impeding visual counting. Furthermore, dissociating zebrafish and counting labeled cancer cells was time consuming and less reliable because some cancer cells were destroyed during processing. Therefore, we decided to develop a quantitative Xt ELISA using a human cell-specific antibody. For these studies, we assessed several human antibodies, including HLA and HLA-DR (Invitrogen, Carlsbad, CA), survivin, XIAP, and Bax (Cell Signaling Technology, Danvers, MA), and Keap1 (Santa Cruz Technology, Santa Cruz, CA). Three dpf zebrafish were microinjected with 1500 CM-DiI-labeled Colo320 cells and fixed at 3 dpx. Whole mount immunostaining was performed using human survivin antibody, followed by staining with FITC-conjugated secondary antibody. Uninjected zebrafish and Xt zebrafish with no primary antibody staining were used as negative controls. CM-DiI-labeled Xt cancer cells were visualized in the rhodamine channel and antibody-stained cells were visualized in the FITC channel using a Zeiss Stemi SV11 Apo upright fluorescence microscope equipped with a Zeiss AxioCam. CM-DiI-labeled cells were visualized in the rhodamine channel and human antibody-stained zebrafish were visualized in the FITC channel using a Zeiss fluorescence microscope. Human survivin antibody stained Colo320 colon cancer cells in zebrafish, but exhibited limited cross-reactivity with whole zebrafish (data not shown).

We then performed the Xt ELISA as described in Section 17.3; signals from both survivin and XIAP were adequate for assay development. Survivin and XIAP, members of the inhibitors of apoptosis family of proteins (IAP), are constitutively expressed in human cells and expression level has been shown to be elevated in many human cancers (Deveraux et al., 1998; Altieri and Marchisio, 1999; Deveraux and Reed, 1999; Tamm et al., 2000; Kasof and Gomes, 2001). In addition, survivin and XIAP have previously been validated as cancer drug targets (Olie et al., 2000; Grossman et al., 2001). Because survivin generated a higher signal/noise ratio than XIAP, we used it for Xt ELISA development.

17.4.3 Linear Relationship Between Chemiluminescence Signal and Number of Xt Zebrafish in Microwells

We next established a linear relationship between chemiluminescence signal and number of Xt zebrafish per well. Two dpf zebrafish were microinjected with 1500 Colo320 cells and fixed 3 days post injection for whole zebrafish ELISA processing. Using relative chemiluminescence units (RCU), a linear relationship was observed between chemiluminescence signal and number of Xt zebrafish per well (data not shown).

Using same stage zebrafish processed with and without primary antibody, we also determined that $S/N = 2.3$, which indicated that the assay discriminated Xt zebrafish signal from background.

17.4.4 Effects of FDA-Approved Drugs on Xenotransplant Colon Cancer Cell Proliferation

To validate Xt ELISA for drug screening, we assessed effects of four FDA-approved colorectal cancer drugs, 5-FU, oxaliplatin, camptothecin, and leucovorin, and one drug combination, 5-FU + leucovorin (Table 17.2), on human colon cancer cell in Xt

Table 17.2 Mechanisms of Test Compounds

Compound	Mechanism of action
5-Fluorouracil	Potent antitumor agent
	Inhibits aminoisobutyrate–pyruvate aminotransferase
	Induces apoptosis in human primary and metastatic colon adenocarcinoma *in vitro*
	Arrests cell cycle at G2
Oxaliplatin	Antitumor platinum compound with activity against colorectal cancer; cytotoxicity follows formation of DNA adducts
	Induces apoptosis
Camptothecin	Binds irreversibly to the DNA topoisomerase I complex, blocking the cell cycle in S phase
	Induces apoptosis in many normal and cancer cell lines
Leucovorin	The active form of the B complex vitamin, folate, used in combination with chemotherapy to maintain folate levels
	Protects normal cells and increases antitumor effects of 5-FU

zebrafish. These drugs have been shown to induce apoptosis in human colon cancer cells. 5-FU, a potent antitumor agent that inhibits aminoisobutyrate–pyruvate aminotransferase, has been used as a chemotherapeutic for more than 40 years and it remains the most widely prescribed colon cancer drug. This drug also induces apoptosis in human primary and metastatic colon adenocarcinoma cell lines and arrests the cell cycle at G2. Oxaliplatin is a platinum compound that was approved by the FDA for colon cancer treatment in 2002. The mechanism of this drug involves formation of DNA adducts, which causes cytotoxicity and induces apoptosis. Camptothecin binds irreversibly to the DNA topoisomerase I complex, blocking the cell cycle in the S phase and inducing apoptosis. Although leucovorin, the active form of the B complex vitamin, folate, is not a cancer drug, it protects normal cells and enhances 5-FU efficacy. If drugs induced similar effects on Xt colon cancer cells, we expected decreased number of colon cancer cells or decreased antigen expression, both resulting in lower chemiluminescence signal. Three concentrations of each drug, determined by LC_{10} experiments, were assessed as described below. Drug treatment began immediately after cell transplant and Xt zebrafish were incubated in drug solution at 35°C continuously for 3 days. At 3 days post treatment (dpt), Xt zebrafish were then processed for ELISA experiments.

17.4.5 Effect of Cancer Drugs on Human Colon Cancer Cells in Xt Zebrafish

We first estimated LC_{10} of each cancer drug by treating nontransplanted, 3dpf zebrafish with test drugs for 3 days (see Section 17.3). LC_{10} for camptothecin was 0.24 µM; however, since no lethality was observed at the highest test concentration,

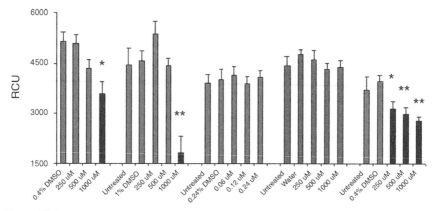

Figure 17.3 Drug effects on Xt Colo320 cells. Significantly lower chemiluminescence signal was observed after treatment with 5-FU, oxaliplatin, and 5-FU + leucovorin, indicating inhibition of Xt Colo320 cell proliferation, whereas no effect was observed after treatment with camptothecin and leucovorin. Data are presented as mean $\pm$ SE ($n = 14$). 0.24%, 0.4%, and 1% DMSO or water were used as vehicle controls. Red bars indicate concentrations that caused statistically significant effects. RCU, relative chemiluminescence units. * indicates $p < 0.05$ and ** indicates $p < 0.01$. (See the color version of this figure in Color Plates section.)

1000 μM, LC_{10} was not determined for 5-FU, oxaliplatin, and leucovorin. Based on LC_{10} estimations, three concentrations, 0.24, 0.12, and 0.06 μM, were assessed for camptothecin, and 1000, 500, and 250 μM were assessed for oxaliplatin, leucovorin, and 5-FU + leucovorin combination.

Test drug was added to fish water immediately after xenotransplant. After treatment for 3 days, zebrafish were processed for Xt ELISA using survivin antibody. For all experiments, drug-treated Xt zebrafish were compared with vehicle control, either DMSO or fish water, and results are shown in Fig. 17.3. Significant inhibition ($P < 0.05$) of Xt cancer cell proliferation was observed for 5-FU (3–43% inhibition), oxaliplatin (3–60% inhibition), and 5-FU + leucovorin (22–31% inhibition), and inhibition was dose dependent, whereas significant inhibition was not observed for camptothecin. As expected, leucovorin alone did not inhibit Xt cancer cell proliferation; however, 5-FU + leucovorin combination caused significant inhibition.

To further validate Xt ELISA, we compared cancer drug results in zebrafish with results in humans. Overall prediction rate in Xt zebrafish was 80% (4/5) for Colo320 cells and 60% (3/5) for SW620 cells.

17.5 CONCLUSIONS

In vitro cell lines, including the NCI-60 human cancer cell lines, are extensively used for primary drug screening; however, results using these *in vitro* assays are significantly different from results observed in complex, *in vivo* physiological environments. Furthermore, xenotransplant rodent cancer models are laborious, time consuming, and costly. Our Xt zebrafish model is designed as a comparatively high-throughput *in vivo* assay that can bridge the gap between cell-based screening and mammalian testing (Li et al., 2009).

REFERENCES

ALTIERI DC and MARCHISIO PC (1999). Survivin apoptosis: an interloper between cell death and cell proliferation in cancer. Lab Invest 79(11): 1327–1333.

DANILOVA N, HOHMAN VS, SACHER F, OTA T, WILLETT CE, and STEINER LA (2004). T cells and the thymus in developing zebrafish. Dev Comp Immunol 28(7–8): 755–767.

DETRICH HW, WESTERFIELD M, and ZON L (1999). The zebrafish biology. Methods Cell Biol 59: 305.

DEVERAUX QL and REED JC (1999). IAP family proteins: suppressors of apoptosis. Genes Dev 13(3): 239–252.

DEVERAUX QL, ROY N, STENNICKE HR, Van ARSDALE T, ZHOU Q, SRINIVASULA SM, ALNEMRI ES, SALVESEN GS, and REED JC (1998). IAPs block apoptotic events induced by caspase-8 and cytochrome *c* by direct inhibition of distinct caspases. EMBO J 17(8): 2215–2223.

GREINER DL, HESSELTON RA, and SHULTZ LD (1998). SCID mouse models of human stem cell engraftment. Stem Cells 16(3): 166–177.

GROSSMAN D, KIM PJ, SCHECHNER JS, and ALTIERI DC (2001). Inhibition of melanoma tumor growth *in vivo* by survivin targeting. Proc Natl Acad Sci USA 98(2): 635–640.

HALDI M, LUO LQ, TANG J, LI CQ, and SENG WL (2009). Zebrafish orthotopic brain cancer model for drug screening. Proceedings of the 100th AACR Annual Meeting, Washington, DC, Abstract 2322, p. 562.

HALDI M, TON C, SENG WL, and McGRATH P (2006). Human melanoma cells transplanted into zebrafish proliferate, migrate, produce melanin, form masses and stimulate angiogenesis in zebrafish. Angiogenesis 9(3): 139–151.

HERLYN M and KOPROWSKI H (1988). Melanoma antigens: immunological and biological characterization and clinical significance. Annu Rev Immunol 6: 283–308.

HO RK and KANE DA (1990). Cell-autonomous action of zebrafish spt-1 mutation in specific mesodermal precursors. Nature 348(6303): 728–730.

KASOF GM and GOMES BC (2001). Livin, a novel inhibitor of apoptosis protein family member. J Biol Chem 276(5): 3238–3246.

KATSANIS E, WEISDORF DJ, and MILLER JS (1998). Activated peripheral blood mononuclear cells from patients receiving subcutaneous interleukin-2 following autologous stem cell transplantation prolong survival of SCID mice bearing human lymphoma. Bone Marrow Transplant 22(2): 185–191.

KELLAND LR (2004). Of mice and men: values and liabilities of the athymic nude mouse model in anticancer drug development. Eur J Cancer 40(6): 827–836.

LEE JT and HERLYN M (2006). Embryogenesis meets tumorigenesis. Nat Med 12(8): 882–884.

LEE LM, SEFTOR EA, BONDE G, CORNELL RA, and HENDRIX MJ (2005). The fate of human malignant melanoma cells transplanted into zebrafish embryos: assessment of migration and cell division in the absence of tumor formation. Dev Dyn 233(4): 1560–1570.

LI C, SENG WL, and McGRATH P (2008). Whole zebrafish cytochrome P450 microplate assays for assessing drug metabolism and drug safety. Society of Toxicology Annual Meeting, Seattle, WA.

LI CQ, LUO LQ, TANG J, TANG S, and SENG WL (2009). Zebrafish xenotransplant colon cancer model for drug screening. Proceedings of the 100th AACR Annual Meeting, Washington, DC, Abstract 2340, p. 567.

McGRATH P and LI CQ (2008). Zebrafish: a predictive model for assessing drug-induced toxicity. Drug Discov Today 13(9–10): 394–401.

McGRATH P and SERBEDZIJA G (2004). Methods for heterologous cell transplant. U.S. Patent 6,761,876.

OLIE RA, SIMOES-WUST AP, BAUMANN B, LEECH SH, FABBRO D, STAHEL RA, and ZANGEMEISTER-WITTKE U (2000). A novel antisense oligonucleotide targeting survivin expression induces apoptosis and sensitizes lung cancer cells to chemotherapy. Cancer Res 60(11): 2805–2809.

SENG WL, ENG K, LEE J, and McGRATH P (2004). Use of a monoclonal antibody specific for activated endothelial cells to quantitate angiogenesis *in vivo* in zebrafish after drug treatment. Angiogenesis 7(3): 243–253.

STOLETOV K, MONTEL V, LESTER RD, GONIAS SL, and KLEMKE R (2007). High-resolution imaging of the dynamic tumor cell vascular interface in transparent zebrafish. Proc Natl Acad Sci USA 104(44): 17406–17411.

TAMM I, KORNBLAU SM, SEGALL H, KRAJEWSKI S, WELSH K, KITADA S, SCUDIERO DA, TUDOR G, QUI YH, MONKS A, ANDREEFF M, and REED JC (2000). Expression and prognostic significance of IAP-family genes in human cancers and myeloid leukemias. Clin Cancer Res 6(5): 1796–1803.

TOPCZEWSKA JM, POSTOVIT LM, MARGARYAN NV, SAM A, HESS AR, WHEATON WW, NICKOLOFF BJ, TOPCZEWSKI J, and HENDRIX MJ (2006). Embryonic and tumorigenic pathways converge via Nodal signaling: role in melanoma aggressiveness. Nat Med 12(8): 925–932.

van WEERDEN WM and ROMIJN JC (2000). Use of nude mouse xenograft models in prostate cancer research. Prostate 43(4): 263–271.

WESTERFIELD M (1993). The Zebrafish Book: A Guide for the Laboratory Use of Zebrafish (*Danio rerio*). University of Oregon Press, Eugene, OR.

WILLETT CE, ZAPATA AG, HOPKINS N, and STEINER LA (1997). Expression of zebrafish rag genes during early development identifies the thymus. Dev Biol 182(2): 331–341.

YANG EB, TANG WY, ZHANG K, CHENG LY, and MACK PO (1997). Norcantharidin inhibits growth of human HepG2 cell-transplanted tumor in nude mice and prolongs host survival. Cancer Lett 117(1): 93–98.

ZHANG JH, CHUNG TD, and OLDENBURG KR (1999). A simple statistical parameter for use in evaluation and validation of high throughput screening assays. J Biomol Screen 4(2): 67–73.

Chapter 18

Zebrafish Assays for Identifying Potential Muscular Dystrophy Drug Candidates

Jian Tang, Susie Tang, Maryann Haldi, and Wen Lin Seng

Phylonix, Cambridge, MA, USA

18.1 INTRODUCTION

The aims of this research were to (1) develop a reproducible zebrafish muscular dystrophy (MD) animal model and (2) optimize a battery of zebrafish assays to identify potential drug candidates. In this research, we used short interfering RNA (siRNA) injection to knockdown dystroglycan, known to be involved in muscle cell integrity and function, and generated phenotypes consistent with those observed in *mdx* mice, a naturally occurring dystrophin mutant. Next, we developed assays to assess myotome length, muscle structure, level of ROS, and motility in MD zebrafish. Using histology, we also confirmed muscular dystrophy disease progression. In addition, we showed that two drugs, prednisone and epigallocatechin-3-gallate (EGCG), which improve muscle strength in both Duchenne muscular dystrophy patients and the *mdx* mouse model (Bonifati et al., 2000; Dorchies et al., 2006), partially inhibited disease progression in MD zebrafish. Although zebrafish is phylogenetically distant from humans, key gene families are conserved and dystroglycan knockdown (KD), performed by injecting siRNA into wild-type animals, is highly specific. This study highlights the convenience of generating zebrafish disease models and describes a variety of complementary whole animal assays for drug screening.

18.1.1 Muscular Dystrophy and Dystrophin–Glycoprotein Complex

Muscular dystrophy refers to a number of clinically and genetically heterogeneous disorders characterized by degeneration of skeletal or voluntary muscles and progressive weakness. In some forms of MD, heart and involuntary muscles are also affected. Major forms of MD, which can occur at any age, include Becker, Duchenne, myotonic, limb-girdle, facioscapulohumeral, congenital, oculopharyngeal, distal, and Emery–Dreifuss (Mathews, 2003). Duchenne is the most common childhood form and myotonic is the most common adult form. Identification of defective dystrophin in Duchenne muscular dystrophy (DMD) and isolation of a number of dystrophin-associated proteins in skeletal muscle have provided valuable clues for MD pathogenesis.

The large oligomeric dystrophin–glycoprotein complex (DGC), which connects the extracellular matrix with the actin cytoskeleton, contains both structural and signal transduction properties (Lapidos et al., 2004) and causes specialization of cardiac and skeletal muscle membranes. Dystrophin binds to filamentous γ-actin through the actin amino-terminal binding domain (Rybakova et al., 2000) and binds to β-dystroglycan through the C-terminal cysteine-rich domain. DGC elements lacking dystrophin are smaller at the sarcolemma and unstable. Dystroglycan, originally isolated from skeletal muscle, was later found to be expressed during development and it is present in nearly all cell types. In vertebrates, dystroglycan is comprised of alpha- and beta-subunits encoded by a single gene (Holt et al., 2000). At the sarcolemma, β-dystroglycan binds intracellularly to dystrophin and extracellularly to α-dystroglycan. α-Dystroglycan binds to extracellular matrix proteins including laminin, neurexin, agrin, and perlecan, completing the link from the cytoskeleton to the basal lamina. In addition to dystrophin and dystroglycan, the DGC contains the sarcoglycan complex and sarcospan. The sarcolemmal DGC interacts with a pair of syntrophins and α-dystrobrevin within the cytosol via dystrophin. DGC contributes to the structural stability of muscle cell membranes and protects muscles from stress-induced damage.

In humans, dystrophin gene mutations have been shown to cause Duchenne and Becker muscular dystrophy. Additional gene mutations in the DGC have been identified in different forms of MD including sarcoglycans (limb-girdle muscular dystrophies, LGMD2C-F; α, β, δ, γ), laminine α2 chain (congenital muscular dystrophy), laminine α7 (congenital muscular dystrophy), and dysferlin (LGMD 2B and Miyoshi myopathy) (Xu et al., 1994; Lim et al., 1995; Araishi et al., 1999; Weiler et al., 1999). Other defective non-DGC molecules have also been found in MD patients, including calpain, an intracellular calcium-activated protease (autosomal recessive limb-girdle muscular dystrophy, LGMD2A), telethonin (TCAP), a muscle-specific protein that localizes to the Z-disc of skeletal muscle (LBMD2G), Trim 32, a potential E3-ubiquitin ligase (LGMD2H), emerin, an integral nuclear membrane protein (Emery Dreifuss syndrome), dystrophia myotonica protein kinase (DMPK) (myotonic dystrophy, DM1), and ZFF9, a zinc finger protein (DM2) (Allamand and Campbell, 2000; Mathews, 2003).

18.1.2   Current Animal Models for Screening Muscular Dystrophy Drugs

Homologs of dystrophin and other DGC components have been identified not only in mammals, birds, and zebrafish, but also in invertebrates, including worms and flies, indicating that this complex is evolutionarily conserved. Current muscular dystrophy models are classified as mammalian and nonmammalian (Collins and Morgan, 2003).

18.1.2.1   Mammalian Models   *Mdx* mice, the most widely used murine MD model, were derived from a naturally occurring mutant found in the C57Bl/10 colony (Bulfield et al., 1984) and were later characterized as dystrophin mutants (Hoffman et al., 1987). Since *mdx* mice are responsive to drugs, they are a useful model for identifying potential protectants (Granchelli et al., 2000; De Luca et al., 2002). Although the *mdx* mouse is the most common DMD model, animals exhibit mild phenotypes, active regeneration, and a normal life span, probably due to effective compensation by utrophin (Khurana and Davies, 2003). *Mdx* pathology can be partially rescued by the homologous protein, utrophin (Rybakova et al., 2002). Double knockout mice, lacking both dystrophin and utrophin (mdx/utrn$^{-/-}$), exhibit more severe phenotypes, including cardiomyopathy (Deconinck et al., 1997; Grady et al., 1997), a phenotype closer to what is observed in DMD patients. The widely studied Golden Retriever Muscular Dystrophic (GRMD) dog (Cooper et al., 1988) more closely resembles human pathogenesis than the *mdx* mouse. Other MD mouse models include chimeric mice with dystroglycan deficiency, chimeric mice with α5 integrin deficiency, integrin α7 null mice, and α-dystrobrevin null mice (Mayer et al., 1997; Taverna et al., 1998; Cote et al., 1999). In addition to these models, cats (Kohn et al., 1993; Gaschen and Burgunder, 2001) and hamsters (Mizuno et al., 1995; Watchko et al., 2002) exhibiting muscular dystrophy phenotypes have also been generated. Although these models have been useful in elucidating MD disease mechanisms and assessing potential therapeutic strategies, none are useful for large-scale drug screening.

18.1.2.2   Nonmammalian Models   Less complex *in vivo* model systems, including *Caenorhabditis elegans* (*dys-1*), have been used for morphological, behavioral, genetic suppressor, and drug screens (Bessou et al., 1998; Baumeister and Ge, 2002) and prednisone has been shown to slow muscle degeneration in dystrophin-deficient animals (Gaud et al., 2004). However, due to their lack of complex metabolism and relatively low correlation to human physiology, invertebrate organisms have been poor predictors of drug toxicity and efficacy in humans.

18.1.3   Zebrafish: A Good Surrogate Model for MD

Zebrafish exhibit many inherent advantages for studying MD. First, zebrafish skeletal muscle is simply organized; single myotomes extend across somites, both ends of which are attached to the extracellular matrix of the transverse myoseptum. Second, loss of dystroglycan and the DGC results in less pleiotropic phenotypes than those observed in mice and dogs. Third, generation of dystroglycan knockdown zebrafish by simple injection of gene specific reagents is straightforward. Fourth, assessment of disease phenotypes and drug effects in zebrafish is faster and cheaper than assessment in other models.

In zebrafish, somites, responsible for controlling movement, are primarily comprised of myotomes that are a significant component of the body plan. Each V-shaped myotome contains muscle cells and myoseptum, a connective tissue that separates dorsal and ventral somites (horizontal myoseptum) from myotomes (vertical myoseptum). The horizontal myoseptum is visible as a black line running through the middle of myotomes (Fig. 18.7). The structure and function of myoseptum are similar to the those of mammalian tendons. Occasionally, disrupted myoseptum in the posterior section of the zebrafish body was also observed. These defects often cause bent notochord and deformed posterior body plan.

Several muscular dystrophy phenotypes have been generated in zebrafish using different methods. *Sapje*, a well-characterized dystrophin gene mutant (Bassett et al., 2003; Bassett and Currie, 2004; Guyon et al., 2007), exhibits defective muscle attachment that causes progressive muscle degeneration and cell death. We note that utrophin, a dystrophin homolog that can substitute for dystrophin in early development, is not present in zebrafish muscle fiber ends (Bassett and Currie, 2003). Therefore, the *sapje* mutant and the dystrophin morphant exhibit severe phenotypes more similar to dystrophin/utrophin double knockout mice (Grady et al., 1997; Janssen et al., 2005; Guyon et al., 2007) than to *mdx* mice, which are only deficient in dystrophin.

18.1.4 Technologies for Gene Knockdown in Zebrafish

Several methods for gene knockdown in zebrafish have been described including antisense morpholino oligonucleotides (MOs), negatively charged peptide nucleic acids (PNAs), ribozymes, and long and short dsRNAs including siRNAs (Skromne and Prince, 2008). MOs have been widely used to block mRNA translation or splicing in zebrafish and phenotypes have been shown to be similar to those observed in the zebrafish *sapje* mutant (Nasevicius and Ekker, 2000). However, sequence-specific "off-target" effects, which are mediated through a p53-dependent cell death pathway, were recently discovered (Robu et al., 2007) and careful design of controls is required. Other modified oligonucleotides, including PNAs, have been shown to function with potency and specificity in zebrafish (Urtishak et al., 2003; Wickstrom et al., 2004) and ribozymes for generating knockdown *no tail* zebrafish have also been assessed (Xie et al., 1997; Pei et al., 2007); however, neither of these methods is widely used.

MOs have also been used to knockdown zebrafish dystrophin genes and MO zebrafish exhibited bent or curved tails and were less active than controls (Guyon et al., 2003). Dystrophin siRNA knockdown animals exhibited similar phenotypes (Dodd et al., 2004). In addition, hooked tail and muscle necrosis were observed in dystroglycan MO knockdown zebrafish (Parsons et al., 2002). Disruption of DGC in KD zebrafish resulted in loss of sarcomere and sarcoplasmic organization, indicating that dystroglycan is required to maintain long-term muscle cell survival (Parsons et al., 2002). Inhibition of dystroglycan protein production in zebrafish has been confirmed by dystroglycan antibody staining. Recently, overexpression of myostatin-2, a member of the TGF-β superfamily and a potent negative regulator of skeletal muscle and growth, was found to decrease dystrophin-associated protein complex (DAPC) expression, resulting in muscle dystrophy (Amali et al., 2008).

RNA interference, which is primarily used for research applications, has received increasing attention as a potential therapeutic strategy and both long ds (double stranded) RNAs (>100) and short (20–80 nucleotides, nt) ds RNAs have been assessed. Long ds siRNAs have been shown to cause nonspecific effects and a short 21-nt ds siRNA has also been used to knockdown the dystrophin gene in zebrafish (Dodd et al., 2004). A recent research report (Zhao et al., 2008) sheds additional light on reducing potential nonspecific siRNA effects and coinjection of preprocessed micro-RNA-430 bases (b) can prevent nonspecific defects associated with siRNA injection. siRNA injection transiently inhibits protein production until siRNA degrades. In our zebrafish MD model, using whole mount dystroglycan antibody staining, we observed elimination of dystroglycan protein ~20 h post fertilization (hpf). In general, siRNA dystroglycan knockdown zebrafish phenotypes were similar to MO phenotypes including developmental delay, truncated bodies, curved tails (data not shown), and decreased and uncoordinated movement. These results are also consistent with data reported by Parsons et al. (2002). In humans, DMD progression leads to a variety of physical symptoms affecting spine, legs, feet, joints, and tendons. Symptoms can include general muscle weakness, overdeveloped calves (pseudohypertrophy), increased muscle volume due to fat deposits, lordosis and scoliosis, curvature of the spine, joint and tendon restriction, speech and mental impairment, and respiratory difficulties. In dystroglycan KD zebrafish, we observed several defects that resemble human congenital myopathies, including rapid, progressive muscular degeneration, immobility, muscle and brain deformities, bent spine (notochord), and cardiac defects.

18.1.5 MD Treatment Options

Previous studies have shown that three- to fourfold increase in utrophin expression in *mdx* mouse muscle was sufficient to prevent or dramatically reduce muscular dystrophy pathology (Tinsley et al., 1998). However, for the majority of DMD patients, steroids, including prednisone, which reduces inflammation, are the only available therapeutic option (Campbell and Jacob, 2003; Manzur et al., 2004, 2008; Beenakker et al., 2005). Use of antibodies that block the action of myostatin (Bogdanovich et al., 2002, 2005), a negative regulator of muscle mass, has resulted in functional improvement of dystrophic muscle in mice. MYO-029, a neutralizing antibody, has also been assessed in clinical trials for various forms of muscular dystrophy. More recently, inhibition of histone deacetylases (HDACs) resulted in upregulation of regeneration-activated genes and formation of hypernucleated, larger than normal myotubes. Efficacy of deacetylase inhibitors, including trichostatin A (TSA) and MS27 (a selective inhibitor of class I deacetylases), injected daily intraperitoneally has been assessed in 3-month-old dystrophin-deficient (*mdx*) mice (Minetti et al., 2006). These drugs increased skeletal muscle size and mitigated morphological and pathological consequences of the primary genetic defects. A beneficial effect on muscle repair has been reported after administering nitric oxide (NO) using a NO-releasing derivative of flurbiprofen (HCT1026), a nonsteroidal anti-inflammatory drug (Minetti et al., 2006). Proteasome inhibitors MG-132, Velcade (bortezomib or PS-341), and MLN273 (PS-273) have been shown to restore both dystrophin and dystrophin-associated protein expression (Bonuccelli et al., 2003, 2007; Assereto et al., 2006). Humans with myotonic dystrophy and mice expressing the muscular dystrophy gene exhibited free radical overload and oxidative stress. In *mdx* mice, epigallocatechin-3-gallate, an antioxidant component of green tea, has been shown to delay MD onset (Nakae et al., 2008) and *N*-acetylcysteine, another antioxidant, protects against dystrophic muscle damage (Whitehead et al., 2008), increased β-dystroglycan and utrophin expression, and decreased expression of NF-κB, the pro-inflammatory cytokine. Although effective therapies for muscular dystrophies are still lacking, several novel strategies are under investigation, including gene and cell therapies briefly described below.

18.1.5.1 Gene Therapy Intramuscular injection of γ-sarcoglycan-expressing adeno-associated viral (AAV) vectors is under investigation for limb-girdle muscular dystrophy. For dystrophin, two strategies are currently undergoing testing in mammals: exon skipping and dystrophin variant expression using AAV vectors. Despite minor efficacy, in addition to viral vectors, nonviral delivery is also undergoing clinical experimentation and direct intramuscular injection of plasmids expressing human dystrophin resulted in protein expression at the site of injection (Romero et al., 2004).

18.1.5.2 Cell Therapy Strategies to replace affected cells (Cossu and Sampaolesi, 2007), including pioneering experiments in mouse DMD models, demonstrated that myoblasts transplanted into dystrophic muscle generated dystrophin-expressing myofibers (Partridge et al., 1989; Skuk et al., 2004, 2006). Several adult-derived stem cells have been isolated, including bone marrow-derived stem cells, blood- and muscle-derived CD133[+] cells, muscle-derived stem cells (MDSCs), side population (SP) cells, and mesoangioblasts (Gussoni et al., 1999; Jiang et al., 2002; Minasi et al., 2002; Torrente et al., 2004; Dezawa et al., 2005; Dellavalle et al., 2007).

18.2 MATERIALS AND METHODS

18.2.1 Compounds and Reagents

Unless specifically described, compounds and reagents were purchased from Sigma (St. Louis, MO).

18.2.2 Embryo Handling

Phylonix AB zebrafish were generated by natural pairwise mating in our aquaculture facility, as described by Westerfield (1993). Four to five pairs were set up for each mating; on average, 50–100 embryos per mating were generated. Embryos were maintained in embryo water (5 g of Instant Ocean Salt with 3 g of $CaSO_4$ in 25 L distilled water) at 28°C for approximately 24 h before sorting for viability. Because embryos receive nourishment from an attached yolk ball, no additional maintenance was required.

18.2.3 Generation of MD Zebrafish Using Dystroglycan-Specific siRNAs

Four smartpool siRNAs specific for zebrafish dystroglycan gene (NM_173274) were designed and synthesized by Dharmacon RNA Technologies/Thermo Fisher Scientific (Lafayette, CO). Target sequences were siRNA1 (SMP-1), GTTATAGGCGATAGCAATA; siRNA2 (SMP-2), GAGGATGGAACTACCGATA; siRNA3 (SMP-3), CGGACAAGGGCGTTCACTA; and siRNA4 (SMP-4), GGAGATGCGCGTACGGTCA. A proprietary siRNA comprised of a nonreacting coding sequence (AS siRNA) (Dharmacon RNA Technologies/Thermo Fisher Scientific) was used as control. siRNA1 was directly labeled with DY547 at the $5'$ sense end. The $5'$ antisense end of each siRNA was modified with a phosphate group.

18.2.4 Microinjection of siRNAs

To reduce variation between batches, randomized embryo samples from three or four independent pairwise matings were used. Fertilized eggs ($\sim$1000) were microinjected at the one-cell stage (Meng and Wolf, 1997). 0.5 nL volume was delivered using an air pressure-controlled microinjector (WPI, Sarasota, FL). Injected zebrafish were cultured and grown as described by Westerfield (1993).

18.2.5 Zebrafish Dechorionation

Embryos hatch from a chorion at $\sim$2.5dpf stage. In order to assess zebrafish earlier than the 2.5dpf stage, we dechorionated zebrafish. Briefly, embryos were placed on an agar gel-coated Petri dish and treated with pronase (1 mg/mL) for 3 min, followed by extensive washing with fish water to remove residual pronase. Treated embryos were then placed in a 28°C incubator for $\sim$30 min and removed from chorions by gentle shaking.

18.2.6 TaqMan Quantitative PCR (qRT-PCR) Analysis

Total RNA was isolated from (a) uninjected control, (b) knockdown control (AS siRNA injected), and (c) MD zebrafish (SMP siRNA injected) at 4, 6, and 20 h post injection (hpi). Real-time quantitative reverse transcriptase PCR (qRT-PCR) was performed at Beth Israel Medical Center qPCR core facility (Boston, MA). Total RNA was isolated from zebrafish using RNAqueous-4PCR kit (Ambion, Austin, TX) and 2 μg of RNA was reverse transcribed to cDNA using oligo dT primer and Omniscript RT kit (Qiagen, Valencia, CA). Real-time PCR was performed in triplicate using the QuantiTect™ probe PCR kit (Qiagen). Thermal cycling conditions included 95°C for 15 min, followed by 40 amplification cycles at 94°C for 15 s, followed by 60°C for 1 min. TaqMan PCR primers against zebrafish dystroglycan gene (NM_173274) were designed by Applied Biosystems (Foster City, CA). β-Actin primers used as an internal control were synthesized by Applied Biosystems. Real-time PCR primer sequences specific for dystroglycan were as follows: forward primer, ACGACAACCAAGCCACCAA, reverse primer, GGGTTACGCAGCTCAGGTTTTATAT, and the probe sequence was 6FAMCTGCAGTTATAGGCGATAGCMGBNFQ. Primer and probe sequences for zebrafish β-actin were 5′-AGGTCATCACCATCGGCAAT-3′, 5′-GATGTCCACGTCGCACTTCAT-3′, and 5′-ViC CTTCCAGCCTTCCTTCCTGGGTATGGA-MGBNFQ-3′. All primers were tested for nonspecific amplicons and primer dimers by visualizing PCR products on 2% agarose gels before performing qRT-PCR and analyzing dissociation curves after performing qRT-PCR.

18.2.7 Whole Mount Antibody Staining

Zebrafish were fixed in 4% paraformaldehyde (PFA) in phosphate-buffered saline (PBS) for 2 h, rinsed, dehydrated, and stored in methanol at −20°C. Whole mount staining using an anti-human dystroglycan monoclonal antibody was then performed on fixed zebrafish samples that were rehydrated using standard methods (Westerfield, 1993).

18.2.8 Fluorescence Microscopy

All fluorescence microscopy studies were performed using a Zeiss M2Bio fluorescence microscope (Zeiss, Thornwood, NY), equipped with a red rhodamine cube, a green FITC filter (excitation: 488 nm, emission: 515 nm), a chilled CCD camera (AxioCam MRm, Zeiss) equipped with 1.6× 10× objective lenses, and 10× eye pieces. Images were processed with AxioVision software Rel 4.6 (Zeiss) and Adobe Photoshop 7.0 software (Adobe, San Jose, CA).

18.2.9 Myotome Length

Length of eight myotomes, two anterior and six posterior to the anal pore, was measured using images of live 3dpf zebrafish. Images were analyzed using ImageJ software (NIH, Bethesda, MD). The region of interest (ROI) was outlined using the software drawing function and relative length was automatically measured.

18.2.10 Reactive Oxygen Species Staining Using H₂DCFDA Respiratory Burst Assay

ROS production was quantitated in live 3dpf zebrafish using the method described by Hermann et al. (2004). Single zebrafish were deposited into each well of black 96-well microplates (Corning Life Sciences, Lowell, MA) in 100 μL of fish water. One hundred microliters of $2',7'$-dihydrodichlorofluorescein diacetate (H_2DCFDA, Invitrogen) (1 μg/mL) was then added to wells containing zebrafish. After incubating at 28°C for 1 h, animals were placed on slides and exposed to UV light for 30 s; fluorescence images were then captured using the same gain and exposure time.

18.2.11 Motility Assay

To quantitate zebrafish movement, we used the VideoTrack system (ViewPoint Life Sciences, Lyons, France). The system uses infrared light to view zebrafish in microwells and to continuously monitor motion. Six dpf zebrafish were placed in 24-well microplates, one animal per well in 500 μL of fish water. Total distance (D) traveled by individual zebrafish was recorded for 60 min in alternating 10 min light and dark photoperiods.

18.2.12 Histology

Histology was performed following standard procedures. Briefly, zebrafish were fixed in 4% PFA in PBS overnight at 4°C, and embedded in JB-4. Cross and sagittal sections were prepared and stained in hematoxylin–eosin (H&E) for examination.

18.2.13 Drug Administration

We assessed effects of five drugs: prednisone, EGCG, MG132, TSA, and dantrolene. Twenty-four hpf MD zebrafish were treated with three compound concentrations (1, 10, and 100 μM) continuously for 48 h. At 3dpf, drug treatment was terminated and zebrafish were processed to assess myotome length and ROS level. Ten embryos were used for each condition. 0.1% dimethyl sulfoxide (DMSO) was used as carrier solvent. To ensure dystroglycan KD specificity, KD control zebrafish were included in each experiment.

18.2.14 Fluorescence Image Analyses

Fluorescence intensity in images after ROS staining was quantified by applying a constant threshold value to ROI (tail region posterior to the anal pore); histograms of highlighted areas were quantitated using the Adobe Photoshop measurement function.

18.2.15 Statistics

For multiple concentration assays, ANOVA was used to analyze drug effects, followed by Dunnett's test to identify concentrations that caused significant differences ($P < 0.05$). For single concentration tests, Student's t-test was used to analyze drug effects.

18.3 RESULTS

In this research, using siRNAs, we generated a zebrafish muscular dystrophy disease model by knocking down dystroglycan gene expression. We confirmed specificity by both immunostaining and qRT-PCR. KD MD zebrafish exhibited small heads, small eyes, pericardial edema, truncated bodies, short tails, and short myotomes. To assess muscle fibers in MD zebrafish, we quantitated myotome length in the tail region. Since mitochondria membrane instability induced by oxidative stress is considered an important mechanism involved in MD, and elevated reactive oxygen species (ROS) level is often observed prior to muscle necrosis in *mdx* mouse, we assessed ROS upregulation in MD zebrafish using a live fluorescent dye, H_2DCFDA. Next, using a quantitative, functional motility assay, we showed that movement decreased in MD zebrafish. We also performed histology on zebrafish tissue sections that clearly showed muscle degeneration during a 5-day time course. Using these methods, we then assessed effects of several compounds from different drug classes shown to partially rescue MD phenotypes in both *mdx* mice and humans. Our results showed that prednisone treatment inhibited shortening of myotomes and increase in ROS level, supporting use of zebrafish to identify potential MD drug candidates.

18.3.1 Reproducible Zebrafish Dystroglycan Knockdown MD Disease Model

To generate a reproducible MD disease model, we first optimized KD conditions. Various KD reagents, including antisense MOs against both dystrophin (Guyon et al., 2003) and dystroglycan (Parsons et al., 2002), and siRNAs against dystrophin (Dodd et al., 2004), have been used to generate MD phenotypes in zebrafish. Since siRNA is currently a major interest of the pharmaceutical industry both for generating disease models and for developing potential therapeutics, we decided to investigate use of this reagent to generate a zebrafish MD disease model. We initially injected varying doses of four siRNAs (SMP1–4) targeted by dystroglycan into one-cell stage zebrafish and determined that SMP1 was the optimum KD reagent for generating reproducible MD phenotypes.

In these initial experiments, we observed that MD phenotypes were sensitive to siRNA dose; higher doses induced toxic effects and lower doses generated mild MD phenotypes. We determined that 0.5 nL of 15 µM SMP1 siRNA (MD zebrafish), which did not induce nonspecific effects, was the optimum concentration. Nonspecific AS siRNA-injected animals (KD control) exhibited morphology similar to

Table 18.1   Survival Rate and Phenotype Percentage for MD and KD Control Zebrafish

Stage (dpf)	MD (SMP) siRNA		KD control (AS) siRNA	
	Survival (%)	MD phenotype (%)	Survival (%)	MD phenotype (%)
1	100	84.8	100	<1
2	98.2	82.3	97.5	<1
3	93.8	73.3	97.5	<1
4	93.8	71.4	97.5	<1
5	82.1	59.8	97.5	<1
6	70.5	53.2	97	<1
7	58.0	53.8	95	<1

uninjected zebrafish. In initial experiments, we followed progression of MD phenotypes for 7 days. Survival rates and phenotype percentages are shown in Table 18.1. We observed that ~93.8% of MD zebrafish survived to 4dpf stage and 71.4% of surviving MD zebrafish exhibited short tails (Fig. 18.1), indicating a high success rate for generating consistent phenotypes. By 5dpf, ~82% of MD zebrafish survived and ~60% exhibited short tails (Fig. 18.1). In contrast, <1% of KD control zebrafish exhibited short tails, indicating that MD phenotypes generated by dystroglycan siRNA KD were not caused by nonspecific toxicity. Since survival of injected

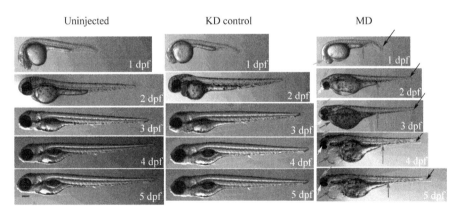

Figure 18.1   Comparison of phenotypes for uninjected, AS siRNA (KD control), and SMP siRNA (MD) injected zebrafish. One-cell stage zebrafish were injected with control AS siRNA (KD control) zebrafish or dystroglycan SMP siRNA (MD). Uninjected zebrafish were used for comparison. After injection, animals were examined under a dissecting microscope; a representative image from each group is shown. KD control animals exhibited similar phenotype as uninjected zebrafish, indicating that control siRNA did not induce abnormal phenotypes. MD zebrafish exhibited short tails (black arrows), truncated body (red arrows), small heads (green arrows), small eyes (yellow arrows), and pericardial edema (blue arrows). (See the color version of this figure in Color Plates section.)

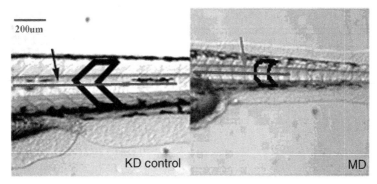

Figure 18.2 Myotome structure. Three dpf zebrafish tails are shown (3.2× magnification).
In KD control zebrafish, myotomes in the tail exhibited the characteristic chevron, V-shape (black arrow;
thick black outline), whereas MD zebrafish exhibited defective myotomes (red arrow; thick black outline).
Blue scale bar indicates 200 μm. Blue lines span myotomes used for quantitation in Fig. 18.4.

zebrafish decreased after 5dpf, we decided to focus on MD phenotype development
from 1 to 5dpf (Fig. 18.1).

Uninjected zebrafish and KD controls exhibited normal body morphology,
whereas MD zebrafish exhibited small heads (green arrows), small eyes (yellow
arrows), pericardial edema (blue arrows), truncated bodies (red arrows), and short or
bent tails (black arrows, Fig. 18.1). At 2, 3, and 4dpf stages, difference in overall body
and tail length, from the anal pore to the tip of the tail (Fig. 18.2), was significant.

18.3.2 Confirmation of Dystroglycan Knockdown by Whole Mount Immunostaining

To confirm that MD phenotypes resulted from dystroglycan knockdown and not from
nonspecific siRNA effects, we performed whole mount immunostaining using an anti-
human dystroglycan antibody shown to cross-react in zebrafish (Parsons et al., 2002).
Staining pattern showed low (Fig. 18.3b) or loss of (Fig. 18.3c) dystroglycan
expression and abnormal sarcolemma.

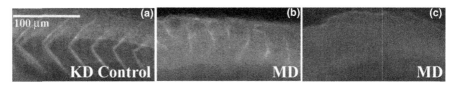

Figure 18.3 Whole mount dystroglycan immunostaining. Twenty hpf zebrafish were stained with a
dystroglycan-specific antibody. (a) KD control zebrafish exhibited dystroglycan staining in the
myoseptum of normal myotomes, whereas MD zebrafish exhibited either (b) weak dystroglycan
staining in irregular shaped myotomes or (c) absence of dystroglycan staining. These results confirmed
KD of dystroglycan expression by siRNA. Fluorescence images were captured at 4× magnification;
anterior, left. White scale bar represents 100 μm.

18.3.3 Confirmation of Dystroglycan Knockdown by qRT-PCR Analysis

To further confirm that siRNA injection specifically knocked down dystroglycan gene expression, we also performed real-time qRT-PCR. Briefly, total RNA was isolated from uninjected control, KD control, and MD zebrafish at 4, 6, and 20 hpi and qRT-PCR was performed using conventional methods. Results showed that dystroglycan gene expression (expressed as percent of uninjected control) decreased in MD zebrafish at 4, 6, and 20 hpi ($85.1 \pm 8.0\%$, $94.2 \pm 8.4\%$, and $59.5 \pm 3.2\%$, respectively), confirming dystroglycan knockdown and results by immunostaining (Fig. 18.3).

18.3.4 MD Zebrafish Exhibit Short, Disorganized Myotomes

MD zebrafish exhibited short myotomes lacking the characteristic chevron, V-shaped structure (Fig. 18.2, red arrow; thick black outline) (Parsons et al., 2002). In contrast, KD control zebrafish exhibited normal V-shaped myotomes (Fig. 18.2, black arrow; thick black outline). In MD zebrafish, some myotomes were disrupted or missing. Myotomes in the tail are primarily comprised of muscle fibers that align along the anterior–posterior axis of the body and tail length reflects muscle fiber length.

Next, we measured eight myotomes along the anterior–posterior body axis (two anterior and six posterior to the anal pore) (blue lines, Fig. 18.2). A significant difference was observed ($P = 7.1 \times 10^{-14}$); mean length for KD control zebrafish was 0.642 ± 0.075 mm and for MD zebrafish was 0.450 ± 0.079 mm ($n = 33$), which was ~30% shorter (($1 - 0.450/0.642) \times 100\% = 29.9\%$) than KD control (Fig. 18.4).

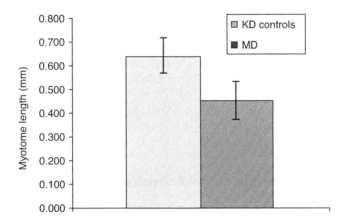

Figure 18.4 Decreased myotome length in MD zebrafish. Myotomes (eight myotomes) in 3dpf MD zebrafish were 39% shorter than KD control ($P = 7.1 \times 10^{-14}$, $n = 33$).

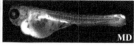

Figure 18.5 ROS staining in MD zebrafish. Live 3dpf zebrafish, stained with H_2DCFDA, are shown at $3.2\times$ magnification; anterior, left. Extensive staining was observed in MD zebrafish. In contrast, no staining was detected in uninjected and KD control animals. Note that fluorescence is detectable in the gut (white arrows) because animals ingested dye. Scale bar indicates $200\,\mu m$.

18.3.5 Increased ROS Level in MD Zebrafish

ROS have been shown to play a critical role in muscular dystrophy pathogenesis (Whitehead et al., 2006). Rando et al. provided evidence that in *mdx* mice, muscle is more susceptible to ROS-induced damage (Rando et al., 1998) and increased ROS production causes lipid peroxidation during the stage preceding necrosis (Disatnik et al., 1998). There is also evidence indicating oxidative stress increases in MD patients (Rodriguez and Tarnopolsky, 2003; Grosso et al., 2008) and elevated levels of ROS indicate oxidative stress, which can lead to inflammation (Whitehead et al., 2006). As shown in Fig. 18.5, at 3dpf, ROS staining in MD zebrafish muscles, specifically at the boundaries of the myotomes, was observed. In comparison, no staining was detected in muscle tissue in uninjected and KD control zebrafish. Note that fluorescence is detectable in the gut (Fig. 18.5, white arrows) because animals ingested the dye.

18.3.6 Decreased Motility in MD Zebrafish

Muscle strength is an important parameter for evaluating MD progression. During development, zebrafish demonstrate stereotypical motility patterns; weak muscle strength can decrease motility. In related studies, we observed that zebrafish movement differs during alternating light and dark photoperiods and the dark period serves as a clear trigger for increased activity (Emran et al., 2007; MacPhail et al., 2009).

   To quantitate zebrafish movement, we used the VideoTrack System. Total distance (D) traveled by individual zebrafish was recorded automatically for 60 min in alternating 10 min light and dark photoperiods. We compared total distance traveled in the dark period by MD, uninjected, and KD control animals (Fig. 18.6). The difference between uninjected (726.8 ± 245.5 mm) and KD (1049.5 ± 187.5 mm) control zebrafish was insignificant ($P = 0.3062$). Distance traveled by MD zebrafish (131.4 ± 48.1 mm) was 18% ($131.4/726.8 \times 100\%$) of uninjected control and 12% ($131.4/1049.5 \times 100\%$) of KD control zebrafish, significantly lower ($P = 0.0310$ and 0.0003, respectively).

18.3.7 Characterization of Zebrafish MD Disease Progression by Histology

Since treatment at different stages of disease progression can result in different therapeutic outcomes, we next elucidated progression of MD pathology in zebrafish

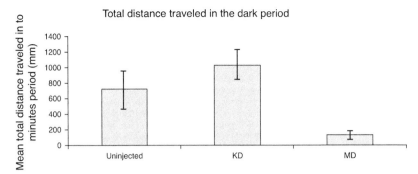

Figure 18.6 Total distance traveled decreased in MD zebrafish. Mean distance traveled in a 10 min dark period for uninjected, KD control, and MD zebrafish. Each bar represents mean ± SE ($n = 15$).

using histology. In *mdx* mice, presence of small muscle fibers exhibiting centralized nuclei, which indicates regeneration of muscle cells subsequent to necrosis (Karpati et al., 1988; Gibson et al., 1995; Rando et al., 2000), is a prominent MD phenotype, characterized by histology (Karpati et al., 1988; Gibson et al., 1995; Rando et al., 2000). As shown in Fig. 18.7, H&E staining of sagittal sections in 2–5dpf zebrafish confirmed MD phenotypes.

By 2dpf, muscle fibers were small and exhibited centralized nuclei (short black arrows). Myotome shape was irregular or boundaries were missing (Fig. 18.7, blue arrows). Although in uninjected and KD control zebrafish, midlines were identifiable at the tip of the V-shaped myotome structure (Fig. 18.7, long black arrows), in MD zebrafish they were unidentifiable. By 3dpf, large vacuole spaces (Fig. 18.7, yellow arrows) between muscle fibers, indicating detached sarcolemma and muscle degeneration, were frequently observed. By 5dpf, vacuole spaces were more prominent and muscle structure was severely disrupted. In contrast, uninjected and KD control zebrafish exhibited normal V-shaped myotomes (Fig. 18.7, white arrows), muscle fibers appeared to be tightly packed, and nuclei were not centralized, indicating normal structure of muscle fiber. Note that staining of uninjected and KD control samples was performed in batches, which contributed to variable color in images. These results indicate that KD control did not exhibit MD phenotypes, whereas MD zebrafish exhibited phenotypes similar to those reported for both *mdx* mice and the zebrafish *sapje* mutant (Bassett and Currie, 2004).

18.3.8 Number of Muscle Fibers Increased in MD Zebrafish

We also analyzed muscle structure on histology slides containing cross sections of KD control and MD zebrafish (data not shown). In 1dpf MD zebrafish, number of muscle cells exhibiting a thin cytoplasm layer and prominent, centered nuclei (regenerated muscle fiber) increased and muscle fibers were smaller. In 2dpf MD zebrafish, muscle fibers remained small with centered nuclei and muscle fibers appeared more abundant,

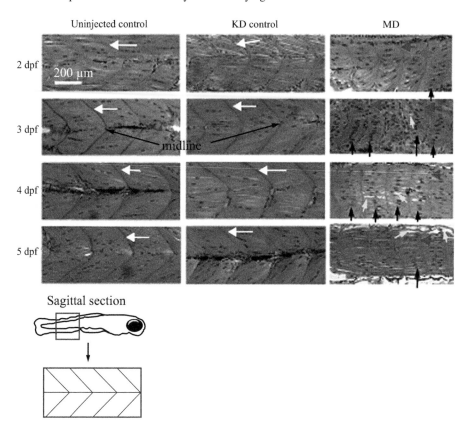

Figure 18.7 Histological assessment of muscle structure in uninjected, KD control, and MD zebrafish. Two to 5dpf uninjected, KD control, and MD zebrafish were processed for JB-4 embedding, sectioning, and H&E staining. Sagittal sections of tail muscles posterior to the anal pore (blue box in cartoon in left panels) are shown at 400× magnification. Uninjected and KD control animals exhibited normal V-shaped myotomes (white arrows) with a clear midline at the tip of the V (long black arrows), whereas MD zebrafish exhibited short, missing, or irregular shaped myotomes (gray black arrows), and no discernable midline. In addition, at 3dpf, centralized nuclei (short black arrows), indicating regeneration subsequent to necrosis, and vacuole spaces between fibers (white arrows), indicating degeneration, were observed in most muscle fibers. In contrast, muscle fibers in control animals exhibited tightly packed, less centralized nuclei; large vacuole spaces were not observed. White scale bar indicates 200 μm; anterior, right. (See the color version of this figure in Color Plates section.)

reflecting regeneration; these phenotypes persisted through 4dpf. On 3 and 4dpf, transverse myoseptum (the connective tissue separating the dorsal and ventral myotomes) was missing in MD zebrafish. In comparison, muscle fibers and transverse myoseptum appeared normal in KD control zebrafish. At 3 and 4dpf, MD zebrafish exhibited 22 and 25 muscle fibers/area, respectively. In contrast, KD control exhibited 8 and 7 muscle fibers/area, respectively; 2.75-fold ($22/8 = 2.75$) and 3.57-fold ($25/7 = 3.57$) increase in number of muscle fibers was observed in MD zebrafish at 3 and 4dpf, respectively, clearly indicating muscle cell regeneration subsequent to necrosis (Karpati et al., 1988; Gibson et al., 1995; Rando et al., 2000).

18.3.9 Drug Effects on MD Zebrafish

To validate use of our MD zebrafish model for drug screening, we next assessed effects of five drugs, prednisone, EGCG, MG132, TSA, and dantrolene, shown to reduce symptoms in MD patients and animal models. To assess drug effects on myotome length, we analyzed animals after treatment with the highest nonlethal concentration. To determine drug effects on ROS level using fluorescence staining and quantitative morphometric analysis, we assessed all three concentrations for each drug and established concentration–response curves. Since myotome length was longer and level of ROS was lower after prednisone treatment, we subsequently performed histology on samples treated with this drug.

18.3.10 Quantitation of Myotome Length After Drug Treatment

We next assessed drug effects on myotome length. Because myotome boundaries in the majority of 4 and 5dpf MD zebrafish were disrupted, using ImageJ software (NIH), we measured myotome length in 3dpf animals. Data are presented as mean ± standard error (SE) of the mean ($n = 10$). Drug effects were normalized to percent of vehicle control (0.1% DMSO-treated MD zebrafish) and results showed that, after prednisone treatment, myotomes were significantly longer ($125.1 \pm 3.8\%$) than vehicle-treated control MD zebrafish ($P < 0.0001$) (Table 18.2). Although higher drug concentrations were tested, increased toxicity often induced severe deformity or death.

18.3.11 Drug Effects on ROS Level

Next, we performed ROS staining as described previously. We focused on assessing ROS level in the tail region, where muscle degeneration was observed. As shown in Fig. 18.8, prednisone and EGCG treatment significantly reduced ROS staining,

Table 18.2 Drug Effect on Myotome Length

Drug	Myotome length (percent of vehicle control) ($n = 10$)	P
Prednisone	125.1 ± 3.8	<0.0001
EGCG	109.7 ± 2.6	0.1136
Dantrolene	105.5 ± 4.3	0.6305
MG132	104.7 ± 3.5	0.3881
TSA	102.4 ± 1.3	0.1485

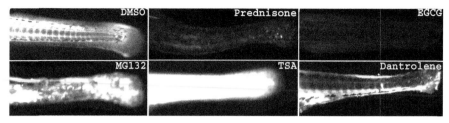

Figure 18.8   ROS staining in drug-treated MD zebrafish. Control and drug-treated 3dpf MD zebrafish were stained with H_2DCFDA; fluorescence staining in the tail region was examined. Prednisone and EGCG treatment significantly reduced ROS-specific fluorescence intensity; TSA and dantrolene treatment increased ROS-specific fluorescence intensity. MG132 treatment did not induce significant changes in fluorescence intensity.

whereas MG132- and dantrolene-treated MD zebrafish exhibited staining intensity similar to vehicle control (DMSO). TSA-treated zebrafish exhibited increased staining intensity. Severe toxicity was observed after treatment with TSA, which may have contributed to a high level of fluorescence. Our results after prednisone treatment were consistent with reports that anti-inflammatory effects of some glucocorticoids, including prednisone and dexamethasone, may be mediated by suppressing ROS (Sanner et al., 2002).

Next, using image-based morphometric analysis, we quantified ROS-specific fluorescence intensity in the tail region after drug treatment. To highlight the area of interest, we applied a consistent threshold value (100) to each tail image. The histogram value (hv) of the highlighted region was quantitated using Photoshop software (Adobe, San Jose, CA); 8–10 animals were used for each condition. Signal intensity for uninjected, KD control, and MD zebrafish was $16,192 \pm 9960$, $18,067 \pm 6315$, and $125,471 \pm 59,988$, respectively. Difference in ROS level for uninjected and KD control zebrafish was insignificant ($P = 0.7314$), indicating that KD control did not increase ROS level in injected animals; however, the difference between KD control and MD zebrafish ($P = 0.0018$) indicated significant ROS induction. 0.1% DMSO-treated MD zebrafish exhibited fluorescence intensity of $135,516 \pm 86,715$, and the difference between MD zebrafish was insignificant ($P = 0.7667$), indicating that carrier solvent did not affect ROS level.

We then assessed effects of varying concentrations of prednisone, EGCG, MG132, TSA, and dantrolene. In order to compare results from different experiments, drug effects were normalized to percent of control (percent of control = hv(drug)/ hv(DMSO control)). Dose response curves were then generated for each drug using level of ROS as percent of control versus drug concentration (Fig. 18.9). Prednisone and EGCG significantly decreased ROS staining ($P = 0.0078$ and $P = 0.0004$, respectively), whereas MG132 and dantrolene did not cause significant effects ($P = 0.0587$ and $P = 0.7739$, respectively); at high concentrations, TSA caused a significant increase in ROS staining ($P < 0.0001$), possibly due to compound-induced toxicity.

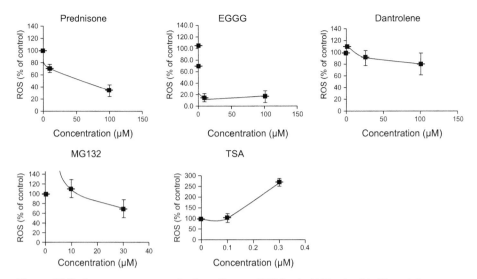

Figure 18.9   Dose–response curves for drug effects on ROS level of MD zebrafish. Three dpf animals were stained with H_2DCFDA and processed for morphometric analysis. Each point represents mean $\pm$ SE ($n = 8$–10).

18.3.12   Confirmation of Prednisone Effects on Muscle Structure by Histology

Since prednisone treatment decreased MD effects on myotome length and ROS level, we next examined effects on muscle structure using histology. As observed in previous experiments, untreated MD zebrafish exhibited MD muscle phenotypes, including small muscle fibers, loss of V-shaped myotomes, centralized nuclei, and disorganized muscle fibers. In contrast, muscle structure in prednisone-treated 3dpf MD zebrafish resembled 3dpf KD control zebrafish, including large muscle fibers, less centralized nuclei, and straight muscle fibers with clear midline; V-shaped myotomes were still absent. These results indicated that prednisone treatment partially inhibited progression of MD phenotypes (data not shown).

18.4   DISCUSSION

Although several different mouse MD models have been developed, a consistent phenotype has not yet been generated, contributing to the difficulty of using mammalian models for drug screening. In contrast to murine dystroglycan knockouts, dystroglycan KD zebrafish can survive up to 10dpf, indicating that dystroglycan is dispensable for basement membrane formation during early development. However, disruption of DGC in dystroglycan KD zebrafish leads to loss of sarcomere and sarcoplasmic organization, indicating that dystroglycan is required to maintain long-term muscle cell survival in zebrafish.

In dystroglycan KD zebrafish, we observed several defects that resemble human congenital myopathies found in both DMD patients and *mdx* mice including (1) rapid, progressive muscular degeneration, (2) immobility, (3) muscle defects, and (4) bent spine (notochord). Although the zebrafish is phylogenically distant from humans, key gene families have been shown to be highly conserved, and dystroglycan knockdown performed by simple injection into wild-type animals is straightforward and highly reproducible. An additional compelling advantage of this model organism is that a battery of disease-specific zebrafish assays can be designed to assess drug effects.

ACKNOWLEDGMENT

This research was supported by a National Institutes of Health grant: 1R43AR055390.

REFERENCES

ALLAMAND V and CAMPBELL KP (2000). Animal models for muscular dystrophy: valuable tools for the development of therapies. Hum Mol Genet 9(16): 2459–2467.

AMALI AA, LIN CJ, CHEN YH, WANG WL, GONG HY, REKHA RD, LU JK, CHEN TT, and WU JL (2008). Overexpression of Myostatin2 in zebrafish reduces the expression of dystrophin associated protein complex (DAPC) which leads to muscle dystrophy. J Biomed Sci 15(5): 595–604.

ARAISHI K, SASAOKA T, IMAMURA M, NOGUCHI S, HAMA H, WAKABAYASHI E, YOSHIDA M, HORI T, and OZAWA E (1999). Loss of the sarcoglycan complex and sarcospan leads to muscular dystrophy in beta-sarcoglycan-deficient mice. Hum Mol Genet 8(9): 1589–1598.

ASSERETO S, STRINGARA S, SOTGIA F, BONUCCELLI G, BROCCOLINI A, PEDEMONTE M, TRAVERSO M, BIANCHERI R, ZARA F, BRUNO C, LISANTI MP, and MINETTI C (2006). Pharmacological rescue of the dystrophin–glycoprotein complex in Duchenne and Becker skeletal muscle explants by proteasome inhibitor treatment. Am J Physiol Cell Physiol 290(2): C577–C582.

BASSETT DI, BRYSON-RICHARDSON RJ, DAGGETT DF, GAUTIER P, KEENAN DG, and CURRIE PD (2003). Dystrophin is required for the formation of stable muscle attachments in the zebrafish embryo. Development 130(23): 5851–5860.

BASSETT DI and CURRIE PD (2003). The zebrafish as a model for muscular dystrophy and congenital myopathy. Hum Mol Genet 12 (Suppl 2): R265–R270.

BASSETT DI and CURRIE PD (2004). Identification of a zebrafish model of muscular dystrophy. Clin Exp Pharmacol Physiol 31(8): 537–540.

BAUMEISTER R and GE L (2002). The worm in us: *Caenorhabditis elegans* as a model of human disease. Trends Biotechnol 20(4): 147–148.

BEENAKKER EA, FOCK JM, VAN TOL MJ, MAURITS NM, KOOPMAN HM, BROUWER OF, and VAN DER HOEVEN JH (2005). Intermittent prednisone therapy in Duchenne muscular dystrophy: a randomized controlled trial. Arch Neurol 62(1): 128–132.

BESSOU C, GIUGIA JB, FRANKS CJ, HOLDEN-DYE L, and SEGALAT L (1998). Mutations in the *Caenorhabditis elegans* dystrophin-like gene dys-1 lead to hyperactivity and suggest a link with cholinergic transmission. Neurogenetics 2(1): 61–72.

BOGDANOVICH S, KRAG TO, BARTON ER, MORRIS LD, WHITTEMORE LA, AHIMA RS, and KHURANA TS (2002). Functional improvement of dystrophic muscle by myostatin blockade. Nature 420(6914): 418–421.

BOGDANOVICH S, PERKINS KJ, KRAG TO, WHITTEMORE LA, and KHURANA TS (2005). Myostatin propeptide-mediated amelioration of dystrophic pathophysiology. FASEB J 19(6): 543–549.

BONIFATI MD, RUZZA G, BONOMETTO P, BERARDINELLI A, GORNI K, ORCESI S, LANZI G, and ANGELINI C (2000). A multicenter, double-blind, randomized trial of deflazacort versus prednisone in Duchenne muscular dystrophy. Muscle Nerve 23(9): 1344–1347.

BONUCCELLI G, SOTGIA F, CAPOZZA F, GAZZERRO E, MINETTI C, and LISANTI MP (2007). Localized treatment with a novel FDA-approved proteasome inhibitor blocks the degradation of dystrophin and dystrophin-associated proteins in mdx mice. Cell Cycle 6(10): 1242–1248.

BONUCCELLI G, SOTGIA F, SCHUBERT W, PARK DS, FRANK PG, WOODMAN SE, INSABATO L, CAMMER M, MINETTI C, and LISANTI MP (2003). Proteasome inhibitor (MG-132) treatment of mdx mice rescues the expression and membrane localization of dystrophin and dystrophin-associated proteins. Am J Pathol 163(4): 1663–1675.

BULFIELD G, SILLER WG, WIGHT PA, and MOORE KJ (1984). X chromosome-linked muscular dystrophy (mdx) in the mouse. Proc Natl Acad Sci USA 81(4): 1189–1192.

CAMPBELL C and JACOB P (2003). Deflazacort for the treatment of Duchenne dystrophy: a systematic review. BMC Neurol 3: 7.

COLLINS CA and MORGAN JE (2003). Duchenne's muscular dystrophy: animal models used to investigate pathogenesis and develop therapeutic strategies. Int J Exp Pathol 84(4): 165–172.

COOPER BJ, VALENTINE BA, WILSON S, PATTERSON DF, and CONCANNON PW (1988). Canine muscular dystrophy: confirmation of X-linked inheritance. J Hered 79(6): 405–408.

COSSU G and SAMPAOLESI M (2007). New therapies for Duchenne muscular dystrophy: challenges, prospects and clinical trials. Trends Mol Med 13(12): 520–526.

COTE PD, MOUKHLES H, LINDENBAUM M, and CARBONETTO S (1999). Chimaeric mice deficient in dystroglycans develop muscular dystrophy and have disrupted myoneural synapses. Nat Genet 23(3): 338–342.

DECONINCK N, TINSLEY J, DE BACKER F, FISHER R, KAHN D, PHELPS S, DAVIES K, and GILLIS JM (1997). Expression of truncated utrophin leads to major functional improvements in dystrophin-deficient muscles of mice. Nat Med 3(11): 1216–1221.

DELLAVALLE A, SAMPAOLESI M, TONLORENZI R, TAGLIAFICO E, SACCHETTI B, PERANI L, INNOCENZI A, GALVEZ BG, MESSINA G, MOROSETTI R, LI S, BELICCHI M, PERETTI G, CHAMBERLAIN JS, WRIGHT WE, TORRENTE Y, FERRARI S, BIANCO P, and COSSU G (2007). Pericytes of human skeletal muscle are myogenic precursors distinct from satellite cells. Nat Cell Biol 9(3): 255–267.

DE LUCA A, PIERNO S, LIANTONIO A, and CONTE CAMERINO D (2002). Pre-clinical trials in Duchenne dystrophy: what animal models can tell us about potential drug effectiveness. Neuromuscul Disord 12 (Suppl 1): S142–S146.

DEZAWA M, ISHIKAWA H, ITOKAZU Y, YOSHIHARA T, HOSHINO M, TAKEDA S, IDE C, and NABESHIMA Y (2005). Bone marrow stromal cells generate muscle cells and repair muscle degeneration. Science 309(5732): 314–317.

DISATNIK MH, DHAWAN J, YU Y, BEAL MF, WHIRL MM, FRANCO AA, and RANDO TA (1998). Evidence of oxidative stress in mdx mouse muscle: studies of the pre-necrotic state. J Neurol Sci 161(1): 77–84.

DODD A, CHAMBERS SP, and LOVE DR (2004). Short interfering RNA-mediated gene targeting in the zebrafish. FEBS Lett 561(1–3): 89–93.

DORCHIES OM, WAGNER S, VUADENS O, WALDHAUSER K, BUETLER TM, KUCERA P, and RUEGG UT (2006). Green tea extract and its major polyphenol (−)-epigallocatechin gallate improve muscle function in a mouse model for Duchenne muscular dystrophy. Am J Physiol Cell Physiol 290(2): C616–C625.

EMRAN F, RIHEL J, ADOLPH AR, WONG KY, KRAVES S, and DOWLING JE (2007). OFF ganglion cells cannot drive the optokinetic reflex in zebrafish. Proc Natl Acad Sci USA 104(48): 19126–19131.

GASCHEN F and BURGUNDER JM (2001). Changes of skeletal muscle in young dystrophin-deficient cats: a morphological and morphometric study. Acta Neuropathol 101(6): 591–600.

GAUD A, SIMON JM, WITZEL T, CARRE-PIERRAT M, WERMUTH CG, and SEGALAT L (2004). Prednisone reduces muscle degeneration in dystrophin-deficient *Caenorhabditis elegans*. Neuromuscul Disord 14(6): 365–370.

GIBSON AJ, KARASINSKI J, RELVAS J, MOSS J, SHERRATT TG, STRONG PN, and WATT DJ (1995). Dermal fibroblasts convert to a myogenic lineage in mdx mouse muscle. J Cell Sci 108(Pt 1): 207–214.

GRADY RM, TENG H, NICHOL MC, CUNNINGHAM JC, WILKINSON RS, and SANES JR (1997). Skeletal and cardiac myopathies in mice lacking utrophin and dystrophin: a model for Duchenne muscular dystrophy. Cell 90(4): 729–738.

GRANCHELLI JA, POLLINA C, and HUDECKI MS (2000). Pre-clinical screening of drugs using the mdx mouse. Neuromuscul Disord 10(4–5): 235–239.

GROSSO S, PERRONE S, LONGINI M, BRUNO C, MINETTI C, GAZZOLO D, BALESTRI P, and BUONOCORE G (2008). Isoprostanes in dystrophinopathy: evidence of increased oxidative stress. Brain Dev 30(6): 391–395.

GUSSONI E, SONEOKA Y, STRICKLAND CD, BUZNEY EA, KHAN MK, FLINT AF, KUNKEL LM, and MULLIGAN RC (1999). Dystrophin expression in the mdx mouse restored by stem cell transplantation. Nature 401(6751): 390–394.

GUYON JR, MOSLEY AN, ZHOU Y, O'BRIEN KF, SHENG X, CHIANG K, DAVIDSON AJ, VOLINSKI JM, ZON LI, and KUNKEL LM (2003). The dystrophin associated protein complex in zebrafish. Hum Mol Genet 12(6): 601–615.

GUYON JR, STEFFEN LS, HOWELL MH, PUSACK TJ, LAWRENCE C, and KUNKEL LM (2007). Modeling human muscle disease in zebrafish. Biochim Biophys Acta 1772(2): 205–215.

HERMANN AC, MILLARD PJ, BLAKE SL, and KIM CH (2004). Development of a respiratory burst assay using zebrafish kidneys and embryos. J Immunol Methods 292(1–2): 119–129.

HOFFMAN EP, BROWN R.H. Jr., and KUNKEL LM (1987). Dystrophin: the protein product of the Duchenne muscular dystrophy locus. Cell 51(6): 919–928.

HOLT KH, CROSBIE RH, VENZKE DP, and CAMPBELL KP (2000). Biosynthesis of dystroglycan: processing of a precursor propeptide. FEBS Lett 468(1): 79–83.

JANSSEN PM, HIRANANDANI N, MAYS TA, and RAFAEL-FORTNEY JA (2005). Utrophin deficiency worsens cardiac contractile dysfunction present in dystrophin-deficient mdx mice. Am J Physiol Heart Circ Physiol 289(6): H2373–H2378.

JIANG Y, JAHAGIRDAR BN, REINHARDT RL, SCHWARTZ RE, KEENE CD, ORTIZ-GONZALEZ XR, REYES M, LENVIK T, LUND T, BLACKSTAD M, DU J, ALDRICH S, LISBERG A, LOW WC, LARGAESPADA DA, and VERFAILLIE CM (2002). Pluripotency of mesenchymal stem cells derived from adult marrow. Nature 418(6893): 41–49.

KARPATI G, CARPENTER S, and PRESCOTT S (1988). Small-caliber skeletal muscle fibers do not suffer necrosis in mdx mouse dystrophy. Muscle Nerve 11(8): 795–803.

KHURANA TS and DAVIES KE (2003). Pharmacological strategies for muscular dystrophy. Nat Rev Drug Discov 2(5): 379–390.

KOHN B, GUSCETTI F, WAXENBERGER M, and AUGSBURGER H (1993). Muscular dystrophy in a cat. Tierarztl Prax 21(5): 451–457.

LAPIDOS KA, KAKKAR R, and MCNALLY EM (2004). The dystrophin glycoprotein complex: signaling strength and integrity for the sarcolemma. Circ Res 94(8): 1023–1031.

LIM LE, DUCLOS F, BROUX O, BOURG N, SUNADA Y, ALLAMAND V, MEYER J, RICHARD I, MOOMAW C, SLAUGHTER C, et al. (1995). Beta-sarcoglycan: characterization and role in limb-girdle muscular dystrophy linked to 4q12. Nat Genet 11(3): 257–265.

MACPHAIL RC, BROOKS J, HUNTER DL, PADNOS B, IRONS TD, and PADILLA S (2009). Locomotion in larval zebrafish: influence of time of day, lighting and ethanol. Neurotoxicology 30(1): 52–58.

MANZUR AY, KUNTZER T, PIKE M, and SWAN A (2004). Glucocorticoid corticosteroids for Duchenne muscular dystrophy. Cochrane Database Syst Rev (2): CD003725.

MANZUR AY, KUNTZER T, PIKE M, and SWAN A (2008). Glucocorticoid corticosteroids for Duchenne muscular dystrophy. Cochrane Database Syst Rev (1): CD003725.

MATHEWS KD (2003). Muscular dystrophy overview: genetics and diagnosis. Neurol Clin 21(4): 795–816.

MAYER U, SAHER G, FASSLER R, BORNEMANN A, ECHTERMEYER F, VON DER MARK H, MIOSGE N, POSCHL E, and VON DER MARK K (1997). Absence of integrin alpha 7 causes a novel form of muscular dystrophy. Nat Genet 17(3): 318–323.

MENG L and WOLF DP (1997). Sperm-induced oocyte activation in the rhesus monkey: nuclear and cytoplasmic changes following intracytoplasmic sperm injection. Hum Reprod 12(5): 1062–1068.

MINASI MG, RIMINUCCI M, DE ANGELIS L, BORELLO U, BERARDUCCI B, INNOCENZI A, CAPRIOLI A, SIRABELLA D, BAIOCCHI M, DE MARIA R, BORATTO R, JAFFREDO T, BROCCOLI V, BIANCO P, and COSSU G (2002). The meso-angioblast: a multipotent, self-renewing cell that originates from the dorsal aorta and differentiates into most mesodermal tissues. Development 129(11): 2773–2783.

MINETTI GC, COLUSSI C, ADAMI R, SERRA C, MOZZETTA C, PARENTE V, FORTUNI S, STRAINO S, SAMPAOLESI M, DI PADOVA M, ILLI B, GALLINARI P, STEINKUHLER C, CAPOGROSSI MC, SARTORELLI V, BOTTINELLI R, GAETANO C, and PURI PL (2006). Functional and morphological recovery of dystrophic muscles in mice treated with deacetylase inhibitors. Nat Med 12(10): 1147–1150.

MIZUNO Y, NOGUCHI S, YAMAMOTO H, YOSHIDA M, NONAKA I, HIRAI S, and OZAWA E (1995). Sarcoglycan complex is selectively lost in dystrophic hamster muscle. Am J Pathol 146(2): 530–536.

NAKAE Y, HIRASAKA K, GOTO J, NIKAWA T, SHONO M, YOSHIDA M, and STOWARD PJ (2008). Subcutaneous injection, from birth, of epigallocatechin-3-gallate, a component of green tea, limits the onset of muscular dystrophy in mdx mice: a quantitative histological, immunohistochemical and electrophysiological study. Histochem Cell Biol 129(4): 489–501.

NASEVICIUS A and EKKER SC (2000). Effective targeted gene 'knockdown' in zebrafish. Nat Genet 26(2): 216–220.

PARSONS MJ, CAMPOS I, HIRST EM, and STEMPLE DL (2002). Removal of dystroglycan causes severe muscular dystrophy in zebrafish embryos. Development 129(14): 3505–3512.

PARTRIDGE TA, MORGAN JE, COULTON GR, HOFFMAN EP, and KUNKEL LM (1989). Conversion of mdx myofibres from dystrophin-negative to -positive by injection of normal myoblasts. Nature 337(6203): 176–179.

PEI DS, SUN YH, LONG Y, and ZHU ZY (2007). Inhibition of no tail (ntl) gene expression in zebrafish by external guide sequence (EGS) technique. Mol Biol Rep 35(2): 139–143.

RANDO TA, DISATNIK MH, YU Y, and FRANCO A (1998). Muscle cells from mdx mice have an increased susceptibility to oxidative stress. Neuromuscul Disord 8(1): 14–21.

RANDO TA, DISATNIK MH, and ZHOU LZ (2000). Rescue of dystrophin expression in mdx mouse muscle by RNA/DNA oligonucleotides. Proc Natl Acad Sci USA 97(10): 5363–5368.

ROBU ME, LARSON JD, NASEVICIUS A, BEIRAGHI S, BRENNER C, FARBER SA, and EKKER SC (2007). p53 activation by knockdown technologies. PLoS Genet 3(5): e78.

RODRIGUEZ MC and TARNOPOLSKY MA (2003). Patients with dystrophinopathy show evidence of increased oxidative stress. Free Radic Biol Med 34(9): 1217–1220.

ROMERO NB, BRAUN S, BENVENISTE O, LETURCQ F, HOGREL JY, MORRIS GE, BAROIS A, EYMARD B, PAYAN C, ORTEGA V, BOCH AL, LEJEAN L, THIOUDELLET C, MOUROT B, ESCOT C, CHOQUEL A, RECAN D, KAPLAN JC, DICKSON G, KLATZMANN D, MOLINIER-FRENCKEL V, GUILLET JG, SQUIBAN P, HERSON S, and FARDEAU M (2004). Phase I study of dystrophin plasmid-based gene therapy in Duchenne/Becker muscular dystrophy. Hum Gene Ther 15(11): 1065–1076.

RYBAKOVA IN, PATEL JR, DAVIES KE, YURCHENCO PD, and ERVASTI JM (2002). Utrophin binds laterally along actin filaments and can couple costameric actin with sarcolemma when overexpressed in dystrophin-deficient muscle. Mol Biol Cell 13(5): 1512–1521.

RYBAKOVA IN, PATEL JR, and ERVASTI JM (2000). The dystrophin complex forms a mechanically strong link between the sarcolemma and costameric actin. J Cell Biol 150(5): 1209–1214.

SANNER BM, MEDER U, ZIDEK W, and TEPEL M (2002). Effects of glucocorticoids on generation of reactive oxygen species in platelets. Steroids 67(8): 715–719.

SKROMNE I and PRINCE VE (2008). Current perspectives in zebrafish reverse genetics: moving forward. Dev Dyn 237(4): 861–882.

SKUK D, GOULET M, ROY B, CHAPDELAINE P, BOUCHARD JP, ROY R, DUGRE FJ, SYLVAIN M, LACHANCE JG, DESCHENES L, SENAY H, and TREMBLAY JP (2006). Dystrophin expression in muscles of Duchenne muscular dystrophy patients after high-density injections of normal myogenic cells. J Neuropathol Exp Neurol 65(4): 371–386.

SKUK D, ROY B, GOULET M, CHAPDELAINE P, BOUCHARD JP, ROY R, DUGRE FJ, LACHANCE JG, DESCHENES L, HELENE S, SYLVAIN M, and TREMBLAY JP (2004). Dystrophin expression in myofibers of Duchenne muscular dystrophy patients following intramuscular injections of normal myogenic cells. Mol Ther 9(3): 475–482.

TAVERNA D, DISATNIK MH, RAYBURN H, BRONSON RT, YANG J, RANDO TA, and HYNES RO (1998). Dystrophic muscle in mice chimeric for expression of alpha5 integrin. J Cell Biol 143(3): 849–859.

Tinsley J, Deconinck N, Fisher R, Kahn D, Phelps S, Gillis JM, and Davies K (1998). Expression of full-length utrophin prevents muscular dystrophy in mdx mice. Nat Med 4(12): 1441–1444.

Torrente Y, Belicchi M, Sampaolesi M, Pisati F, Meregalli M, D'Antona G, Tonlorenzi R, Porretti L, Gavina M, Mamchaoui K, Pellegrino MA, Furling D, Mouly V, Butler-Browne GS, Bottinelli R, Cossu G, and Bresolin N (2004). Human circulating AC133$^+$ stem cells restore dystrophin expression and ameliorate function in dystrophic skeletal muscle. J Clin Invest 114(2): 182–195.

Urtishak KA, Choob M, Tian X, Sternheim N, Talbot WS, Wickstrom E, and Farber SA (2003). Targeted gene knockdown in zebrafish using negatively charged peptide nucleic acid mimics. Dev Dyn 228(3): 405–413.

Watchko JF, O'Day TL, and Hoffman EP (2002). Functional characteristics of dystrophic skeletal muscle: insights from animal models. J Appl Physiol 93(2): 407–417.

Weiler T, Bashir R, Anderson LV, Davison K, Moss JA, Britton S, Nylen E, Keers S, Vafiadaki E, Greenberg CR, Bushby CR, and Wrogemann K (1999). Identical mutation in patients with limb girdle muscular dystrophy type 2B or Miyoshi myopathy suggests a role for modifier gene(s). Hum Mol Genet 8(5): 871–877.

Westerfield M (1993). The Zebrafish Book: A Guide for the Laboratory Use of Zebrafish. University of Oregon Press, Eugene, OR.

Whitehead NP, Pham C, Gervasio OL, and Allen DG (2008). N-Acetylcysteine ameliorates skeletal muscle pathophysiology in mdx mice. J Physiol 586(7): 2003–2014.

Whitehead NP, Yeung EW, and Allen DG (2006). Muscle damage in mdx (dystrophic) mice: role of calcium and reactive oxygen species. Clin Exp Pharmacol Physiol 33(7): 657–662.

Wickstrom E, Choob M, Urtishak KA, Tian X, Sternheim N, Talbot S, Archdeacon J, Efimov VA, and Farber SA (2004). Sequence specificity of alternating hydroxyprolyl/phosphono peptide nucleic acids against zebrafish embryo mRNAs. J Drug Target 12(6): 363–372.

Xie Y, Chen X, and Wagner TE (1997). A ribozyme-mediated, gene "knockdown" strategy for the identification of gene function in zebrafish. Proc Natl Acad Sci USA 94(25): 13777–13781.

Xu H, Wu XR, Wewer UM, and Engvall E (1994). Murine muscular dystrophy caused by a mutation in the laminin alpha 2 (Lama2) gene. Nat Genet 8(3): 297–302.

Zhao XF, Fjose A, Larsen N, Helvik JV, and Drivenes O (2008). Treatment with small interfering RNA affects the microRNA pathway and causes unspecific defects in zebrafish embryos. FEBS J. 275(9): 2177–2184.

Chapter 19

Cytoprotective Activities of Water-Soluble Fullerenes in Zebrafish Models*

Florian Beuerle[1], Patrick Witte[1], Uwe Hartnagel[1], Russell Lebovitz[2], Chuenlei Parng[3], and Andreas Hirsch[1,2]

[1]*The Institut für Organische Chemie, Universität Erlangen-Nürnberg, Erlangen, Germany*
[2]*C-Sixty Inc., Houston, TX, USA*
[3]*Phylonix, Cambridge, MA, USA*

19.1 INTRODUCTION

Water-soluble fullerenes have been shown to be effective antioxidants against reactive oxygen species (Chueh et al., 1999; Guldi and Asmus, 1999; Lin et al., 1999, 2002, 2004; Puhaca, 1999; Bensasson et al., 2000; Lai et al., 2000; Monti et al., 2000; Okuda et al., 2000; Huang et al., 2001a, 200b; Bosi et al., 2003; Chen et al., 2004), including superoxide (Ali et al., 2004) and hydroxyl radical, and appear to protect cells and tissues from oxidative injury associated with chemical oxidants (Tsai et al., 1997), as well as UV, X-rays, and gamma irradiation (Straface et al., 1999; Fumelli et al., 2000; Lin et al., 2001). Two different mechanisms through which fullerenes provide this protection have been proposed. The first involves covalent attachment of oxygen radicals to the fullerene carbon framework, resulting in fullerene radical adducts (stoichiometric process), which could in some cases after a sequence of subsequent addition and elimination steps convert back to the parent water-soluble derivative (catalytic process). The second mechanism appears to involve a sequence of electron transfer processes between the

*Reprinted with permission: Beuerle F, Witte P, Hartnagel U, Lebovitz R, and Parng C (2007). Cytoprotective activities of water-soluble fullerenes in zebrafish models. J Exp Nanosci 2(3): 147–170.

fullerene and reactive oxygen species. The concentrations required for effective biological protection in mammalian cell cultures appear to be in the 10–100 µM range (Yamago et al., 1995; Dugan et al., 1996, 1997, 2001; Huang et al., 1998, 2001a; Lotharius et al., 1999; Fumelli et al., 2000; Lin et al., 2004). Data using antibodies against fullerenes indicate that water-soluble fullerenes readily enter many cell types, and may be located preferentially with mitochondria (Foley et al., 2002), although it is not clear whether they can enter the mitochondria to any significant extent.

One of the leading candidates for a pharmacologically active form of fullerenes is the e,e,e-trismalonic acid (**1**) (so-called C3) (Fig. 19.1). C3 (**1**) has been shown to protect against oxidative injury and cell death in a variety of cell culture and animal models (Dugan et al., 1996, 1997, 2001; Djojo and Hirsch, 1998; Straface et al., 1999; Bisaglia et al., 2000; Fumelli et al., 2000; Lin et al., 2000; Monti et al., 2000; Foley et al., 2002; Reuther et al., 2002; Ali et al., 2004). Moreover, C3 fullerenes appear to cross the blood–brain barrier in detectable amounts and are eliminated almost entirely from the body via the liver and kidneys.

We have recently synthesized and characterized various new families of water-soluble fullerenes that can be prepared in scalable amounts and have compared their activities against C3 (**1**) (Witte et al., 2007). We observed broad differences among the various modified fullerenes with respect to both superoxide quenching and interaction with cytochrome c. One factor linked to the overall antioxidant capabilities was net charge of the modified fullerene. Anionic fullerenes give rise to stronger binding of cytochrome c. On the other hand, monoadducts (e.g., dendrofullerenes **2–7**) tend to have enhanced activity of superoxide quenching compared with higher adducts such as C3-like fullerenes (**1** and **8–10**) or oxo-aminofullerenes **11** and **12** (Fig. 19.1 and Table 19.1) (Witte et al., 2007). However, other, as yet unknown, factors appear to play a role as well.

We have extended our observations on the cytoprotective antioxidant effects of a subset of our water-soluble fullerenes **1–12** (Fig. 19.1) to zebrafish (*Danio rerio*), a

Table 19.1 IC$_{50}$ Values for the Superoxide Quenching Activities of Fullerene Derivatives **1–12** in Xanthine/Xanthine Oxidase Assays

Compound	Class	IC$_{50}$ superoxide (µM)
1	C3-like fullerenes	18.5
2	Dendrofullerenes	6.2
3	Dendrofullerenes	11.0
4	Dendrofullerenes	15.4
5	Dendrofullerenes	14.7
6	Dendrofullerenes	24.0
7	Dendrofullerenes	26.1
8	C3-like fullerenes	56.0
9	C3-like fullerenes	35.0
10	C3-like fullerenes	202.0
11	Oxo-aminofullerenes	35.0
12	Oxo-aminofullerenes	45.4

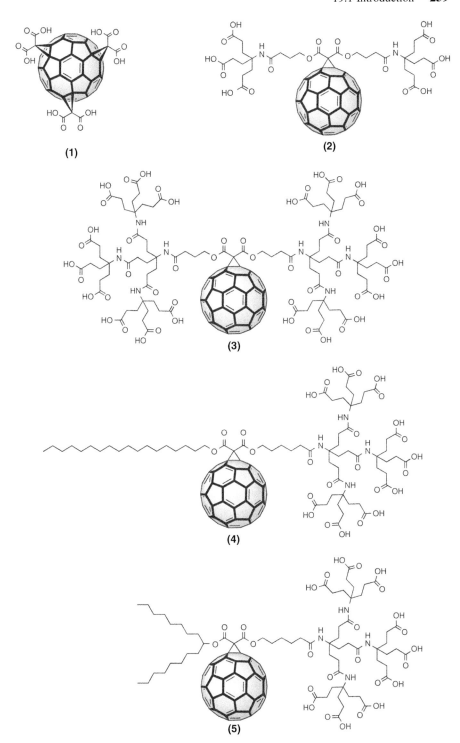

Figure 19.1 Structures of various water-soluble fullerenes **1–12**.

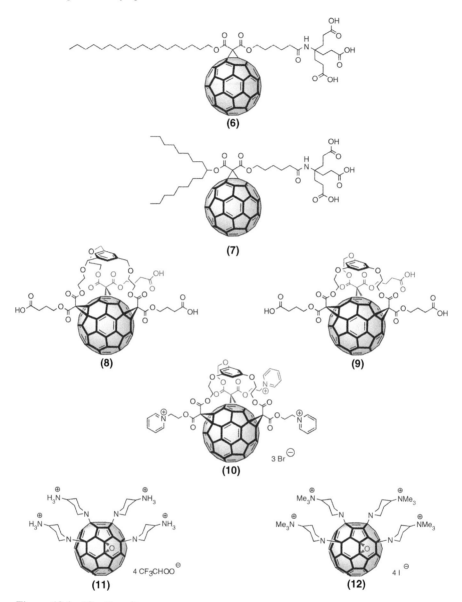

Figure 19.1 *(Continued)*

well-characterized vertebrate animal model for studying general and specific cyto-toxicity of small molecules and biologicals. Zebrafish embryos develop easily recognizable and functional organs, including heart, CNS, and eye, within 24 h after fertilization, and toxicity to these organs from small molecules and radiation mimics similar effects observed in many mammals, including humans. Moreover, since zebrafish embryos are small (they can be studied qualitatively and quantitatively in a single well of a 96-well plate), and transparent up to 10 days post fertilization (dpf), the toxic and cytoprotective effects of small molecules can be studied *in situ* during this period. Finally, since the morphological structure of zebrafish embryos has been mapped to the single-cell level, the toxic, apoptotic, and cytoprotective effects on various organs can be tracked to approximately the single-cell level *in situ* (Rodriguez and Driever, 1997; Cole and Ross, 2001; Langheinrich et al., 2002; Parng et al., 2002).

Our results indicate that many of our anionic water-soluble fullerenes generally exhibit little overall toxicity at concentrations of 400 μM or greater, while cationic fullerenes exhibit significant toxicity at 100 μM or less. We also observed that anionic fullerenes provide significant protection against apoptosis induced by three known chemical cytotoxins, cisplatinum, gentamicin, and 6-hydroxydopamine, which produce similar patterns of cell injury and death in zebrafish and humans.

19.2 MATERIALS AND METHODS

19.2.1 Compounds

Water-soluble fullerenes **1–12** were prepared according to our previously reported synthesis protocols (Witte et al., 2007, and references cited therein). In order to carry out the biological tests, 25–100 mM stock solutions in 10% DMSO and 0.1 N HCl of **1–12** were prepared. All stocks were stored at −20°C and diluted before use.

19.2.2 Standard Procedures for Embryo Collection

Embryos were generated by natural pairwise mating, as described in *The Zebrafish Book* (Westerfield, 1993). Four to five pairs of adult zebrafish were set up for each mating, and, on average, 100–150 embryos per pair were generated. Embryos were maintained at 28°C in fish water (200 mg Instant Ocean Salt per liter of deionized water; pH 6.6–7.0 maintained with 2.5 mg/L Jungle pH Stabilizer (Jungle Laboratories Corporation, Cibolo, TX); conductivity 670–760 μS). Embryos were cleaned (dead embryos removed) and sorted by developmental stage (Kimmel et al., 1995) at 6 and 24 h post fertilization (hpf). Because the embryo receives nourishment from an attached yolk sac, no feeding was required for 7dpf.

19.2.3 Determination of LC$_{50}$ and Organ Toxicity

19.2.3.1 Treatment of Embryos with 1–12   Zebrafish embryos at 24hpf were distributed into 96-well cell culture plates, one embryo per well in 100 μL fish water containing the compound. Embryos were exposed from day 1 to day 5 post fertilization. Typically, the fish water contains PBS at a final concentration of 10%. Treated embryos were compared with untreated and 0.1% DMSO-treated controls.

19.2.3.2 Measurement of LC$_{50}$ and Lethality Curves   Mortality was recorded every 24 h. At 120hpf, total mortality was used to generate the concentration–response curves. The data were averaged from multiple experiments. Best-fit concentration–response curves were generated using KaleidaGraph software (Synergy Software, Reading, PA) using the equation $Y = M_1 + \{(M_2 - M_1)/[1 + (X/M_3)\wedge M_4]\}$, where $M_1 = $ maximum Y value (100% in this case), $M_2 = $ minimum Y value (0% in this case), $M_3 = $ the concentration corresponding to the value midway between M_1 and M_2 (LC_{50}), $M_4 = $ slope of the curve at M_3 (best fit, generating R^2 closest to 1), $X = $ concentration of compound, and $Y = $ percent lethality.

19.2.3.3 *Visual Toxicity Assessment of Developing Embryos*   At 5dpf, organs in five randomly selected embryos were inspected by light microscopy. Body morphology, liver, intestine, and heart were assessed. Since drug treatment was initiated at 24hpf, which is after circulation and heartbeat are present, but before the liver and intestine are developed, observations are for effects on developing organs.

19.2.3.4 *Body Morphology*   Defects in the development of midline structures, including notochord and floor plate, often result in abnormal body shape in zebrafish. To assess the potential drug effects on midline development, abnormalities in body shape, including small body size and bent or missing tail, were examined.

19.2.4 Determination of Otoprotective Activity

19.2.4.1 *Hair Cell Assessment*   To damage inner hair cells, 5-day zebrafish were treated with gentamicin (2.5 µg/mL) or cisplatinum (10 µM) for 24 h to induce apoptosis in inner hair cells. To test drug effects on protection of hair cell damage, test compounds were administered at 0, 0.1, 1, 10, 100, and 250 µM concentrations with either the gentamicin or cisplatinum treatment. The inner hair cells in the lateral neuromasts were examined by 2-[4-(dimethylamino)styryl]-*N*-ethylpyridinium iodide (DASPEI or 2,4-Di-Asp) staining.

19.2.4.2 *DASPEI Staining*   Zebrafish were incubated with DASPEI solution (1 mM) for 2 h and rinsed thoroughly in fish water. Zebrafish were anaesthetized with MESAB (0.5 mM 3-aminobenzoic acid ethyl ester, 2 mM Na_2HPO_4), and mounted in methylcellulose in depression slide for observation.

19.2.4.3 *Morphometric Analysis*   Fluorescent signals (SS) were quantified as SS = staining area × staining intensity of neuromasts by particle analysis (Scion Image, Scion Corporation, Frederick, MD). Images of lateral sides in each animal were obtained by the same exposure time and fluorescent gain (anterior on the left; posterior on the right; dorsal on the top). The size of neuromasts was defined and

specified for particle analysis. Five animals for each treatment were quantified and intensity of fluorescent signals was averaged.

19.2.5 Statistics

All data were presented as mean $\pm$ standard deviation (SD). Student's t-test was used to compare vehicle-treated and drug-treated zebrafish. Significance was defined as $P < 0.05$, $n = 5$.

19.2.6 Determination of General CNS Neuroprotective Activity Against 6-OHDA-Induced Neuronal Apoptosis

19.2.6.1 Microinjection of 6-OHDA 6-Hydroxdopamine (6-OHDA) is extremely unstable in solution; it was prepared fresh for each injection. Three-day zebrafish were anesthetized in MESAB and 500 mM 6-OHDA was microinjected into the midbrain region. The estimated amount of injection was about 1–2 pM per compound. Zebrafish were rapidly transferred to fresh fish water after injection, allowed to recover for 30 min, and incubated with neuroprotective antioxidants. Water alone was injected into zebrafish brain region as a vehicle control.

19.2.6.2 Treatment of Embryos with 1–12 Embryos were exposed to water-soluble fullerenes at 1, 10, 100, and 250 μM for 24 h, and then microinjected with 6-OHDA. Twenty embryos were treated with each concentration of each compound.

19.2.6.3 Morphometric Analysis Fluorescent signals (SS) were quantified as $SS = $ staining area $\times$ staining intensity of neuromasts by particle analysis (Scion Image, Scion Corporation, Frederick, MD). Images of dorsal sides in each animal were obtained by the same exposure time and fluorescent gain. The positive staining was defined by size and fluorescent intensity evaluation, and specified for particle analysis. Five animals for each treatment were quantified and intensity of fluorescent signals was averaged.

19.2.6.4 Fluorescence Microscopy and Image Analyses All fluorescence microscopy studies were performed using a Zeiss M2Bio fluorescence microscope (Carl Zeiss Microimaging Inc., Thornwood, NY) equipped with a rhodamine cube, a green FITC filter (excitation: 488 nm, emission: 515 nm), and a chilled CCD camera (AxioCam MRm, Carl Zeiss Microimaging Inc., Thornwood, NY). Screens were routinely done using 1.6×, 10×, and 20× objective archromats and 10× eye piece. The system was also equipped with a z-motorized stage, deconvolution software, and 4D reconstruction software (AxioVision, Carl Zeiss Microimaging Inc.), which permits reconstruction of 3D objects and analyzes Z-stacks. Images were analyzed with AxioVision software Rel 4.0 (Carl Zeiss Microimaging Inc.), the Adobe Photoshop 6.0 computer program (Adobe, San Jose, CA), and NIH image software (Bethesda, MD). Patterns and intensity of staining were recorded and quantitated.

19.2.6.5 Acridine Orange Staining At 48, 72, and 120 hpf, five embryos were immersed in 0.5 μg/mL acridine orange (acridinium chloride hemi(zinc chloride)) in PBS for 60 min and rinsed thoroughly twice in 10 mL of fresh fish water. Stained embryos were anesthetized with MESAB and mounted in methylcellulose in a depression slide for observation using fluorescent microscopy. Effects of water-soluble fullerenes on apoptosis in the hatching glands, retina, lateral neuromasts, and olfactory pits were examined.

19.2.7 Determination of Dopaminergic CNS Neuroprotective Activity Against 6-Hydroxydopamine-Induced Apoptosis

19.2.7.1 Treatment of Embryos with 1–12 Two-day zebrafish were treated with 1% DMSO for vehicle control and treated with 1% DMSO + 250 μM 6-OHDA for DA-loss control. For compound testing, 2-day embryos were exposed to a mixture of 250 μM 6-OHDA and fullerenes for 72 h.

19.2.7.2 Antibody Staining Embryos were fixed in 4% paraformaldehyde overnight at 4°C. Fixed embryos were permeabilized with cold acetone at −20°C for 20 min and rehydrated in stepwise descending ethanol/PBS solutions (95, 79, 50, 25, and 0% ethanol/PBS mix, 10 min for each step). Embryos were then stained with anti-tyrosine hydroxylase antibody (mouse anti-human, Sigma, St. Louis, MO) at 4°C overnight. The next day, samples were washed with PBS-T (0.1% Tween 20), incubated with secondary antibody (goat anti-mouse), and color was developed using ABC reagent (Vector Labs, Burlingame, CA) according to manufacturer's instructions. Stained embryos were further flat-mounted on a glass slide and five randomly selected embryos were examined for DA neuron loss.

19.2.7.3 Light Microscopy and Image Analyses All microscopy studies were performed using a Zeiss light microscope (Carl Zeiss Microimaging Inc.) equipped with a SPOT camera (Diagnostic Instruments, Sterling, MI). The patterns and the intensity of staining were recorded and quantified.

19.2.7.4 Fluorescence Microscopy All fluorescence microscopy studies were performed using a Nikon Eclipse E600 fluorescence microscope (Nikon Inc., Melville, NY) equipped with a 200 W mercury/xenon lamp, a green FITC filter (excitation: 488 nm, emission: 515 nm), and a C5985 chilled CCD camera

(Hamamatsu Photonics, Hamamatsu City, Japan). Images were analyzed with the Adobe Photoshop 6.0 computer program (Adobe, San Jose, CA).

19.2.8 Vertebrate Animal Care and Safety

The Office of Laboratory Animal Welfare (OLAW), National Institutes of Health (NIH), has approved Phylonix' Animal Welfare Assurance #A4191-01, effective July 14, 2003, for a 5-year approval period. We euthanize zebrafish of all ages by overexposure to tricaine methanesulfonate; adult fish are then rapidly frozen. These procedures are consistent with the American Veterinary Medical Association's (AVMA) Panel on Euthanasia. Approximately 1500 animals were used in this study.

19.3 RESULTS

19.3.1 General Remarks and *In Vitro* Properties

Three different classes of water-soluble fullerenes, dendrofullerenes (**2–7**), C3-like fullerenes (**1** and **8–10**), and oxo-aminofullerenes (**11** and **12**) (Fig. 19.1), were tested in five different zebrafish assays to assess both overall toxicity (measured by LC_{50}) and efficacy in protecting against three different tissue-specific chemical cytotoxins: cisplatinum, gentamicin, and 6-hydroxydopamine. Cisplatinum is a widely used and highly effective anticancer drug and also causes apoptosis in noncancerous hair cells of auditory system and renal tubular cells in mammals. A similar pattern of cytotoxicity was observed in lateral-line neuromast hair cells of zebrafish embryos after exposure to cisplatinum. Gentamicin and related amino-glycoside antibiotics induce apoptosis and cell loss in auditory hair cells in mammals and dorsal mechanoreceptor hair cells in zebrafish and represent a major cause of acquired deafness in children (Yoshikawa, 1980; Stavroulaki et al., 1999). 6-Hydroxydopamine is taken up by CNS neurons, particularly by tyrosine hydroxylase-containing dopaminergic neurons in the central nervous system, and causes oxidative injury and apoptotic death (Nagatsu et al., 2000). The preferential destruction of dopaminergic neurons in mammals has led to widespread use of 6-hydroxydopamine as a model system for induction and study of Parkinson's-like syndromes in experimental animals. The data in Table 19.1 indicate that both dendrofullerenes **2–7** and C3-like anionic fullerenes **8** and **9** show substantial antioxidant activity against superoxide, whereas the activity of the cationic deri-vatives **10–12** is considerably lower (Witte et al., 2007). The results in Table 19.1 were obtained using xanthine/xanthine oxidase-generated superoxide. Similar results have been obtained with potassium superoxide and direct spectroscopic detection of superoxide (Ivanovic et al., unpublished data). The fact that anionic fullerenes interact much stronger with the positively charged cytochrome c than cationic derivatives demonstrates that electrostatic interactions are the major driving force for this binding process (Witte et al., 2007).

19.3.2 Stability of the C3 Fullerene (1)

The e,e,e-trismalonic acid (**1**) (C3) was the first water-soluble fullerene derivative with a defined structure whose antioxidant properties were studied in detail. However, when comparing the *in vitro* and *in vivo* antioxidant, neuroprotective, and toxic properties of C3 (**1**) with **2–12** and other water-soluble fullerenes, we recognized the C3 can easily degrade and that the degradation products can be comparatively toxic. In a systematic investigation, we studied the stability of purified C3 (**1**) in crystalline form as well as after suspension in several different buffers. The results are summarized in Fig. 19.2. In crystalline powder form, C3 (**1**)

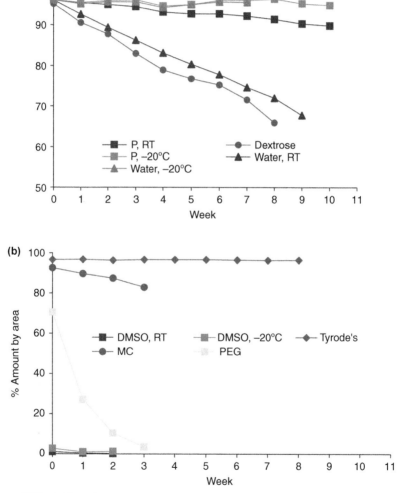

Figure 19.2   Stability of C3 (**1**): (a) as powder (P) and in water and dextrose; (b) in DMSO, Tyrode's solution, and MC. The amount of **1** was determined by the corresponding peak areas in the HPLC profiles.

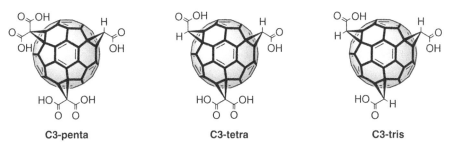

Figure 19.3 Schematic representation of the C3 decarboxylation products C3-penta, C-tetra, and C-tris. Only one stereoisomer each is represented although they were formed as mixtures of isomers (NMR).

degrades approximately 0.5% per week at room temperature, but is stable at −20°C. In solution, the stability of C3 (**1**) is highly variable, depending upon the suspending medium. In DMSO at room temperature, C3 (**1**) is degraded completely within a few minutes; the same is true in DMSO with immediate freezing at −20°C. C3 solutions of distilled water, phosphate-buffered Tyrode's solution, and 5% glucose exhibit a degradation rate of approximately 4% per week at room temperature, but all three solutions are completely stable at −20°C for at least 10 weeks. C3 solutions in polyethylene glycol (PEG) at room temperature exhibit rapid degradation rates of approximately 40% per week.

HPLC analysis of C3 degradation products indicates that decarboxylation reactions of the malonyl adducts represent the major pathway for degradation of C3 (**1**). Three major decomposition products, namely, the mono-, bis-, and tris-decarboxylation products C3-penta, C3-tetra, and C3-tris, could be identified (Fig. 19.3). The initial breakdown product is mostly C3-penta, but with continued degradation, significant amounts of C3-tetra and C3-tris are observed as well. In the presence of DMSO, complete degradation to C3-tris appears within 1–2 min at room temperature.

19.3.3 Toxicity of Water-Soluble Fullerenes in Zebrafish Embryos

We compared the overall toxicity of our water-soluble fullerenes in zebrafish embryos, and the results are summarized in Table 19.2. Fullerenes were added to the water in single wells of a 96-well plate, each containing a single zebrafish embryo at 24hpf, and lethality along with any observed morphologic abnormalities was scored at 120hpf. In most cases, fullerenes were tested at varying concentrations up to 500 μM, but in some cases the maximum concentration tested was 250 μM due to solubility issues at higher concentrations. LC_{50} was calculated as the fullerene concentration at which 50% lethality is observed at 120hpf. For the three cationic fullerenes **10–12** tested, the LC_{50} was less than 120 μM, and in the case of **12**, as low as 30 μM. For the anionic fullerenes, the LC_{50} values were considerably higher reaching

Table 19.2 LC_{50} of Fullerenes **1–12** in Zebrafish

Compound	Class	LC_{50} zebrafish (μM)
1	C3-like fullerenes	596
2	Dendrofullerenes	>500 (30% lethality at 500 μM)
3	Dendrofullerenes	>500 (20% lethality at 500 μM)
4	Dendrofullerenes	ND
5	Dendrofullerenes	ND
6	Dendrofullerenes	≫250 (0% lethality at 250 μM)
7	Dendrofullerenes	ND
8	C3-like fullerenes	≫500 (0% lethality at 250 μM)
9	C3-like fullerenes	≫500 (0% lethality at 250 μM)
10	C3-like fullerenes	117
11	Oxo-aminofullerenes	104
12	Oxo-aminofullerenes	30

values up to 0% lethality at 500 μM for **8** and **9**. In several cases, morphological abnormalities such as shortened body length and abnormal body curvature were detected in embryos treated with fullerene concentrations at or near the LC_{50}. For example, we observed shortened body length after treatment with C3 (**1**) at 250 μM (Fig. 19.4a, panel B) and slight curvature of the body after treatment with **3** at 500 μM (Fig. 19.4a, panel C). The majority of fullerenes tested showed no observable body length or curvature abnormalities (see, for example, Fig. 19.4b). Higher resolution microscopy of individual organs within developing zebrafish embryos showed abnormal cardiac chambers in the presence of 250 μM C3 (**1**) (Fig. 19.5b), enlargement of liver and intestines in the presence of **3** at 500 μM (Fig. 19.5c), and underdevelopment of liver and intestines in the presence of **2** at 500 μM (Fig. 19.5f). In addition, embryos treated with **8** (500 μM) showed slight enlargement of the heart (Fig. 19.5h). These results suggest that at high concentrations, some water-soluble fullerenes may exhibit teratogenic effects, but this has not yet been tested in mammals.

The decarboxylation breakdown products of C3 (**1**) represent an important exception with respect to our generalization of low toxicity for anionic fullerenes. With progressive decarboxylation steps, each resulting C3 breakdown product shows increasing toxicity with respect to the parent C3 (**1**) (Table 19.3). The LC_{50} for C3 (**1**) is 596 μM, followed by C3-penta at 373 μM, C3-tetra at 134 μM, and C3-tris at 10 μM. Interestingly, the increasing toxicity of the progressive C3 decarboxylation products was not observed with cultured murine endothelial cells, at least for C3-tris (data not shown), suggesting that the toxic effects of C3 decarboxylation products may be limited to a specific cell, tissue, or organ type that is critical for overall survival. To further investigate this possibility, we initiated a series of experiments to study the effects of C3 (**1**) and its decarboxylated

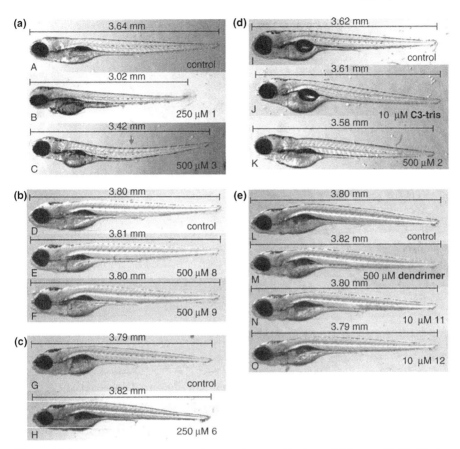

Figure 19.4 Morphological body length abnormalities caused by doses of fullerenes C3 (**1**), C3-tris, **2**, **3**, **6**, **8**, **9**, **11**, and **12**. For comparison, the corresponding free G2 dendron was investigated (e, panel M).

breakdown products on cardiac conduction in zebrafish, since previous work had shown that native fullerenes can dramatically effect cardiac conduction by binding to the ZERG potassium channel in fish (analogous to the HERG potassium channel in humans). We were unable to measure the ECG directly in fish and therefore chose as a surrogate induction of bradycardia by various water-soluble fullerenes, including C3 (**1**), **3**, and C3-tris. Figure 19.6 indicates that C3-tris induces bradycardia at very low doses of 1 μM or less. In contrast, no significant brady-cardia is observed at **3** or C3 concentrations of 100 μM. The extent of bradycardia and associated cardiac abnormalities is probably sufficient to account for the increased toxicity of C3 decarboxylation products in the LC_{50} assay. Figure 19.6 also indicates that cardiac injury is not characteristic of water-soluble fullerenes in general.

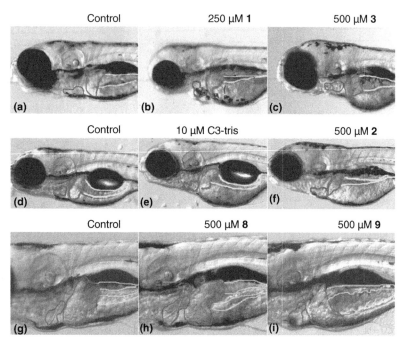

Figure 19.5 Morphological organ abnormalities in the development of liver, intestine, and heart caused by doses of fullerenes C3 (**1**), C3-tris, **2**, **3**, **8**, and **9** near LC_{50}. The liver, intestine, and heart are outlined by magenta, yellow, and red lines, respectively, for easy visual assessment. Compared to controls (a, b, and g), 250 μM C3-treated zebrafish exhibited unfolded cardiac chambers, and underdeveloped liver and intestine (b), 500 μM **3**-treated zebrafish exhibited enlarged liver and intestine (c), and 500 μM **2**-treated zebrafish exhibited underdeveloped liver and intestinal tract (f). No obvious defects on liver, intestine, and heart were observed for 10 μM C3-tris (e), 500 μM **8** (h), or 500 μM **9** (i). However, slightly enlarged cardiac chambers were observed for **8**. No organ necrosis was observed with any of the treatments. (See the color version of this figure in Color Plates section.)

19.3.4 Protection of Zebrafish Embryos from CNS Injury Due to 6-Hydroxydopamine

Previous reports in the literature (Lin et al., 1999, 2002, 2004; Lotharius et al., 1999; Bisaglia et al., 2000; Jin et al., 2000; Dugan et al., 2001; Yang et al., 2001; Tzeng et al., 2002; Simpson et al., 2003) have suggested that C3 (**1**) and related fullerenes

Table 19.3 Toxicity of C3 Decarboxylation Products in Zebrafish and Cell Culture

	1	C3-penta	C3-tetra	C3-tris
Antioxidant activity *in vitro* (superoxide)	174 μM	62 μM	48 μM	44 μM
Zebrafish toxicity	596 μM	373 μM	134 μM	10 μM
Cell culture toxicity	350 μM	ND	ND	325 μM

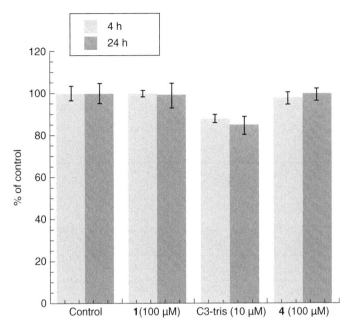

Figure 19.6 Cardiac bradycardia induced by C3-tris compared with C3 (**1**) and percent of control (heart rate) was plotted for control (0.1% DMSO), C3 (**1**) (100 μM), C3-tris (10 μM), and **4** (100 μM) at 4 h (light gray) and 24 h (dark gray).

may be useful in the treatment of Parkinson's disease as well as other CNS degenerative disease related to neuronal apoptosis. CNS dopaminergic neurons are particularly sensitive to the toxic effects of 6-OHDA, and we therefore tested the ability of various fullerenes to protect from 6-OHDA-induced apoptosis of CNS dopaminergic neurons in developing zebrafish. Dopaminergic CNS neurons are easily identified by the presence of tyrosine hydroxylase, which can be detected by immunohistochemistry on whole zebrafish embryos.

Fullerenes **1–4** were tested in various combinations for their ability to protect zebrafish embryos against damage to both general CNS neurons and tyrosine hydroxylase CNS neurons against 6-OHDA. 6-OHDA crosses the blood–brain barrier and is taken up preferentially by dopaminergic CNS neurons, and to a lesser extent by nondopaminergic CNS neurons. Intracellular 6-OHDA induces cellular apoptosis and death, at least in part through oxidative injury (Silva et al., 2005). Dendrofullerene **3** showed the highest levels of general CNS neuroprotection against 6-OHDA injury, and was able to protect 47% of total CNS neurons, while C3 (**1**) and **2** were able to block 42% and 23%, respectively, of 6-OHDA-induced total CNS neuronal death (Table 19.4). With respect to specific protection of tyrosine hydroxylase positive dopaminergic neurons in the diencephalon, **3** (100 μM) protected 100% of dopaminergic neurons, while C3 (**1**) (100 μM) and **4** (250 μM) each protected 60% of dopaminergic neurons. Figure 19.7 and Table 19.5 show results obtained with **4**, in protecting CNS dopaminergic neurons against 6-OHDA-induced cell death.

Table 19.4 Protection Provided by Fullerenes **1–4** Against General CNS Apoptosis Induced by 6-OHDA

Compound	Maximum protection against CNS apoptosis	Maximum protection against dopaminergic CNS apoptosis
1	23%	ND
2	47%	100%
3	ND	60%
4	42%	60%

Table 19.5 Protection Provided by **4** Against Dopaminergic CNS Apoptosis Induced by 6-OHDA

Treatment	Animal 1	Animal 2	Animal 3	Animal 4	Animal 5
Control	28	29	30	26	32
6-OHDA	7	8	10	1	3
6-OHDA + 250 μM **4**	22	14	21	18	26

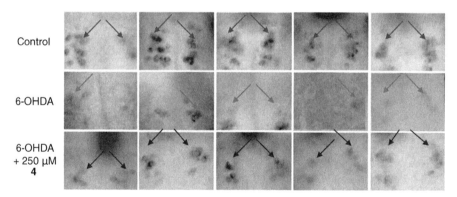

Figure 19.7 Protection of CNS dopaminergic neurons from 6-OHDA-induced apoptosis by **4**. Untreated (top panels), 6-OHDA plus vehicle treatment (middle panels), and 6-OHDA plus 250 μM **4** (lower panels) embryos were stained with anti-tyrosine hydroxylase antibody. Five images of each group were examined; the number of TH-immunoreactive cells clustered in diencephalons (between the eyes) was examined. In untreated animals, 20–40 TH cells were observed (dark gray arrows). After 6-OHDA treatment, decrease in the number of TH-immunoreactive cells was observed (1–17) (light gray arrows). After 250 μM **4** treatment, three of the five **4**-treated animals showed a normal number of TH-immunoreactive cells with normal morphology (black arrows).

Table 19.6 Ability of C3 (**1**), C3-tris, **2**, and **3** to Block Apoptosis During Normal Development of Zebrafish Embryos

Compound	c (μM)	Reduction of apoptosis			
		Hatching glands	Olfactory pits	Retina	Lateral neuromasts
1	250	N	N	N	N
C3-tris	10	Y	N	N	Y
2	500	Y	Y	Y	Y
3	500	N	N	N	N

19.3.5 Blocking Developmental Apoptosis in Zebrafish Embryos

Water-soluble fullerenes appear to be highly effective at blocking apoptotic cell death in the presence of known chemical toxins and electromagnetic radiation (Lin et al., 1999, 2004; Lotharius et al., 1999; Fumelli et al., 2000; Daroczi et al., 2006). We wanted to explore whether they could also block apoptosis that occurs naturally during embryonic development. Zebrafish embryos were exposed to fullerenes (C3 (**1**), C3-tris, **2**, and **3**) from day 1 to day 5 post fertilization, and whole embryos were subsequently stained with acridine orange to detect apoptotic cells in control and fullerene-treated embryos. Table 19.6 shows that **3** and C3-tris block embryogenesis-associated apoptosis, while no effects were observed in these experiments with C3 (**1**) or **2**. Figure 19.8 shows acridine orange staining for C3-tris and **3**, demonstrating reduction in number of positively stained cells in hatching glands and olfactory pits for C3-tris and in retina and neuromast cells for **3**. It is notable that zebrafish embryos appear viable after treatment with these agents based on lethality, although developmental abnormalities in body and organ development have been noted in preceding sections.

19.3.6 Protection of Zebrafish Embryos from Mechanoreceptor Hair Cell Injury and Death Induced by Cisplatinum and Gentamicin

In zebrafish, the dorsal mechanoreceptor hair cells are responsible for sensing position, orientation, balance, and movement and are biochemically and morphologically similar to outer hair cells of the inner ear in mammals. Using DASPEI, a vital dye that stains mechanoreceptor cells specifically, the fate of each lateral-line neuromast cell can be tracked in zebrafish embryos after exposure to chemical toxins that induce apoptosis in these cells (Ton and Parng, 2005). Cisplatinum is a widely used and highly effective chemotherapy agent. The major side effects are damage to renal tubular cells and outer hair cells of the inner ear, leading to renal failure and loss

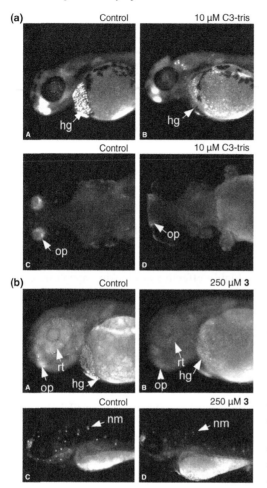

Figure 19.8 Inhibition of normal apoptosis by **3** and C3-tris in zebrafish embryos. (a) Decreased apoptosis after C3-tris treatment. Compared to controls (A and C), 10 µM C3-tris caused decreased apoptosis in the hatching gland (hg) at 2dpf (B) and olfactory pits (op) at 4dpf (D). (b) Decreased apoptosis after treatment with **3**. Compared to controls (A and C), decreased apoptosis in the hatching gland (hg), olfactory pits (op), and retina (rt) was observed at 2dpf (B). Decreased apoptosis in neuromasts (nm) was observed at 5dpf (D).

of hearing acuity or deafness, respectively. The mechanism by which cisplatinum induces cell death is believed to be primarily direct DNA damage, although perturbation of cellular redox pathways may also play a role (Schweitzer, 1993). In zebrafish embryos, cisplatinum rapidly induces apoptotic cell death in all of the dorsal mechanoreceptor hair cells detectable by DASPEI staining. We assessed the comparative ability of our water-soluble fullerenes to block cisplatinum-induced mechanoreceptor cell apoptosis in zebrafish embryos. The results, displayed in Fig. 19.9, indicate an unexpectedly wide variation among various anionic fullerenes. Fullerene **3** displayed the highest efficacy, blocking 50% of cisplatinum-induced apoptosis at low concentrations (29 µM), followed by **4** (194 µM), **7** (285 µM), and **5** (438 µM). C3 (**1**), though highly effective against the superoxide assay (Table 19.1), was almost completely unable to block apoptosis induced by cisplatinum, achieving only 5% protection at 500 µM.

EC_{50} (µM)

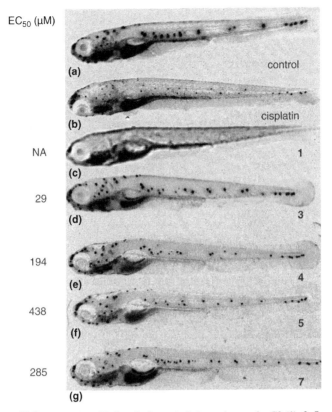

Figure 19.9 Protection of hair cells from cisplatinum damage by C3 (**1**), **3–5**, and 7.

Gentamicin and related aminoglycoside antibiotics are also widely used against a variety of serious infections in both adult and pediatric populations. High-dose gentamicin exposure is occasionally required with life-threatening infections, but may lead to permanent damage to outer hair cells of the inner ear and result in hearing loss or deafness (Yoshikawa, 1980; Matz, 1993; Stavroulaki et al., 1999; Unal et al., 2005). In zebrafish embryos, exposure to gentamicin or the related antibiotic neomycin also induces apoptosis and complete loss of DASPEI-stained dorsal mechanoreceptor hair cells in a pattern indistinguishable from that described for cisplatinum. We therefore assessed the comparative ability of our water-soluble fullerenes to block gentamicin-induced hair cell apoptosis and cell loss. Surprisingly, our results contrast substantially with those observed in the same cells using cisplatinum-induced cell loss (Fig. 19.10). C3 (**1**), **4**, and **6** are most effective in protecting against gentamicin-induced hair cell loss, with EC_{50} values of 34, 29, and 35 µM, respectively, followed closely by **5** (47 µM), **8** (60.7 µM), and **9** (78 µM). Dendrofullerene **3**, the most effective fullerene against cisplatinum-induced hair cell loss, ranks far behind in protection of the same cells from gentamicin-induced apoptosis (128 µM), along with **2** (97.5 µM) and **7** (124 µM) µM. It should be

EC$_{50}$ (μM)

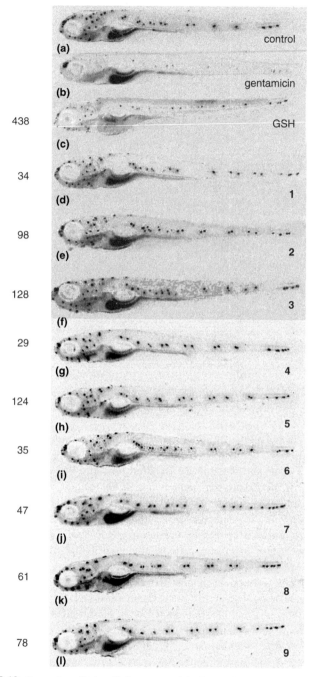

438

34

98

128

29

124

35

47

61

78

Figure 19.10 Protection of hair cells from gentamicin damage by various fullerenes **1–9**.

mentioned that even the least effective fullerenes in this group are three- to fourfold more effective than glutathione ($EC_{50} = 438 \, \mu M$) (Garetz et al., 1994; Lautermann et al., 1995; Pourbakht and Yamasoba, 2003; Ton and Parng, 2005), which has been used successfully in animal models to protect against aminoglycoside-induced hearing loss and renal injury.

19.4 DISCUSSION

Anionic water-soluble fullerenes have previously been shown to exhibit cytoprotective and antioxidant effects in both cell culture and various animal models (Dugan et al., 1996, 1997, 2001; Lai and Chiang, 1997; Tsai et al., 1997; Chi et al., 1998; Huang et al., 1998, 2001a, 2001b; Lai et al., 1998, 2000; Chueh et al., 1999; Lin et al., 1999, 2000, 2001, 2002, 2004; Puhaca, 1999; Straface et al., 1999; Bisaglia et al., 2000; Fumelli et al., 2000; Lee et al., 2000; Monti et al., 2000; Tagmatarchis and Shinohara, 2001; Yang et al., 2001; Foley et al., 2002; Murugan et al., 2002; Tzeng et al., 2002; Bosi et al., 2003; Ali et al., 2004; Chen et al., 2004; Witte et al., 2007). Our current results described above indicate that at least in a highly accessible zebrafish embryo model system, water-soluble anionic fullerenes can block apoptosis of specific cell types induced by chemical toxins and commonly used drugs whose toxicity is well characterized in mammalian systems, including in some cases, humans. However, our results also indicate that the mechanisms by which various modified fullerenes can protect cells from apoptosis are not monolithic and depend upon the specific mechanism of toxicity, even in a single cell type. Clearly, cisplatinum and gentamicin have very different binding and chemical activities, and the differential ability of fullerenes to protect against these two toxins may reflect the intracellular localization of each fullerene, the chemical reactivity against unknown reactive compounds or intermediates, or the ability to interact with proteins or other biological molecules involved in apoptosis signaling and regulation. In this context, the ability of some anionic fullerenes to bind to cytochrome c is of particular interest. Cytochrome c is located within mitochondria and is released to the cytosol after apoptosis-inducing signals are activated either at the cell surface or within the cell. Cytoplasmic cytochrome c facilitates the activation of cytoplasmic caspase cascades, triggering irreversible steps in the apoptosis pathway.

Our current results do not address whether or not fullerenes actually interact directly or indirectly with cytochrome c or related proteins nor do they elucidate whether this could occur in mitochondria, cytoplasm, or both. They do, however, indicate that fullerenes are capable of interacting with cytochrome c *in vitro* (Witte et al., 2007). Moreover, our results suggest that the ability of various fullerenes to interact with cytochrome c is somehow linked to the ability to protect against apoptosis induced by cisplatinum, but not gentamicin, in mechanoreceptor hair cells. Some of our water-soluble fullerenes, such as **4**, exhibit cytoprotective activity against both cisplatinum- and gentamicin-induced apoptosis and may prove to be useful as candidates for antiapoptosis drugs.

While some water-soluble fullerenes can exhibit cytoprotective activities, others show predominantly toxic effects. For example, C3 (**1**) is largely protective and relatively nontoxic, but C3 decarboxylation products show increasing lethality in zebrafish with increasing extent of decarboxylation. Much of this toxicity appears due to effects mediated primarily at the level of the cardiovascular system, including rhythm disturbances in the zebrafish heart. Native fullerenes and unmodified single-walled carbon nanotubes have been shown to inhibit the ZERG potassium channel in zebrafish by binding to the hydrophobic external channel and blocking conduction through the channel (Langheinrich et al., 2003; Park et al., 2003). It is possible that decarboxylated forms of C3 (**1**) are small enough to fit into the 1 nm ZERG channel, while fullerenes with more bulky adducts such as C3 (**1**) and dendritic fullerenes are prevented from binding to the hydrophobic ZERG channel by steric hindrance. However, the current study does not directly measure fullerene binding to the ZERG channel, and the proposed mechanism of toxicity remains speculative.

19.5 CONCLUSIONS

We have investigated the toxicity and the cytoprotective activities of three different families of water-soluble fullerenes. Using zebrafish (*D. rerio*) embryos as a model system, we assessed the ability of each fullerene to protect against toxicity induced by known chemical toxins in this system. We discovered that (i) positively charged water-soluble fullerenes tend to exhibit greater toxicity than negatively charged fullerenes with similar structures; (ii) toxicity varies considerably among negatively charged fullerenes from very low to moderate, depending on structural features; (iii) dendrofullerenes **2–7** (monoadducts of C_{60}) show stronger protection against cisplatinum toxicity in neuromast hair cells while the e,e,e-trismalonic acid (**1**) (so-called C3) shows stronger protection against gentamicin-induced cytotoxicity in the same cells; (iv) C3 (**1**) is relatively unstable in all aqueous solvents tested and breaks down mainly through decarboxylation reactions to form penta-, tetra-, and tris-carboxylated forms, which exhibit increased toxicity *in vivo* compared with C3 (**1**). These findings demonstrate that water-soluble fullerenes can protect against chemical toxin-induced apoptotic cell death in a vertebrate, whole animal model that may be useful for predicting the efficacy and toxicity of these compounds in mammals. Since we have established straightforward synthesis protocols for various families of water-soluble fullerenes allowing for the fine-tuning of pharmacological properties and bioavailability by using suitable combinations of building blocks as exohedral addends, we believe that water-soluble fullerenes represent a promising platform for the development of new potent antioxidant drugs.

19.6 ACKNOWLEDGMENTS

This work was supported by the Deutsche Forschungsgemeinschaft (DFG) and CNI, Houston, TX.

REFERENCES

ALI SS, HARDT JI, QUICK KL, KIM-HAN JS, ERLANGER BF, HUANG TT, EPSTEIN CJ, and DUGAN LL (2004). A biologically effective fullerene (C_{60}) derivative with superoxide dismutase mimetic properties. Free Radic Biol Med 37(8): 1191–1202.

BENSASSON RV, BRETTREICH M, FREDERIKSEN J, GOTTINGER H, HIRSCH A, LAND EJ, LEACH S, MCGARVEY DJ, and SCHONBERGER H (2000). Reactions of e^-_{aq}, CO_2^{*-}, HO^*, O_2^{*-} and $O_2(^1\Delta_g)$ with a dendro[60]fullerene and $C_{60}[C(COOH)_2]_n$ ($n = 2$–6). Free Radic Biol Med 29(1): 26–33.

BISAGLIA M, NATALINI B, PELLICCIARI R, STRAFACE E, MALORNI W, MONTI D, FRANCESCHI C, and SCHETTINI G (2000). C3-fullero-tris-methanodicarboxylic acid protects cerebellar granule cells from apoptosis. J Neurochem 74(3): 1197–1204.

BOSI S, DA ROS T, SPALLUTO G, and PRATO M (2003). Fullerene derivatives: an attractive tool for biological applications. Eur J Med Chem 38(11–12): 913–923.

CHEN YW, HWANG KC, YEN CC, and LAI YL (2004). Fullerene derivatives protect against oxidative stress in RAW 264.7 cells and ischemia-reperfused lungs. Am J Physiol Regul Integr Comp Physiol 287(1): R21–R26.

CHI Y, BHONSLE JB, CANTEENWALA T, HUANG JP, SHIEN J, CHEN BJ, and CHIANG LY (1998). Novel water-soluble hexa(sulfobutyl)fullerenes as potent free radical scavengers. Chem Lett 27: 465.

CHUEH SC, LAI MK, LEE MS, CHIANG LY, HO TI, and CHEN SC (1999). Decrease of free radical level in organ perfusate by a novel water-soluble carbon-sixty, hexa(sulfobutyl)fullerenes. Transplant Proc 31(5): 1976–1977.

COLE LK and ROSS LS (2001). Apoptosis in the developing zebrafish embryo. Dev Biol 240(1): 123–142.

DAROCZI B, KARI G, MCALEER MF, WOLF JC, RODECK U, and DICKER AP (2006). *In vivo* radioprotection by the fullerene nanoparticle DF-1 as assessed in a zebrafish model. Clin Cancer Res 12(23): 7086–7091.

DJOJO F and HIRSCH A (1998). Synthesis and chiropotical properties of enantiomerically pure bis- and trisadducts of C_{60} with and inherent chiral addition pattern. Chem Eur J 4(2): 344–356.

DUGAN LL, GABRIELSEN JK, YU SP, LIN TS, and CHOI DW (1996). Buckminsterfullerenol free radical scavengers reduce excitotoxic and apoptotic death of cultured cortical neurons. Neurobiol Dis 3(2): 129–135.

DUGAN LL, LOVETT EG, QUICK KL, LOTHARIUS J, LIN TT, and O'MALLEY KL (2001). Fullerene-based antioxidants and neurodegenerative disorders. Parkinsonism Relat Disord 7(3): 243–246.

DUGAN LL, TURETSKY DM, DU C, LOBNER D, WHEELER M, ALMLI CR, SHEN CK, LUH TY, CHOI DW, and LIN TS (1997). Carboxyfullerenes as neuroprotective agents. Proc Natl Acad Sci USA 94(17): 9434–9439.

FOLEY S, CROWLEY C, SMAIHI M, BONFILS C, ERLANGER BF, SETA P, and LARROQUE C (2002). Cellular localisation of a water-soluble fullerene derivative. Biochem Biophys Res Commun 294(1): 116–119.

FUMELLI C, MARCONI A, SALVIOLI S, STRAFACE E, MALORNI W, OFFIDANI AM, PELLICCIARI R, SCHETTINI G, GIANNETTI A, MONTI D, FRANCESCHI C, and PINCELLI C (2000). Carboxyfullerenes protect human keratinocytes from ultraviolet-B-induced apoptosis. J Invest Dermatol 115(5): 835–841.

GARETZ SL, ALTSCHULER RA, and SCHACHT J (1994). Attenuation of gentamicin ototoxicity by glutathione in the guinea pig *in vivo*. Hear Res 77(1–2): 81–87.

GULDI DM and ASMUS KD (1999). Activity of water-soluble fullerenes towards OH-radicals and molecular oxygen. Radiat Phys Chem 56(4): 449–456.

HUANG SS, TSAI SK, CHIANG LY, CHIH LH, and TSAI MC (2001a). Cardioprotective effects of hexasulfo-butylated C_{60} (FC_4S) in anesthetized rats during coronary occlusion/reperfusion injury. Drug Devel Res 53(4): 244–253.

HUANG SS, TSAI SK, CHIH CL, CHIANG LY, HSIEH HM, TENG CM, and TSAI MC (2001b). Neuroprotective effect of hexasulfobutylated C_{60} on rats subjected to focal cerebral ischemia. Free Radic Biol Med 30(6): 643–649.

HUANG YL, SHEN CK, LUH TY, YANG HC, HWANG KC, and CHOU CK (1998). Blockage of apoptotic signaling of transforming growth factor-beta in human hepatoma cells by carboxyfullerene. Eur J Biochem 254(1): 38–43.

JIN H, CHEN WQ, TANG XW, CHIANG LY, YANG CY, SCHLOSS JV, and WU JY (2000). Polyhydroxylated C_{60}, fullerenols, as glutamate receptor antagonists and neuroprotective agents. J Neurosci Res 62(4): 600–607.

KIMMEL CB, BALLARD WW, KIMMEL SR, ULLMANN B, and SCHILLING TF (1995). Stages of embryonic development of the zebrafish. Dev Dyn 203(3): 253–310.

LAI HS, CHEN WJ, and CHIANG LY (2000). Free radical scavenging activity of fullerenol on the ischemia–reperfusion intestine in dogs. World J Surg 24(4): 450–454.

LAI YL and CHIANG LY (1997). Water-soluble fullerene derivatives attenuate exsanguination-induced bronchoconstriction of guinea pigs. J Auton Pharmacol 17(4): 229–235.

LAI YL, WU HD, and CHEN CF (1998). Antioxidants attenuate chronic hypoxic pulmonary hypertension. J Cardiovasc Pharmacol 32(5): 714–720.

LANGHEINRICH U, HENNEN E, STOTT G, and VACUN G (2002). Zebrafish as a model organism for the identification and characterization of drugs and genes affecting p53 signaling. Curr Biol 12(23): 2023–2028.

LANGHEINRICH U, VACUN G, and WAGNER T (2003). Zebrafish embryos express an orthologue of HERG and are sensitive toward a range of QT-prolonging drugs inducing severe arrhythmia. Toxicol Appl Pharmacol 193(3): 370–382.

LAUTERMANN J, MCLAREN J, and SCHACHT J (1995). Glutathione protection against gentamicin ototoxicity depends on nutritional status. Hear Res 86(1–2): 15–24.

LEE YT, CHIANG LY, CHEN WJ, and HSU HC (2000). Water-soluble hexasulfobutyl[60]fullerene inhibit low-density lipoprotein oxidation in aqueous and lipophilic phases. Proc Soc Exp Biol Med 224(2): 69–75.

LIN AM, CHYI BY, WANG SD, YU HH, KANAKAMMA PP, LUH TY, CHOU CK, and HO LT (1999). Carboxyfullerene prevents iron-induced oxidative stress in rat brain. J Neurochem 72(4): 1634–1640.

LIN AM, FANG SF, LIN SZ, CHOU CK, LUH TY, and HO LT (2002). Local carboxyfullerene protects cortical infarction in rat brain. Neurosci Res 43(4): 317–321.

LIN AM, YANG CH, UENG YF, LUH TY, LIU TY, LAY YP, and HO LT (2004). Differential effects of carboxyfullerene on MPP$^+$/MPTP-induced neurotoxicity. Neurochem Int 44(2): 99–105.

LIN HS, LIN TS, LAI RS, D'ROSARIO T, and LUH TY (2001). Fullerenes as a new class of radioprotectors. Int J Radiat Biol 77(2): 235–239.

LIN YL, LEI HY, WEN YY, LUH TY, CHOU CK, and LIU HS (2000). Light-independent inactivation of dengue-2 virus by carboxyfullerene C3 isomer. Virology 275(2): 258–262.

LOTHARIUS J, DUGAN LL, and O'MALLEY KL (1999). Distinct mechanisms underlie neurotoxin-mediated cell death in cultured dopaminergic neurons. J Neurosci 19(4): 1284–1293.

MATZ GJ (1993). Aminoglycoside cochlear ototoxicity. Otolaryngol Clin North Am 26(5): 705–712.

MONTI D, MORETTI L, SALVIOLI S, STRAFACE E, MALORNI W, PELLICCIARI R, SCHETTINI G, BISAGLIA M, PINCELLI C, FUMELLI C, BONAFE M, and FRANCESCHI C (2000). C_{60} carboxyfullerene exerts a protective activity against oxidative stress-induced apoptosis in human peripheral blood mononuclear cells. Biochem Biophys Res Commun 277(3): 711–717.

MURUGAN MA, GANGADHARAN B, and MATHUR PP (2002). Antioxidative effect of fullerenol on goat epididymal spermatozoa. Asian J Androl 4(2): 149–152.

NAGATSU T, MOGI M, ICHINOSE H, and TOGARI A (2000). Changes in cytokines and neurotrophins in Parkinson's disease. J Neural Transm Suppl (60): 277–290.

OKUDA K, HIROTA T, HIROBE M, NAGANO T, MOCHIZUKI M, and MASHINO T (2000). Synthesis of various water-soluble C_{60} derivatives and their superoxide-quenching activity. Fullerene Sci Technol 8(1–2): 89–104.

PARK KH, CHHOWALLA M, IQBAL Z, and SESTI F (2003). Single-walled carbon nanotubes are a new class of ion channel blockers. J Biol Chem 278(50): 50212–50216.

PARNG C, SENG WL, SEMINO C, and MCGRATH P (2002). Zebrafish: a preclinical model for drug screening. Assay Drug Dev Technol 1 (1 Pt 1): 41–48.

POURBAKHT A and YAMASOBA T (2003). Ebselen attenuates cochlear damage caused by acoustic trauma. Hear Res 181(1–2): 100–108.

PUHACA B (1999). *In vitro* modulation of adriamycin-induced cytotoxicity of fulerol $C_{60}(OH)_{24}$. Med Pregl 52(11–12): 521–526.

REUTHER U, BRANDMULLER T, DONAUBAUER W, HAMPEL F, and HIRSCH A (2002). A highly regioselective approach to multiple adducts of C_{60} governed by strain minimization of macrocyclic malonate addends. Chemistry 8(10): 2261–2273.

RODRIGUEZ M and DRIEVER W (1997). Mutations resulting in transient and localized degeneration in the developing zebrafish brain. Biochem Cell Biol 75(5): 579–600.

SCHWEITZER VG (1993). Cisplatin-induced ototoxicity: the effect of pigmentation and inhibitory agents. Laryngoscope 103 (4 Pt 2): 1–52.

SILVA RM, RIES V, OO TF, YARYGINA O, JACKSON-LEWIS V, RYU EJ, LU PD, MARCINIAK SJ, RON D, PRZEDBORSKI S, KHOLODILOV N, GREENE LA, and BURKE RE (2005). CHOP/GADD153 is a mediator of apoptotic death in substantia nigra dopamine neurons in an *in vivo* neurotoxin model of parkinsonism. J Neurochem 95(4): 974–986.

SIMPSON EP, YEN AA, and APPEL SH (2003). Oxidative stress: a common denominator in the pathogenesis of amyotrophic lateral sclerosis. Curr Opin Rheumatol 15(6): 730–736.

STAVROULAKI P, APOSTOLOPOULOS N, DINOPOULOU D, VOSSINAKIS I, TSAKANIKOS M, and DOUNIADAKIS D (1999). Otoacoustic emissions: an approach for monitoring aminoglycoside induced ototoxicity in children. Int J Pediatr Otorhinolaryngol 50(3): 177–184.

STRAFACE E, NATALINI B, MONTI D, FRANCESCHI C, SCHETTINI G, BISAGLIA M, FUMELLI C, PINCELLI C, PELLICCIARI R, and MALORNI W (1999). C3-fullero-tris-methanodicarboxylic acid protects epithelial cells from radiation-induced anoikia by influencing cell adhesion ability. FEBS Lett 454(3): 335–340.

TAGMATARCHIS N and SHINOHARA H (2001). Fullerenes in medicinal chemistry and their biological applications. Mini Rev Med Chem 1(4): 339–348.

TON C and PARNG C (2005). The use of zebrafish for assessing ototoxic and otoprotective agents. Hear Res 208(1–2): 79–88.

TSAI MC, CHEN YH, and CHIANG LY (1997). Polyhydroxylated C_{60}, fullerenol, a novel free-radical trapper, prevented hydrogen peroxide- and cumene hydroperoxide-elicited changes in rat hippocampus *in-vitro*. J Pharm Pharmacol 49(4): 438–445.

TZENG SF, LEE JL, KUO JS, YANG CS, MURUGAN P, AI TAI L, and CHU HWANG K (2002). Effects of malonate C_{60} derivatives on activated microglia. Brain Res 940(1–2): 61–68.

UNAL OF, GHOREISHI SM, ATAS A, AKYUREK N, AKYOL G, and GURSEL B (2005). Prevention of gentamicin induced ototoxicity by trimetazidine in animal model. Int J Pediatr Otorhinolaryngol 69(2): 193–199.

WESTERFIELD M (1993). The Zebrafish Book: A Guide for the Laboratory Use of Zebrafish. University of Oregon Press, Eugene, OR.

WITTE P, BEUERLE F, HARTNAGEL U, LEBOVITZ R, SAVOUCHKINA A, SALI S, GULDI D, CHRONAKIS N, and HIRSCH A (2007). Water solubility, antioxidant activity and cytochrome C binding of four families of exohedral adducts of C_{60} and C_{70}. Org Biomol Chem 5(22): 3599–3613.

YAMAGO S, TOKUYAMA H, NAKAMURA E, KIKUCHI K, KANANISHI S, SUEKI K, NAKAHARA H, ENOMOTO S, and AMBE F (1995). *In vivo* biological behavior of a water-miscible fullerene: ^{14}C labeling, absorption, distribution, excretion and acute toxicity. Chem Biol 2(6): 385–389.

YANG DY, WANG MF, CHEN IL, CHAN YC, LEE MS, and CHENG FC (2001). Systemic administration of a water-soluble hexasulfonated C_{60} (FC_4S) reduces cerebral ischemia-induced infarct volume in gerbils. Neurosci Lett 311(2): 121–124.

YOSHIKAWA TT (1980). Proper use of aminoglycosides. Am Fam Physician 21(5): 125–130.

Chapter 20

Fishing to Design Inherently Safer Nanoparticles

Lisa Truong, Michael T. Simonich, Katerine S. Saili, and Robert L. Tanguay

Department of Environmental and Molecular Toxicology, Environmental Health Sciences Center, Oregon State University, Corvallis, OR, USA

20.1 INTRODUCTION

Nanotechnology is a rapidly growing field with a broad range of applications in the electronics, healthcare, cosmetics, technologies, and engineering industries (Forrest, 2001; Okamoto, 2001; Lecoanet et al., 2004; Lecoanet and Wiesner, 2004; Sun, 2005). By manipulating matter at the atomic level, nanoparticles are precisely engineered to exhibit desired physicochemical properties (Lecoanet et al., 2004). These unique properties can be exploited to improve targeted drug delivery, diagnostics systems, therapeutics, and biocompatibility leading to advances in the biomedical sciences (e.g., prosthetics, regenerative medicine, etc.). With so many positive attributes, there are few studies addressing how or why nanoparticles interact with biological systems. Without this knowledge, the full potential of nanotechnology will not be realized.

Applications of nanotechnology will rapidly increase as nanomaterial innovations occur. This will ultimately result in increased environmental and human exposure to nanoparticles. At the forefront of any risk assessment is a need for basic toxicological information, including characterizing how specific properties of nanoparticles govern biological responses. Given the variety of new nanoparticles being developed and marketed, this task requires systematic, collaborative scientific investigations. Voluntary testing adopted while the industry is still young might well avoid introduction of dangerous nanoparticles into the marketplace and environment such as occurred with chemicals in previous decades (i.e., chlorofluorocarbons, commercial use of DDT, asbestos, or lead in gasoline and paint products).

Zebrafish: Methods for Assessing Drug Safety and Toxicity, Edited by Patricia McGrath.
© 2012 John Wiley & Sons, Inc. Published 2012 by John Wiley & Sons, Inc.

Addressing potential risks of nanoparticles should be a priority concern not only for regulatory agencies, but also for the researchers and companies producing the particles. Collaborations between scientific investigators and manufacturers will preemptively minimize negative consequences and allow for development of the most environmentally benign nanochemistries and manufacturing methods. Greener nanotechnology is a practice pioneered at the University of Oregon to include replacing or minimizing usage of hazardous chemicals. Greener nanoscience also seeks to alter nanoparticles to render them nontoxic (e.g., via new reaction mechanisms, controlling physical properties, or surface functionalization).

It is now clear that rapid, relevant, and efficient testing methods must be developed to assess emerging nanoparticles. The investigation of nanoparticle interactions with biological systems must be conducted at multiple levels of biological organization (i.e., molecular, cellular, and organismal levels). There are many models that could be used to assess nano-biological interactions, but due to the rapidly increasing number of manufactured nano-particles, the ideal model must also offer high-throughput capabilities. For example, although *in vitro* techniques (cell-based systems) are preferred for cost and time efficiency, direct translation to whole organisms or humans is difficult. Challenges of *in vitro* studies also include contradictory effects from nano-biological interactions depending on the cell type, organ system, or developmental stage of the cells being used (Nakamura and Isobe, 2003; Bosi et al., 2004; Sayes et al., 2005; Isakovic et al., 2006). *In vivo* models (typically rodents) are more comprehensive and perhaps predictive, but the cost, labor, time, and infrastructure required to conduct studies in rodents make these species less than ideal testing platforms. The embryonic zebrafish model is ideally suited for rapid throughput nanoparticle testing because this is a genetically tractable vertebrate model, with rapid generation time, high sensitivity to environmental insult, and relatively low cost.

20.2 APPLICATION OF EMBRYONIC ZEBRAFISH

Numerous advantages have contributed to the rise in use of the zebrafish model over the past decade (Dodd et al., 2000; Wixon, 2000; Rubinstein, 2003). The embryonic zebrafish is a powerful model for rapidly evaluating nanoparticle toxicity at multiple biological levels while also defining structural properties that lead to adverse biological consequences. The embryonic zebrafish is uniquely sensitive to environmental insult due to the highly complex signaling events that underlie the cellular differentiation, proliferation, and migration events required to complete the developmental process (Truong et al., 2008; Usenko et al., 2008; Harper, 2008a, 2008b). Toxic responses often result from the disruption of these molecular signals.

To fully exploit the embryonic zebrafish model for nanoparticle testing, a three-tier *in vivo* approach was devised to define structural properties that lead to adverse biological consequences (Fig. 20.1). Tier 1 consists of a rapid screening experiment that assesses embryonic zebrafish for morphological changes caused by structurally well-characterized nanoparticles. In addition, exposed embryos are analyzed to determine nanoparticle uptake. All data gathered in each tier are entered into the

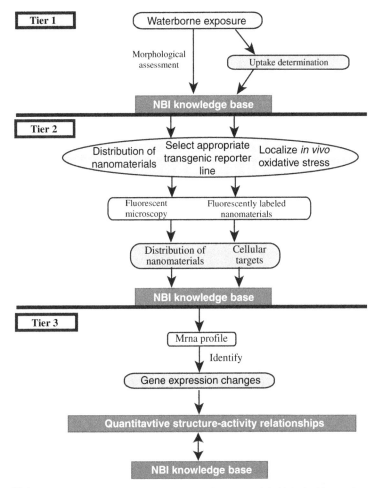

Figure 20.1 Overview of tiered design scheme to probe nanoparticle biological interactions. Tier 1: rapid toxicity screening. Tier 2: cellular toxicity and distribution. Tier 3: molecular expression.

NBI knowledge base. Tier 2 involves further testing of those nanoparticles identified in tier 1 as eliciting significant adverse effects. This involves probing for potential cellular targets and modes of action utilizing transgenic fluorescent zebrafish, assays to evaluate the oxidative environment, cell death, and biodistribution of the nanoparticles. Nanoparticles are then grouped by structural indices in tier 3, where gene expression profiles are defined. These data are also reported to the NBI knowledge base. Ultimately, the combination of data obtained from each tier of discovery will provide a path toward defining structure–activity relationships (SARs).

A key component necessary for defining a structure–activity relationship is to begin with well-characterized nanoparticles. There are currently many methods to create nanoparticles, but there exists no standard guiding the collection or characterization of data (i.e., purity, concentration, ligand size, core size, particle size, etc.). Without knowing the basic structural and purity information, it is not possible to define the structure variable in SAR. Therefore, insufficiently characterized materials are of no value for SAR studies.

20.3 TIER 1: RAPID TOXICITY SCREENING

20.3.1 Waterborne Exposure

Initially, all nanoparticles are processed through a rapid screening assay. The first step of this assay is to remove the embryonic chorion, an acellular envelope that acts as a protective barrier, at 6 h post fertilization (hpf) via enzymatic digestion. Removing the chorion eliminates a potential barrier to uptake from the surrounding water. Embryos are then exposed to a standard range of seven nanoparticle concentrations (fivefold serial dilutions) plus a vehicle/water control with the highest concentration defined by the limit of solubility. If a solvent is used, dimethyl sulfoxide (DMSO) is preferred based on previous studies indicating that DMSO is compatible with embryonic development (Harper, 2007; Isaacson et al., 2007; Usenko et al., 2007, 2008; Harper, 2008b). All concentrations utilizing a solvent are prepared with less than 1% DMSO in embryo media. Embryos are exposed to appropriate nanoparticle concentrations at 8hpf to ensure coverage of gastrulation and organogenesis, the periods of development that are most conserved among vertebrates. A total of 12 animals ($n = 12$) are exposed to each concentration, with one individual per well in a 96-well plate with 100 µL of exposure solution. Thus, for any given nanoparticle, the first step of waterborne exposure will require one 96-well plate.

After waterborne exposure, each embryo is evaluated at two time points: 24 and 120 hpf. At 24 hpf, embryos are evaluated for viability, developmental progression, and spontaneous movement (the earliest behavior in zebrafish). At 120 hpf, larval morphology (body, axis, eye, snout, jaw, otic vesicle, notochord, heart, brain, somites, fin, morphometrics) and behavioral end points (motility and tactile response) are evaluated and scored in a binary fashion (i.e., for each developmental end point, a score of normal or abnormal is recorded) according to well-established methods (Andreasen et al., 2002; Tanguay, 2003; Brent, 2004; Haendel et al., 2004). If more than 10% mortality occurs in the control group, then the assessment for that particular nanoparticle must be repeated. Any nanoparticle that induces adverse effects will proceed to tier 2, in which the potential cause of observed effects is investigated.

20.3.2 Statistical Analysis

Embryos are scored in a binary fashion and a Fisher's exact test is used to provide a proportional statistical analysis of differences between control and nanoparticle-exposed groups. The one-sided Fisher's exact test is used to calculate the sample size of 12 embryos per treatment, and since embryos are exposed individually, a well represents an individual test subject. This provides 80% power to detect differences between control and nanoparticle-exposed groups with a 0.10 significance level. Control and nanoparticle-exposed groups are then statistically compared using one-way analysis of variance (ANOVA) (SigmaStat, SPSS Inc., Chicago, IL). An alternative statistical test, Kruskal–Wallis analysis of variance on ranks, is used when several different groups are analyzed that have received different treatments or if the raw data violate normality or equal variance assumptions. In addition, the lethal concentration causing 50% embryonic mortality (LC_{50}) is calculated using probit and sigmoidal regression analyses. The lowest concentration to elicit significant effects (LOAEL) and the concentration at which no adverse effects are observed (NOAEL) are determined using one-way ANOVA (SPSS Inc., Chicago, IL).

20.3.3 Uptake Determination

All embryos exposed to nanoparticles undergo uptake analysis whether or not the nanoparticles induce significant effects (morphological or behavioral). After uptake efficiency is determined, it can be correlated with the observed effects to determine how dose metric relates to observed responses. These nanoparticles continue onto tier 2. To achieve this, the amount of nanoparticles in individual whole animals must be determined. In many cases, nanoparticles consist of metal cores. In these situations, two types of quantification methods can be employed: instrumental neutron activation analysis (INAA) or inductively coupled plasma optical emission (ICP-OES)/mass spectrometry (ICP-MS). Both analytical instruments are able to detect any element on the periodic table at trace levels. Regardless of the instrumentation, three biological replicates are used, but for either of the ICP instruments, additional three technical replicates are required due to instrument variabilities.

The same volume and concentration required to produce significant abnormal effects via waterborne exposure is used to determine actual uptake concentration via INAA. Each instrument has its advantages and disadvantages. INAA is able to detect the quantity of nanoparticles in individual embryonic zebrafish. Although this method is accurate, technically easy to perform, and does not require pooling of animals, it is also time consuming, expensive, and requires a radiation source, which is not common at academic or commercial settings. In contrast, ICP-OES and ICP-MS may not be as sensitive for certain metals and require embryo pooling in order to get a significant signal (varies depending on the core element and noise to background ratio). ICP-MS is three orders of magnitude more sensitive than ICP-OES. However, it is labor intensive and requires a reference standard for the nanoparticle core. In addition, the matrix of choice for ICP instruments is HNO_3 to minimize background signal interference. If ICP instruments are utilized, the analytical method must be validated by completing a spike and recover assay (i.e., use unexposed embryos and "spike" with the standards to ensure that the results can be replicated). In either INAA or any ICP method, the remaining challenge is determining if the quantified nanoparticles are integrated within the zebrafish or if it they are simply tightly associated with the outside of the animal.

20.4 TIER 2: CELLULAR TOXICITY AND DISTRIBUTION

20.4.1 Cellular Level Responses

Cellular targets can be identified *in vivo* by using (1) transgenic zebrafish that express fluorescent protein in specific cell populations and (2) fluorescent dyes that respond to changes in the cellular oxidation environment.

20.4.2 Transgenic Zebrafish

There are many transgenic zebrafish available, most of which can be found on the Zebrafish Model Organism Database (ZFIN) web site (www.zfin.org). The morphological effects observed in tier 1 are used as a guide to the selection of the reporter transgenic line used to assess target toxicity. Nanoparticle-induced disruption of specific cell populations can be easily investigated using this approach (Lele and Krone, 1996; Blechinger et al., 2002). For example, if nervous system end points are significantly impacted (e.g., brain malformation or abnormal behavior), then specific neuronal effects will be identified using transgenic fish such as *huc*-GFP (GFP produced in most neurons), *islet1*-GFP (GFP produced only in secondary motor neurons), *nbt1*-GFP (GFP expressed in primary spinal neurons), and/or *neurog1*-GFP (GFP produced in Rohon–Beard cells, and later in dorsal root ganglion cell bodies and axons).

After the most appropriate transgenic lines have been selected, 8hpf embryos are exposed as described in tier 1. After exposure, embryos are rinsed, anesthetized, and mounted in 1% agarose on a glass slide for microscopic analysis. Images are captured from live embryos expressing cell-specific fluorescent proteins via digital microscopy. There are multiple software programs available to quantify the amount of GFP positive cells, but image analysis using Image ProPlus 5.1 and tools in Axiovert 4.0 (Zeiss) has been successfully used. If more precise images are necessary, confocal laser scanning microscopy can be used.

20.4.3 Oxidative Stress

Nanoparticles have been implicated in disrupting the oxidative environment of cells (Tsuchiya et al., 1996; Hussain, 2001; Gharbi et al., 2004; Nemmar et al., 2004; Oberdorster, 2004). Excessive quantities of reactive oxygen species can cause oxidation of lipids, proteins, and DNA, cellular events that have been implicated in cardiovascular diseases (hypertension) and neurodegenerative diseases (Parkinson's and Alzheimer's). To directly measure oxidative stress *in vivo*, whole mount oxidative stress assays have been developed to identify temporal and spatial localization. To conduct these assays, we use embryonic exposures identical to that described for tier 1, except a positive control for oxidative stress (hydrogen peroxide) is also included.

20.4.4 Cellular Death

The embryonic zebrafish model is ideal for mapping cellular death in the whole animal and identifying the mechanism of cell death (i.e., necrosis versus apoptosis). To determine overall cellular death, embryos are exposed to nanoparticle solutions as described for tier 1, and then stained with acridine orange to label cells that have lost cell membrane integrity. The terminal deoxynucleotidyl transferase (TdT)-mediated dUTP nick end labeling (TUNEL) assay is used specifically to quantify apoptosis.

20.4.5 *In Vivo* Nanoparticle Distribution

Determining nanoparticle localization, even in an organism as small as larval zebrafish, is technically challenging. Distribution of nanoparticles assessed in tier 1 and 2 testing is currently tracked in two ways. The simpler approach is to fluorescently label the nanoparticle. In the case of a waterborne exposure, fluorescent microscopy can then be used to determine whether the nanoparticle was absorbed or was simply stuck to the epithelium. This simple discrimination cannot be determined using INAA, ICP-OES, and ICP-MS methods described in tier 1. A caveat to the fluorescently labeled approach is that the label could alter the nanoparticle properties altering the uptake profile compared to the native nanoparticle.

The second method for detecting nanoparticle distribution is to use high-resolution imaging scanning emission microscopy (SEM) or transmission electron microscopy (TEM) to scan throughout the animal. To do this, nanoscale slices are cut through mounted embryos and images are acquired using the electron microscopes. This is followed by the use of software to overlay adjacent images, thereby recreating a three-dimensional picture of the whole animal. Distribution data are critical to understanding the mechanism of how each nanoparticle interacts with the biological system. Developing nanoparticles for biomedical application must be accompanied by careful assessment of their biodistribution for the product to be both efficacious and safe.

Data gathered on the localization of oxidative stress, cell death, and distribution of nanoparticles in tier 2 are processed into the NBI knowledge base and combined with the data gathered from tier 1. This accumulation of data for each nanoparticle enables conclusions to be drawn about structure–activity relationships in tier 3.

20.5 TIER 3: MOLECULAR EXPRESSION

Once distribution is localized, the nanoparticle is moved to tier 3 where global gene expression profiling is conducted following waterborne exposures that elicit a tier 1 effect in 100% of the subjects (i.e., EC_{100}). Gene expression patterns, coupled with the nanoparticle distribution data, can then be used to help identify cellular targets. For example, if the nanoparticles are located in the liver, then one might reasonably expect to see altered expression of liver-specific genes as an activity resulting from the exposure. There are various aspects that must be considered to conduct a useful global gene expression analysis for the grouped structural indices and potential structure–activity relationships.

20.5.1 Selecting an End Point

The most critical step is to identify an end point for investigating structure–activity effects. This end point must occur for all nanoparticles sharing the same structure pattern (i.e., core size, core material, or surface functionalization) in 100% of exposed animals (EC_{100}) at a concentration that does not cause high mortality.

Once the end point is characterized, the "window of sensitivity" must be determined. This is a period during the development of the zebrafish at which a nanoparticle disrupts gene or protein expression prior to the visibility of the end point (EC_{100}). The goal is to define the gene expression changes that may underlie or drive the toxicity end point. The window may range from a few hours to several days. The goal is to identify the narrowest window of exposure capable of inducing the effect. To accomplish this, 6hpf embryos are dechorionated and exposed as described in tier 1, except the embryos are kept in the static nanoparticle solution for various durations and rinsed thoroughly with embryo media to remove agglomeration of the nanoparticle on the outside of the embryo. The typical experimental design involves exposure beginning at 8hpf and removal from the solution at 24, 48, and 72hpf and placement into fresh nanoparticle-free solution. The 72hpf time point is selected as the last end point since organogenesis is essentially complete and for most nanoparticle exposure experiments the morphological effects are first evident at this time point. Twenty-four-hour increments ensure that one end of the time window will be sampled, but to pinpoint the beginning and the end of the sensitivity window, shorter nanoparticle exposures are needed, ideally down to a few hours resolution.

Identifying critical windows of exposure is essential for efficient and cost-effective tier 3 gene expression analysis. Analysis of gene expression after the end point has occurred offers little, if any, mechanistic insight to the nanoparticle structure–activity relationship. An additional benefit of the microarray analyses is that they provide valuable data on how perturbation of one pathway may affect another, offering mechanistic insight into yet a higher level of biological interaction.

Nanoparticles have various physiochemical properties, all potentially inducing differential effects, making direct comparison between any two nanoparticles difficult. Effects can be a result of a synergism between the core size and material or the functional group and core size, and so on. This highlights the importance of grouping nanoparticles by similar physicochemical properties. For comparative gene expression analysis, relevant concentrations must be used. There is no value in using the no observable adverse effect level (NOAEL) or lowest observable adverse effect level (LOAEL) for end points that do not occur in 100% of the animals since the majority of the exposed animals in the pool will be indistinguishable from the nonexposed controls. The concentration should rather be just below the EC_{100} for a nonlethal end point to meaningfully link measured gene expression changes to the observed end point.

20.6 EMBRYONIC ZEBRAFISH DATA TO DESIGN "SAFER" NANOPARTICLES

The embryonic zebrafish model offers advantages for testing the exponentially increasing number of nanoparticles in development and already in the marketplace. However, the data obtained in this model or any model will be of little value unless a database repository of experimental data and conclusions is available. Such a knowledge base, the Nanomaterial–Biological Interactions (NBI) created at Oregon State University, is now in use as a clearinghouse for information gathered from all three-tier levels of testing (http://nbi.oregonstate.edu). We believe that a searchable platform such as NBI will allow for a better connection between the manufacturer of these nanoparticles and the researchers that determine the biological response potential for each nanoparticle. This will in turn facilitate industrial/regulatory cooperation toward safer nanoparticles.

The NBI is not restricted to the zebrafish model. Data obtained from other *in vitro* or *in vivo* models are encouraged. A greater wealth of structure–activity relationships, characterized in a diversity of models, will only facilitate better nanoparticle chemistry. A database of structure–activity relationships that is readily available to manufacturers is an invaluable resource where R&D budget constraints may not allow for extensive in-house safety testing of promising nanochemistries. A comprehensive knowledge database of structure–activity relationships in high-throughput models, including the embryonic zebrafish, will enable tuning of nanoparticles to maximize performance and safety before they reach commercial production. Nanochemistries in need of tuning can be placed into the safer nanoparticle design flow (Fig. 20.2): a continuum of criteria for safer design. The NBI will also limit the blind duplication

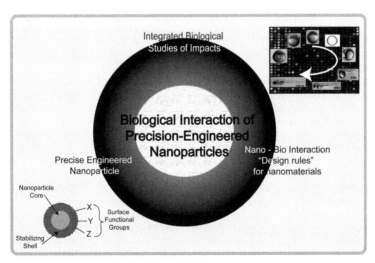

Figure 20.2   Using biological interactions of engineered nanoparticles to understand design rules for nanoparticles with enhanced biological properties. (See the color version of this figure in Color Plates section.)

of effort in testing newly synthesized nanoparticles that may have already been produced and tested.

20.7 CONCLUSIONS

Without a realistic means to test the massive numbers of new nanoparticles, some of these materials will be developed for public use before potential health risks are identified. Proactive assessments performed in appropriate test models can help to avoid this scenario and maintain consumer confidence. The platforms are in place and the application of the rapid throughput embryonic zebrafish model will maximize testing efficiency, minimizing the barriers to research, development, and marketing of safer nanoparticles by design.

REFERENCES

ANDREASEN EA, HAHN ME, HEIDEMAN W, PETERSON RE, and TANGUAY RL (2002). The zebrafish (*Danio rerio*) aryl hydrocarbon receptor type 1 is a novel vertebrate receptor. Mol Pharmacol 62(2): 234–249.

BLECHINGER SR, WARREN JT JR., KUWADA JY, and KRONE PH (2002). Developmental toxicology of cadmium in living embryos of a stable transgenic zebrafish line. Environ Health Perspect 110(10): 1041–1046.

BOSI S, FERUGLIO L, DA ROS T, SPALLUTO G, GREGORETTI B, TERDOSLAVICH M, DECORTI G, PASSAMONTI S, MORO S, and PRATO M (2004). Hemolytic effects of water-soluble fullerene derivatives. J Med Chem 47(27): 6711–6715.

BRENT RL (2004). Utilization of juvenile animal studies to determine the human effects and risks of environmental toxicants during postnatal developmental stages. Birth Defects Res B Dev Reprod Toxicol 71(5): 303–320.

DODD A, CURTIS PM, WILLIAMS LC, and LOVE DR (2000). Zebrafish: bridging the gap between development and disease. Hum Mol Genet 9(16): 2443–2449.

FORREST DR (2001). Molecular Nanotechnology. IEEE, pp. 11–20.

GHARBI K, SEMPLE JW, FERGUSON MM, SCHULTE PM, and DANZMANN RG (2004). Linkage arrangement of Na, K-ATPase genes in the tetraploid-derived genome of the rainbow trout (*Oncorhynchus mykiss*). Anim Genet 35(4): 321–325.

HAENDEL MA, TILTON F, BAILEY GS, and TANGUAY RL (2004). Developmental toxicity of the dithiocarbamate pesticide sodium metam in zebrafish. Toxicol Sci 81(2): 390–400.

HARPER SL (2007). Biodistribution and toxicity of nanomaterials *in vivo*: effects of composition, size, surface functionalization and route of exposure. NSTI-NanoTech 1: 666–669.

HARPER SL (2008a). Proactively designing nanomaterials to enhance performance and minimize hazard. J Nanotechnol 5(1): 124–142.

HARPER SL (2008b). *In vivo* biodistribution and toxicity depends on nanomaterial composition, size, surface functionalization and route of exposure. J Exp Nanosci 3: 195–206.

HUSSAIN N. (2001). Fluorometric method for the simultaneous quantitation of differently-sized nanoparticles in rodent tissue. Int J Pharm 214(1–2): 55–61.

ISAACSON CW, USENKO CY, TANGUAY RL, and FIELD JA (2007). Quantification of fullerenes by LC/ESI-MS and its application to *in vivo* toxicity assays. Anal Chem 79(23): 9091–9097.

ISAKOVIC A, MARKOVIC Z, TODOROVIC-MARKOVIC B, NIKOLIC N, VRANJES-DJURIC S, MIRKOVIC M, DRAMICANIN M, HARHAJI L, RAICEVIC N, NIKOLIC Z, and TRAJKOVIC V (2006). Distinct cytotoxic mechanisms of pristine versus hydroxylated fullerene. Toxicol Sci 91(1): 173–183.

LECOANET HF, BOTTERO JY, and WIESNER MR (2004). Laboratory assessment of the mobility of nanomaterials in porous media. Environ Sci Technol 38(19): 5164–5169.

Lecoanet HF and Wiesner MR (2004). Velocity effects on fullerene and oxide nanoparticle deposition in porous media. Environ Sci Technol 38(16): 4377–4382.

Lele Z and Krone PH (1996). The zebrafish as a model system in developmental, toxicological and transgenic research. Biotechnol Adv 14(1): 57–72.

Nakamura E and Isobe H (2003). Functionalized fullerenes in water. The first 10 years of their chemistry, biology, and nanoscience. Acc Chem Res 36(11): 807–815.

Nemmar A, Hoet PH, Vermylen J, Nemery B, and Hoylaerts MF (2004). Pharmacological stabilization of mast cells abrogates late thrombotic events induced by diesel exhaust particles in hamsters. Circulation 110(12): 1670–1677.

Oberdorster E. (2004). Manufactured nanomaterials (fullerenes, C_{60}) induce oxidative stress in the brain of juvenile largemouth bass. Environ Health Perspect 112(10): 1058–1062.

Okamoto Y. (2001). *Ab initio* investigation of hydrogenation of C_{60}. J Phys Chem A 105(32): 7634–7637.

Rubinstein AL (2003). Zebrafish: from disease modeling to drug discovery. Curr Opin Drug Discov Dev 6(2): 218–223.

Sayes CM, Gobin AM, Ausman KD, Mendez J, West JL, and Colvin VL (2005). Nano-C_{60} cytotoxicity is due to lipid peroxidation. Biomaterials 26(36): 7587–7595.

Sun O (2005). Clustering of Ti on a C_{60} surface and its effect on hydrogen storage. J Am Chem Soc 127(42): 14582–14583.

Tanguay RL (2003). Dioxin toxicity and aryl hydrocarbon receptor signaling in fish. In: Schecter A and Gasiewicz TA (Eds.), Dioxins and Health. Plenum Press, New York, pp. 603–628.

Truong L, Harper SL, and Tanguay RL (2008). Evaluation of embryotoxicity using the zebrafish model. Methods Mol Med 691: 271–279.

Tsuchiya T, Oguri I, Yamakoshi YN, and Miyata N (1996). Novel harmful effects of [60]fullerene on mouse embryos *in vitro* and *in vivo*. FEBS Lett 393(1): 139–145.

Usenko CY, Harper SL, and Tanguay RL (2007). *In vivo* evaluation of carbon fullerene toxicity using embryonic zebrafish. Carbon N Y 45(9): 1891–1898.

Usenko CY, Harper SL, and Tanguay RL (2008). Fullerene C_{60} exposure elicits an oxidative stress response in embryonic zebrafish. Toxicol Appl Pharmacol 229(1): 44–55.

Wixon J (2000). Featured organism: *Danio rerio*, the zebrafish. Yeast 17(3): 225–231.

Chapter 21

Radiation-Induced Toxicity and Radiation Response Modifiers in Zebrafish

Adam P. Dicker[1], Gabor Kari[1], and Ulrich Rodeck[1,2]

[1]Department of Radiation Oncology, Thomas Jefferson University, Philadelphia, PA, USA
[2]Department of Dermatology, Thomas Jefferson University, Philadelphia, PA, USA

21.1 INTRODUCTION

Radiation therapy (RT) remains a key therapeutic option for a wide range of neoplastic diseases, either alone or in combination with other treatment modalities. However, the radiation doses that can be safely administered are often lower than those required to eradicate tumor cells. This is due in large part to "collateral damage" by radiation, that is, toxicity to normal cells and tissues in the radiation field. The normal tissues most relevent to radiation toxicity typically contain epithelial cells with high rates of turnover including the lining of the gastrointestinal tract and the skin. Existing radiation protectors including the FDA-approved amifostine (Brizel, 2007), sucralfate (Hovdenak et al., 2005), and mesalazine (Sanguineti et al., 2003) are of limited utility in protecting these normal tissues against radiation effects primarily due to their own toxicity.

The identification and functional exploration of novel radioprotective compounds critically depend on performing studies in whole animals because the interplay of several cell types is likely to contribute to both radiation toxicity and radiation protection (Stone et al., 2003; Burdelya et al., 2006). In recognition of this requirement, many studies have been done in higher vertebrates such as mice. In the following, we will describe the use of zebrafish embryos as a complementary system to identify targeted agents that interfere with select signaling intermediaries and radioprotect the whole organism and/or specific organ systems.

Zebrafish: Methods for Assessing Drug Safety and Toxicity, Edited by Patricia McGrath.
© 2012 John Wiley & Sons, Inc. Published 2012 by John Wiley & Sons, Inc.

21.2 MATERIALS AND METHODS

21.2.1 Experimental Animals and Husbandry

Embryos are generated by natural pairwise mating in embryo collection tanks, as described in *The Zebrafish Book* (Westerfield, 1993). Four to five pairs of adult zebrafish are set up for each mating, and, on average, 100–150 embryos per pair are generated. Because the embryo receives nourishment from an attached yolk sac, no feeding is required for 7 days post fertilization (dpf). Viable embryos are washed and sorted (10 per well in six-well plates) at the one- to two-cell developmental stage and maintained under normoxic conditions at 28.5°C. Embryo medium is changed at 24, 72, and 120 h post fertilization (hpf). In select experiments, embryos (24 hpf) are dechorionated by placing in embryo medium supplemented with 1–5 µg/mL pronase (Sigma) for approximately 10 min at room temperature, and then gently agitating with a plastic pipette until the embryos are liberated from the chorions. After dechorionation, embryos are rinsed thoroughly with embryo medium, and placed in fresh embryo medium.

21.2.2 Radiation Protocol

Embryos are exposed to ionizing radiation ranging in dose from 0 to 80 Gy at 24 hpf using 250 kVp X-rays (PanTak) with 50 cm source-to-skin distance with a 2 mm aluminum filter. After irradiation, zebrafish embryos are maintained at 28.5°C for up to 7 dpf to monitor effects of treatments on survival, morphology, and organ-specific toxicity.

21.2.3 Effects on Zebrafish Morphology and Survival

Dechorionated embryos at 72 hpf are anesthetized with 1:100 dilution of 4 mg/mL tricaine and immobilized by placing them on 3% methylcellulose on a glass depression slide. Morphology is assessed visually using a light transmission microscope (Olympus BX51, Olympus, Melville, NY) at 40–100× magnification, and representative images are recorded using a QImaging camera using the iVision software (BioVision Technologies, Inc.). Similarly, survival of embryos is assessed visually at 24 h intervals up to 144 hpf by light microscopy. The criterion for embryonic survival is the presence of cardiac contractility.

21.2.4 Dextran Clearance Assay

Clearance of tetramethylrhodamine-labeled, 10 kDa dextran from the cardiac area is determined as described previously (Hentschel et al., 2005) with minor modifications. Briefly, zebrafish embryos are exposed at 24 hpf to ionizing radiation and maintained in embryo medium. At 72 hpf, embryos are anesthetized using a 1:100 dilution of

4 mg/mL tricaine methanesulfonate (Sigma) and dorsally positioned on a 3% methylcellulose gel. Tetramethylrhodamine-labeled, 10 kDa dextran (Molecular Probes) is injected into the cardiac venous sinus and embryos imaged 1, 5, and 24 h following injection. The average fluorescence emission at 590 nm following excitation at 570 nm is detected over the cardiac area, and the relative intensity measured using a Leica microscope (Leica Mikroskopie and Systeme GmbH). Images are transformed into grayscale and evaluated with NIH ImageJ software as described (Hentschel et al., 2005).

21.2.5 Assessment of Gastrointestinal Morphology and Function

The functional and morphological integrity of the developing gastrointestinal system is assessed in zebrafish embryos using PED6, a fluorescent reporter of phospholipase A2 (PLA2) activity. PED6 is a fluorogenic substrate for PLA2, which contains a BODIPY FL dye-labeled acyl chain and a dinitrophenyl quencher group (Farber et al., 2001). The cleavage of the dye-labeled acyl chain by phospholipase A2 within cells lining the intestine unquenches the dye and leads to the fluorescent labeling of the gall bladder and the lumen of the developing gastrointestinal tract. PED6 is added to zebrafish embryos at a stage when feeding and swallowing has commenced (5 dpf) followed by imaging the fish at 6dpf with the average fluorescence emission at 540 nm and excitation at 505 nm. Images are taken using a Leica microscope and analyzed using ImageJ software.

21.2.6 Ototoxicity Assay

Neuromast development is assessed as described by Harris et al. (2003), using the fluorescent vital dye 2-[4-(dimethylamino)styryl]-N-ethylpyridinium iodide (DASPEI, Molecular Probes, Carlsbad, CA). Zebrafish embryos at 5 dpf are exposed to 80 Gy. Neuromast staining was done 24 h later by incubation with 0.005% DASPEI in embryo medium for 15 min. Embryos are rinsed once with embryo medium and anesthetized for 5 min. The fluorescent emission at 515 nm following excitation at 450–490 nm is detected using a fluorescence microscope (Olympus BX51, Olympus). Neuromast staining was evaluated as follows: each neuromast on one side of the body is given a DESPEI score of +2 for normal staining, +1 for reduced staining, and +0 for no staining. Values are normalized to the maximum possible score of 54 (27 neuromasts).

21.2.7 Determination of Apoptosis by Acridine Orange Staining

Six hours after radiation exposure (20 Gy at 24 hpf), embryos are dechorionated, stained for 15 min using 5 μg/mL of acridine orange dye (Sigma), and rinsed five times with embryo medium as described previously (Westerfield, 1993). Zebrafish embryos are imaged with QImaging camera and iVision software; the images are analyzed using ImageJ software.

21.2.8 Drug Treatments

Stock solutions of the agents described here are typically prepared in DMSO followed by further dilution in embryo medium. DMSO concentrations in final solutions are typically <0.1% and nontoxic. In all experiments, embryo medium containing no more than 0.1% DMSO is used as a vehicle control. Amifostine (MedImmune Oncology, Inc., Gaithersburg, MD) serves as a positive control for radiation protectors and is used at 4 mmol as described by us previously. DF-1 was provided by C-Sixty Inc. and CDDO-TFEA by Reata Pharmaceuticals. The IKK inhibitor 2 (wedelolactone), IKK inhibitor 3 (BMS-345541), IKK-2 inhibitor 4, and IKK-2 inhibitor 5 (IMD-0354) are from Calbiochem as are azakenpaullone, SB361540, and SB361549 (Calbiochem). LiCl is obtained from Sigma and SB216763 from Selleck Chemicals LLC. All compounds are used at nontoxic concentrations arrived at by toxicity testing using log dilutions. When possible, target inhibition is assessed by measuring effects of drug *in vivo* in zebrafish. Current efforts focus on accomplishing this by monitoring the activity of reporter plasmids that contain specific promoters responsive to the molecular pathways targeted by the agents under investigation and injected into zebrafish embryos at the two- to four-cell stage (Kari et al., in preparation). To observe drug effects in the presence and absence of radiation, zebrafish embryos are maintained at 28.5°C for up to 7 dpf and the time-dependent effects of treatments on survival and morphology are scored.

21.3 VALIDATION OF ZEBRAFISH EMBRYOS AS A MODEL SYSTEM FOR RADIATION PROTECTORS/SENSITIZERS

Early work focused on proof-of-principle experiments to validate the potential of zebrafish to identify modulators of the radiation response (Traver et al., 2004; McAleer et al., 2005). While hematopoietic stem cell transplantation was shown to protect adult fish against hematopoietic radiation syndrome (Traver et al., 2004), much of the subsequent work has focused on embryos. In this work, viable zebrafish embryos were exposed to 0–10 Gy single-fraction 250 kVp X-rays at defined developmental stages at 1–24 hpf and scored for survival. These experiments revealed increasing radiation resistance at later stages of development and radiation protection of 4–8 hpf embryos by the FDA-approved amifostine. In contrast, pharmacological inhibition of the epidermal growth factor receptor radiosensitized zebrafish embryos consistent with earlier results *in vitro* and in mice (for review see Brizel, 2007).

Subsequent studies were done by irradiating at 24 hpf for the following reasons: (i) at this developmental stage most major organs have started to form, (ii) the effects of morpholino oligonucleotides (MOs) to downregulate target gene expression are well established, and (iii) pharmacological agents can be easily administered at defined time points before and after radiation to assess radiation protection and mitigation, respectively. For practical reasons, the LD_{50} of ionizing radiation (IR; 20 Gy single-dose whole body exposure) was defined by determining survival at 7 dpf of embryos irradiated at 1 dpf. Principally, this was due to minimal husbandry requirements in the first 6–7 days of development. However, when embryos irradiated at 24 hpf were scored for survival at 3 months, the LD_{50} decreased to 6 Gy and thus was close to the LD_{50} of humans exposed to whole body radiation (4 Gy).

21.4 GROSS MORPHOLOGICAL ALTERATIONS ASSOCIATED WITH RADIATION EXPOSURE

Irradiation of zebrafish embryos at or before 4 hpf produces multiple morphological defects, including microcephaly, microophthalmia, micrognathia, distal notochord and segmental abnormalities, pericardial edema, and inhibition of yolk resorption. As outlined in the preceding paragraph, organogenesis has commenced by 24 hpf rendering this the preferred developmental stage to study radiation-induced effects on major organ systems. When compared to younger embryos, the range of known organ-specific functional impairments in embryos irradiated at 24 hpf is more limited (see below). However, embryos irradiated at 24 hpf with 10–40 Gy demonstrate a characteristic morphological phenotype readily apparent within 1–2 days following ionizing radiation exposure and observable without visual aids. This phenotype consists of dorsal curvature of the body axis previously described as "curly-up" or CUP and ascribed to defects in midline development of zebrafish embryos (Fig. 21.1a) (Brand et al., 1996). The incidence of the CUP phenotype is approximately 60% after 20 Gy at 24 hpf. Although the mechanism causing CUP after IR is presently unknown, the ease with which this phenotype can be observed makes it an attractive parameter to score radiation toxicity in zebrafish embryos (Daroczi et al., 2006).

21.5 RADIATION-ASSOCIATED APOPTOSIS INCIDENCE

Radiation induces apoptosis in zebrafish, which can be easily visualized and scored using acridine orange (AO) staining of live fish. An example of increased AO staining after 20 Gy and assayed 6 h after radiation exposure is shown in Fig. 21.1b. Although several tissues reveal increased AO staining after radiation exposure, the central nervous system, the eyes, and cells in the spinal cord/notochord region along the body axis preferentially show marked increases in punctate AO staining and lend themselves to quantification of this phenomenon by measuring fluorescence in a defined area of the developing embryo using ImageJ software. In addition to AO staining, terminal deoxyribonucleotidyl transferase-mediated dUTP nick end labeling (TUNEL) and determination of caspase activity by use of fluorescent substrates have been used to determine radiation-associated apoptosis in zebrafish embryos. Compared to AO staining, these methods are presently more cumbersome (TUNEL) or have low signal-to-noise ratios (our unpublished results).

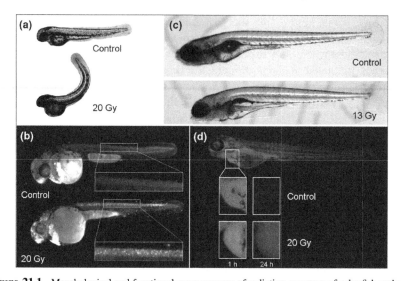

Figure 21.1 Morphological and functional consequences of radiation exposure of zebrafish embryos. (a) Characteristic dorsal curvature of body axis ("curly-up") observed at 48hpf. (b) AO staining of irradiated fish to detect apoptotic cells. The area used for quantitative analysis of AO staining at 30hpf (6 h post radiation) is boxed and magnified. (c) PED6 staining of the gastrointestinal tract. Reduced fluorescence and impaired gastrointestinal lumen formation is apparent in irradiated fish.
(d) Fluorescent dextran retention upon radiation exposure of zebrafish embryos. The dye is injected intracardially at 72hpf and the decrease in signal over time (1 and 24 h post radiation) is monitored in the cardiac area (as highlighted). Representative examples of control and irradiated fish showing dye retention are shown. All experiments are typically performed on embryos irradiated at 24hpf. Note that different radiation doses are used in different functional testing scenarios to enhance the sensitivity of identifying potential radiation protectors.

21.6 RADIATION-ASSOCIATED GASTROINTESTINAL TOXICITY

Damage to the small and large intestines is a major concern for the treatment of many abdominal malignancies with radiation therapy. To investigate gastrointestinal toxicity and the effects of radiation protectors on this phenomenon in zebrafish embryos, several methods have proven informative. Perhaps the simplest way to assess overall gastrointestinal function is to score "long-term" (up to 15 dpf) survival of fish larvae irradiated with 15 Gy. Fish larvae become dependent on external food sources at approximately 6 dpf when the contents of the yolk sac are depleted. Significant functional damage to the gastrointestinal system will thus lead to death by starvation within 10 dpf. Conversely, survival of fish beyond 2 weeks indicates establishment of a functionally adequate gastrointestinal system among other organ sites. Fish irradiated with 15 Gy do not survive beyond 14 dpf.

The effect of radiation on gastrointestinal function and lumen formation can be further monitored by using the fluorescent reporter PED6 (Ho et al., 2004) that is metabolized and excreted through the gastrointestinal system. Fluorescence microscopy enables a view of the gastrointestinal lumen filled with fluorescent PED6 and un-quenched by the activity of phospholipase A2 present in the mucosal lining of the gastrointestinal tract (Fig. 21.1c). IR (15 Gy) severely impairs fluorescence intensity and blunts lumen formation consistent with significant damage to the GI tract of higher vertebrates exposed to 10–15 Gy IR (Bhanja et al., 2009). The histological appearance of the gastrointestinal mucosa 5 days after radiation exposure reveals widespread damage caused by a sublethal IR dose (12 Gy). Specifically, the mucosal epithelium of the hindgut proximal to the cloaca is shaped irregularly with redistribution of nuclei away from the basal lamina and decreased goblet cell numbers (Daroczi et al., 2009).

21.7 RADIATION-ASSOCIATED NEPHROTOXICITY

Impairment of renal function has been recognized as a result of radiation therapy in the clinical setting. In zebrafish embryos, radiation exposure induced extensive edema formation raising the question whether kidney damage leads to water retention. However, edema may also be the result of damage to organs other than the kidneys including the heart and vessels. To assess kidney function, a technique based on renal clearance of fluorescently labeled dextran originally described by Hentschel et al. (2005) was employed. The labeled dextran is injected at 72 hpf into the cardiac venous sinus and the relative fluorescent intensity of the cardiac area measured within 24 h. Compared with nonirradiated controls, exposure to 20 Gy markedly reduces clearance of this agent from the heart in the absence of gross morphological alterations of the intracardiac lumen (Fig. 21.1d). This result is consistent with retention of the labeled contrast agent due to reduced renal excretion and points to compro-mised renal function associated with radiation exposure. Time lapse microscopy of cardiac contractility (Incardona et al., 2004) represents a useful tool to distinguish whether edema formation in zebrafish embryos is secondary to reduced cardiac function. In the case of irradiated embryos, this analysis revealed only marginal effects of 20 Gy ionizing radiation on heart rate and blood flow of sham-irradiated and irradiated fish (unpublished results). Collectively, these results demonstrate the utility of using organ function assays to evaluate radiation toxicity even when the cell or tissue type responsible for the observed toxicity is unknown (Daroczi et al., 2006).

21.8 OTOTOXICITY IN IRRADIATED ZEBRAFISH

Neurotoxicity represents a dose-limiting toxicity associated with radiation therapy of the central nervous system. While short-term assays to ascertain nerve cell damage have yet to be adapted to the radiation setting, an assay for ototoxicity has been used to monitor radiation-induced effects on these specialized nerve cells. This is done by using the vital dye DASPEI (Molecular Probes) that exclusively stains mechanore-ceptive hair cells (neuromasts) that are comparable to inner ear nerve hair cells in higher vertebrates. When embryos are exposed to 80 Gy IR at 5 dpf, staining of neuromasts is markedly reduced at 6 dpf whereas lower doses (20–60 Gy) do not exert measurable effect on neuromast development (Daroczi et al., 2006).

Table 21.1   Comparative Efficacy of Radiation Protectors in Zebrafish and Mammals

Target	Compound	Zebrafish	Effective dose	Mammals	Effective dose
NF-κB	Ethyl pyruvate	+	1 mM	+	0.6 mM
	CDDO	+	1 μM	+	1 μM
	IKK inhibitor 2 (wedelolactone)	+	0.1 μM		
	IKK inhibitor 3 (BMS-345541)	+	5 μM		
	IKK2 inhibitor 4	+	0.25 μM		
	IKK2 inhibitor 5 (IMD-0354)	+	0.05 μM		
GSK-3	LiCl	+	20 mM	+	18.9 mM
	SB216763	+	5 μM	+	1.6 μM
	Azakenpaullone	+	0.3 μM		
	TDZD-8	+	1 μM		
	Inhibitor VIII	+	10 μM		
ROS	Amifostine	+	4 mM	+	1.9 mM
	DF-1	+	100 μM	+	~100 μM

Protection against IR-induced lethality is indicated by "+". Only a select number of drugs have been tested in mice or rats to date. All compounds used in zebrafish were from Calbiochem/EMD. Note that protection was achieved in zebrafish and mice at roughly equimolar concentrations, where data are available in both model systems. Radioprotection of the gastrointestinal system in zebrafish was selectively tested and observed for EP and CDDO as well as for LiCl, SB216763, and azakenpaullone. Animal testing data were compiled using the following references: Daroczi et al. 2006, 2009, Meyer et al. (2006), Epperly et al. (2007), Thotala et al. (2008), and Wang et al. (2009). Testing in mammals was performed using mice and, in the case of CDDO, rats. Zebrafish GSK3 inhibitor data from Kari et al. (in preparation). Inhibition of the molecular targets as indicated is described in the following references: ethyl pyruvate (Han et al., 2005), CDDO (Ahmad et al., 2006), wedelolactone (Kobori et al., 2004), BMS-345541 (Burke et al., 2003), IKK2 inhibitor 4 (Podolin et al., 2005), IMD-0354 (Tanaka et al., 2005), LiCl (Zhang et al., 2003), SB216763 (Coghlan et al., 2000), azakenpaullone (Kunick et al., 2004), TDZD-8 (Martinez et al., 2002), Inhibitor VIII (Bhat et al., 2003), amifostine (Andreassen et al., 2003), and DF-1 (Daroczi et al., 2006).

21.9 RADIATION PROTECTORS IN ZEBRAFISH

The results described above represent a cross section of reproducible "phenotypes" and functional impairments of zebrafish embryos due to radiation exposure. These results set the stage for testing the effects of modulators of the radiation response in this system. Table 21.1 provides a summary of recent results from different studies across molecular targets and species (zebrafish, mice, and rats) that aptly demonstrate the utility of zebrafish for this purpose. The results in this table summarize beneficial effects of the agents tested on overall survival and include analysis of GI protection. In addition to ROS scavengers, agents that target nuclear factor (NF)-κB and glycogen synthase kinase 3 (GSK3) have been identified as promising radiation protectors. The focus on NF-κB was due to earlier observations that irradiated normal epithelial tissues are frequently characterized by excessive inflammation and that this inflammatory response limits the dose of radiation that can be safely administered. Since

NF-κB is a central mediator of inflammatory responses to radiation exposure (Li and Karin, 1998; Beetz et al., 2000; Linard et al., 2004), it was hypothesized that downmodulating NF-κB activity by pharmacological means could be of benefit in the radiation setting. As shown recently, this is indeed the case in zebrafish embryos (Daroczi et al., 2009), in mice (Epperly et al., 2007), and in rats (Linard et al., 2004). In all three species, a survival advantage was found by using different pharmacological NF-κB inhibitors. That NF-κB is the relevant target for radiation protection is underlined by the finding in zebrafish that multiple inhibitors of canonical NF-κB activation were effective whereas several proteasome inhibitors including PS-341 (bortezomib) were not (Daroczi et al., 2009).

A similar argument can be made for GSK3 inhibitors, which were first tested as radiation protectors in the cranial radiation setting (Thotala et al., 2008). Recent unpublished work showed that a wide variety of GSK3 inhibitors with different mechanisms of action afford radiation protection to zebrafish measured by extended survival as well as improved gastrointestinal morphology and function (Table 21.1). While it is presently unknown how attenuating GSK3 activity exerts radioprotection, these results underline that it is likely that GSK3 is a relevant target for radiation protection.

21.10 SUMMARY

During the past 20 years, the molecular mechanisms of the radiation and genotoxic stress responses have been mapped in exquisite detail. While this work has revealed complex, interconnected molecular networks, less has been learned about the integration of these networks *in vivo* in tissues and organisms. As recognized by a recent NCI workshop Stone et al., 2003, the tissue context is indispensable to accurately gauge the functional importance of specific molecular events in the whole organism consistent with the concept of emergent properties of complex systems. For example, inflammatory processes rely on the interplay of multiple cell types rather than homogeneous cell populations as studied in culture. Recent work has shown that screening for radiation modulators in zebrafish is a promising tool to identify suitable targets and select appropriate agents for pharmacological intervention. Effects of ionizing radiation on both overall survival and organ function have been defined as described in more detail above. It has been shown that zebrafish embryos are protected by known radiation protectors (ROS scavengers). In addition, pharmacological modulation of NF-κB and of GSK3 signaling has been observed to radioprotect both zebrafish and mice and this protection extends to the gastrointestinal system, a radiosensitive organ site of high clinical relevance. An important advantage of using zebrafish in studying radiation protectors is the capacity to rapidly test multiple agents affecting a given pathway. This circumstance helps to distinguish the contribution of a specific molecular target to radioprotection as opposed to off-target effects associated with any individual agent. In addition, antisense techniques based on morpholino-mediated inhibition of translation can be employed to further probe the contribution of a given pathway or drug target in the radiation response. An example for the use of this technique to characterize off-target drug effects is the identification of p73 as a target of Pifithrin-α, which was originally developed as a p53 inhibitor with radioprotective properties (Komarov et al., 1999; Davidson et al., 2008). Collectively, these considerations illustrate that studies in zebrafish may provide useful and novel information to complement the characterization of modifiers of the radiation response in higher vertebrates.

REFERENCES

AHMAD R, RAINA D, MEYER C, KHARBANDA S, and KUFE D (2006). Triterpenoid CDDO-Me blocks the NF-kappaB pathway by direct inhibition of IKKbeta on Cys-179. J Biol Chem 281: 35764–35769.

ANDREASSEN CN, GRAU C, and LINDEGAARD JC (2003). Chemical radioprotection: a critical review of amifostine as a cytoprotector in radiotherapy. Semin Radiat Oncol 13: 62–72.

BEETZ A, PETER RU, OPPEL T, KAFFENBERGER W, RUPEC RA, MEYER M, VAN BEUNINGEN D, KIND P, and MESSER G (2000). NF-kappaB and AP-1 are responsible for inducibility of the IL-6 promoter by ionizing radiation in HeLa cells. Int J Radiat Biol 76(11): 1443–1453.

BHANJA P, SAHA S, KABARRITI R, LIU L, ROY-CHOWDHURY N, ROY-CHOWDHURY J, SELLERS RS, ALFIERI AA, and GUHA C (2009). Protective role of R-spondin1, an intestinal stem cell growth factor, against radiation-induced gastrointestinal syndrome in mice. PLoS One 4(11): e8014.

BHAT R, XUE Y, BERG S, HELLBERG S, ORMO M, NILSSON Y, RADESATER AC, JERNING E, MARKGREN PO, BORGEGARD T, NYLOF M, GIMENEZ-CASSINA A, HERNANDEZ F, LUCAS JJ, DIAZ-NIDO J, and AVILA J (2003). Structural insights and biological effects of glycogen synthase kinase 3-specific inhibitor AR-A014418. J Biol Chem 278: 45937–45945.

BRAND M, HEISENBERG CP, WARGA RM, PELEGRI F, KARLSTROM RO, BEUCHLE D, PICKER A, JIANG YJ, FURUTANI-SEIKI M, VAN EEDEN FJ, GRANATO M, HAFFTER P, HAMMERSCHMIDT M, KANE DA, KELSH RN, MULLINS MC, ODENTHAL J, and NUSSLEIN-VOLHARD C (1996). Mutations affecting development of the midline and general body shape during zebrafish embryogenesis. Development 123: 129–142.

BRIZEL DM (2007). Pharmacologic approaches to radiation protection. J Clin Oncol 25(26): 4084–4089.

BURDELYA LG, KOMAROVA EA, HILL JE, BROWDER T, TARAROVA ND, MAVRAKIS L, DICORLETO PE, FOLKMAN J, and GUDKOV AV (2006). Inhibition of p53 response in tumor stroma improves efficacy of anticancer treatment by increasing antiangiogenic effects of chemotherapy and radiotherapy in mice. Cancer Res 66 (19): 9356–9361.

BURKE JR, PATTOLI MA, GREGOR KR, BRASSIL PJ, MACMASTER JF, MCINTYRE KW, YANG X, IOTZOVA VS, CLARKE W, STRNAD J, QIU T, and ZUSI FC (2003). BMS-345541 is a highly selective inhibitor of I kappa B kinase that binds at an allosteric site of the enzyme and blocks NF-kappa B-dependent transcription in mice. J Biol Chem 278: 1450–1456.

COGHLAN MP, CULBERT AA, CROSS DA, CORCORAN SL, YATES JW, PEARCE NJ, RAUSCH OL, MURPHY GJ, CARTER PS, ROXBEE COX L, MILLS D, BROWN MJ, HAIGH D, WARD RW, SMITH DG, MURRAY KJ, REITH AD, and HOLDER JC (2000). Selective small molecule inhibitors of glycogen synthase kinase-3 modulate glycogen metabolism and gene transcription. Chem Biol 7: 793–803.

DAROCZI B, KARI G, MCALEER MF, WOLF JC, RODECK U, and DICKER AP (2006). In vivo radioprotection by the fullerene nanoparticle DF-1 as assessed in a zebrafish model. Clin Cancer Res 12(23): 7086–7091.

DAROCZI B, KARI G, REN Q, DICKER AP, and RODECK U (2009). Nuclear factor kappaB inhibitors alleviate and the proteasome inhibitor PS-341 exacerbates radiation toxicity in zebrafish embryos. Mol Cancer Ther 8 (9): 2625–2634.

DAVIDSON W, REN Q, KARI G, KASHI O, DICKER AP, and RODECK U (2008). Inhibition of p73 function by Pifithrin-alpha as revealed by studies in zebrafish embryos. Cell Cycle 7(9): 1224–1230.

EPPERLY M, JIN S, NIE S, CAO S, ZHANG X, FRANICOLA D, WANG H, FINK MP, and GREENBERGER JS (2007). Ethyl pyruvate, a potentially effective mitigator of damage after total-body irradiation. Radiat Res 168 (5): 552–559.

FARBER SA, PACK M, HO SY, JOHNSON ID, WAGNER DS, DOSCH R, MULLINS MC, HENDRICKSON HS, HENDRICKSON EK, and HALPERN ME (2001). Genetic analysis of digestive physiology using fluorescent phospholipid reporters. Science 292(5520): 1385–1388.

HAN Y, ENGLERT JA, YANG R, DELUDE RL, and FINK MP (2005). Ethyl pyruvate inhibits nuclear factor-kappaB-dependent signaling by directly targeting p65. J Pharmacol Exp Ther 312: 1097–1105.

HARRIS JA, CHENG AG, CUNNINGHAM LL, MACDONALD G, RAIBLE DW, and RUBEL EW (2003). Neomycin-induced hair cell death and rapid regeneration in the lateral line of zebrafish (Danio rerio). J Assoc Res Otolaryngol 4(2): 219–234.

HENTSCHEL DM, PARK KM, CILENTI L, ZERVOS AS, DRUMMOND I, and BONVENTRE JV (2005). Acute renal failure in zebrafish: a novel system to study a complex disease. Am J Physiol Renal Physiol 288(5): F923–F929.

HO SY, THORPE JL, DENG Y, SANTANA E, DEROSE RA, and FARBER SA (2004). Lipid metabolism in zebrafish. Methods Cell Biol 76: 87–108.

HOVDENAK N, SORBYE H, and DAHL O (2005). Sucralfate does not ameliorate acute radiation proctitis: randomised study and meta-analysis. Clin Oncol (R Coll Radiol) 17(6): 485–491.

INCARDONA JP, COLLIER TK, and SCHOLZ NL (2004). Defects in cardiac function precede morphological abnormalities in fish embryos exposed to polycyclic aromatic hydrocarbons. Toxicol Appl Pharmacol 196(2): 191–205.

KOBORI M, YANG Z, GONG D, HEISSMEYER V, ZHU H, JUNG YK, GAKIDIS MA, RAO A, SEKINE T, IKEGAMI F, YUAN C, and YUAN J (2004). Wedelolactone suppresses LPS-induced caspase-11 expression by directly inhibiting the IKK complex. Cell Death Differ 11: 123–130.

KOMAROV PG, KOMAROVA EA, KONDRATOV RV, CHRISTOV-TSELKOV K, COON JS, CHERNOV MV, and GUDKOV AV (1999). A chemical inhibitor of p53 that protects mice from the side effects of cancer therapy. Science 285 (5434): 1733–1737.

KUNICK C, LAUENROTH K, LEOST M, MEIJER L, and LEMCKE T (2004). 1-Azakenpaullone is a selective inhibitor of glycogen synthase kinase-3 beta. Bioorg Med Chem Lett 14: 413–416.

LI N and KARIN M (1998). Ionizing radiation and short wavelength UV activate NF-kappaB through two distinct mechanisms. Proc Natl Acad Sci USA 95(22): 13012–13017.

LINARD C, MARQUETTE C, MATHIEU J, PENNEQUIN A, CLARENCON D, and MATHE D (2004). Acute induction of inflammatory cytokine expression after gamma-irradiation in the rat: effect of an NF-kappaB inhibitor. Int J Radiat Oncol Biol Phys 58(2): 427–434.

MARTINEZ A, ALONSO M, CASTRO A, PEREZ C, and MORENO FJ (2002). First non-ATP competitive glycogen synthase kinase 3 beta (GSK-3beta) inhibitors: thiadiazolidinones (TDZD) as potential drugs for the treatment of Alzheimer's disease. J Med Chem 45: 1292–1299.

MCALEER MF, DAVIDSON C, DAVIDSON WR, YENTZER B, FARBER SA, RODECK U, and DICKER AP (2005). Novel use of zebrafish as a vertebrate model to screen radiation protectors and sensitizers. Int J Radiat Oncol Biol Phys 61(1): 10–13.

MEYER CJ, SPORN MB, WIGLEY WC, and SONIS ST (2006). RAT 402 (CDDO-Me) suppresses tumor and treatment induced inflammation, sensitizing tumors to and protecting normal tissue from radiation. European Journal of Cancer (Suppl.) 4: 162.

PODOLIN PL, CALLAHAN JF, BOLOGNESE BJ, LI YH, CARLSON K, DAVIS TG, MELLOR GW, EVANS C, and ROSHAK AK (2005). Attenuation of murine collagen-induced arthritis by a novel, potent, selective small molecule inhibitor of IkappaB Kinase 2, TPCA-1 (2-[(aminocarbonyl)amino]-5-(4-fluorophenyl)-3-thiophene-carboxamide), occurs via reduction of proinflammatory cytokines and antigen-induced T cell Proliferation. J Pharmacol Exp Ther 312: 373–381.

SANGUINETI G, FRANZONE P, MARCENARO M, FOPPIANO F, and VITALE V (2003). Sucralfate versus mesalazine versus hydrocortisone in the prevention of acute radiation proctitis during conformal radiotherapy for prostate carcinoma. A randomized study. Strahlenther Onkol 179(7): 464–470.

STONE HB, COLEMAN CN, ANSCHER MS, and MCBRIDE WH (2003). Effects of radiation on normal tissue: consequences and mechanisms. Lancet Oncol 4(9): 529–536.

TANAKA A, KONNO M, MUTO S, KAMBE N, MORII E, NAKAHATA T, ITAI A, and MATSUDA H (2005). A novel NF-kappaB inhibitor, IMD-0354, suppresses neoplastic proliferation of human mast cells with constitutively activated c-kit receptors. Blood 105: 2324–2331.

THOTALA DK, HALLAHAN DE, and YAZLOVITSKAYA EM (2008). Inhibition of glycogen synthase kinase 3 beta attenuates neurocognitive dysfunction resulting from cranial irradiation. Cancer Res 68(14): 5859–5868.

TRAVER D, WINZELER A, STERN HM, MAYHALL EA, LANGENAU DM, KUTOK JL, LOOK AT, and ZON LI (2004). Effects of lethal irradiation in zebrafish and rescue by hematopoietic cell transplantation. Blood 104(5): 1298–1305.

WANG Y, HUANG WC, WANG CY, TSAI CC, CHEN CL, CHANG YT, KAI JI, and LIN CF (2009). Inhibiting glycogen synthase kinase-3 reduces endotoxaemic acute renal failure by down-regulating inflammation and renal cell apoptosis. Br J Pharmacol 157: 1004–1013.

WESTERFIELD M (1993). The Zebrafish Book: A Guide for the Laboratory Use of Zebrafish (*Danio rerio*). University of Oregon Press, Eugene, OR.

ZHANG F, PHIEL CJ, SPECE L, GURVICH N, and KLEIN PS (2003). Inhibitory phosphorylation of glycogen synthase kinase-3 (GSK-3) in response to lithium. Evidence for autoregulation of GSK-3. J Biol Chem 278: 33067–33077.

Chapter 22

Caudal Fin Regeneration in Zebrafish

Tamara Tal, Sumitra Sengupta, and Robert L. Tanguay

Department of Environmental and Molecular Toxicology, Environmental Health Sciences Center, Oregon State University, Corvallis, OR, USA

22.1 INTRODUCTION

The field of regenerative medicine is aimed at awaking our latent ability to replace damaged tissues and structures by reactivating evolutionarily conserved signaling pathways that promote regeneration. Mammals can regenerate a limited number of structures including skin, digits, skeletal muscle, hair cells, gut epithelium, liver, pancreas, and blood. In stark comparison, mammalian CNS tissues and structures (brain, retina, and spinal cord) and heart, limb, and kidney all fail to regenerate. Therefore, numerous human conditions resulting from injury, aging, and disease could be significantly improved if therapies that encourage tissue regeneration were available.

The field of regenerative medicine is aimed at developing strategies to restore individual cell types, complex tissues, or structures that are lost or damaged. This emerging field is approached from two distinct angles. In recent years, stem cell-based models have been developed to generate a suite of differentiated cells for therapeutic applications. The alternative approach employs the inherent regenerative capacity of nonmammalian models to define the molecular events that permit tissue regeneration. While there are several regenerative animal models including salamanders, newts, hydra, and flatworms that are established to evaluate tissue regeneration, zebrafish have emerged as a powerful vertebrate regenerative model because of advances in zebrafish genetics, rapid regeneration time, and the ability to obtain vast quantities of externally fertilized eggs.

Zebrafish: Methods for Assessing Drug Safety and Toxicity, Edited by Patricia McGrath.
© 2012 John Wiley & Sons, Inc. Published 2012 by John Wiley & Sons, Inc.

Zebrafish have the remarkable capability to regenerate their heart muscle (Poss et al., 2002; Raya et al., 2004), retina, optic nerve (Becker and Becker, 2002), liver (Sadler et al., 2007), spinal cord (Becker et al., 2004), and sensory hair cells (Lopez-Schier and Hudspeth, 2006). Understanding the molecular and genetic pathways that coordinately function to accomplish regeneration in this model, may explain why mammals fail to respond to tissue injury with a regenerative response and to the generation of therapeutic interventions designed to overcome limitations in mammalian regenerative abilities.

22.2   SIGNALING AND EPIMORPHIC REGENERATION

The central goal of regeneration research is to elucidate the molecular signaling networks that coordinately function to promote regeneration. There has been significant progress in identifying genes and signaling pathways that function during early fin regeneration through the use of mutagenic screens, loss-of-function and gain-of-function experiments, gene expression studies, and chemical genetic approaches. Both FGF and Wnt signaling have emerged as two main signaling pathways that regulate virtually all aspects of epimorphic fin regeneration. These signaling networks have also been shown to be influential regulators of heart (Lepilina et al., 2006; Stoick-Cooper et al., 2007b), retina (Del Rio-Tsonis et al., 1997; Osakada et al., 2007; Petersen and Reddien, 2008), and skeletal muscle (Floss et al., 1997; Otto et al., 2008) regeneration, and in the case of Wnt, liver (Monga et al., 2001; Goessling et al., 2009) and sensory hair cell regeneration. The ability of cells to determine their position in three dimensions is critical for the establishment of proper patterning during regeneration (Stoick-Cooper et al., 2007a) and activin, hedgehog, and retinoic acid signaling have all been implicated in regulating cellular patterning during epimorphic regeneration (Ferretti and Geraudie, 1995; Quint et al., 2002). Ultimately, a better understanding of the complexity of signaling events that choreograph epimorphic regeneration will provide new avenues for comparative studies in mammalian models with the hope of developing novel therapeutics to slow or prevent tissue loss from aging, injury, and disease.

22.3   CAUDAL FIN ARCHITECTURE

Zebrafish caudal fins are complex structures that contain 16–18 lepidotrichia (fin rays) connected by soft tissue interrays that lack skeletal elements. The fin rays are a series of bony segments comprised of a pair of concave hemirays surrounded by a monolayer of scleroblasts (bone secreting cells). The hemirays function to protect an intraray core consisting of mesenchymal cells, blood vessels, nerves, melanocytes, and fibroblasts. Mesenchymal compartmental components are also present in the interray space. The entire multiray fin is covered by an epithelial cell layer. The fin displays an indeterminate ontogenetic growth pattern meaning that fin growth occurs by the gradual addition of bony ray segments to the distal tip of the fin.

22.4 STAGES OF EPIMORPHIC REGENERATION

Upon caudal fin amputation, an impressive series of regenerative stages are initiated that result in the complete restoration of lost bone, epidermis, blood vessels, nerves, connective tissue, and pigmentation (Fig. 22.1). This complex process, termed epimorphic regeneration, is completed in approximately 3 days in larval fish (Fig. 22.2) and ~2 weeks in adult fish (reviewed in Lovine, 2007). Epimorphic regeneration is a term used to describe a regenerative process involving the formation of a mass of undifferentiated proliferative multipotent mesenchymal cells called a blastema (Akimenko et al., 2003). In the adult model, an initial wound healing stage, characterized by nonproliferative lateral epithelial cell migration over the wound and subsequent formation of the apical epidermal cap (AEC), is initiated immediately following surgical removal of caudal fin tissue (Nechiporuk and Keating, 2002). Second, the wound epithelium thickens and mesenchymal tissue proximal to the amputation plane begins to disorganize. Cellular disorganization is thought to occur as a result of growth factors that originate from the mature wound epidermis and stimulate mesenchymal cells to dedifferentiate and proliferate as they migrate distally toward the area directly proximal to the AEC (Nechiporuk and Keating, 2002). In the third stage, a series of blastemas form at the severed portion of each amputated fin ray. Blastema formation is the main event that distinguishes regeneration from limb development. The blastema is an accumulated mass of progenitor cells that are thought to be pluripotent or able to produce daughter cells capable of differentiating into a variety of cell types required to populate the regenerating tissue. Although pluripotency is inferred from aforementioned migratory and proliferative qualities, lineage tracing studies have not yet been completed. Therefore, it is possible that resident stem cells or some combination of mesenchymal dedifferentiation and stem cell proliferation gives rise to new cells in the regenerating tissue.

The immature blastema proliferates slowly with a median G2 cell cycle time of >6 h (Nechiporuk and Keating, 2002). Twenty-four hours following caudal fin amputation, blastemal cells segregate into two morphologically identical, but functionally distinct, subpopulations. The distal blastema, located proximally to the AEC, contains msxb positive cells, proliferates extremely slowly, and is hypothesized to specify the regenerating boundary and direct regenerative growth (Nechiporuk and Keating, 2002). In sharp contrast, the proximal blastema contains a rapidly proliferative mass of msxb negative cells with a mean G2 time of <60 min. Together, the proximal and distal blastemas form a proliferation gradient with a 50-fold difference in proliferation across an approximate distance of 50 µM or 10 cell diameters (Nechiporuk and Keating, 2002).

The next phase of regenerative outgrowth is marked by intense proliferation in the proximal blastema. A moderately proliferative patterning zone is located immediately proximal to the proximal blastema. The patterning zone contains newly divided cells that migrate to appropriate locations and differentiate to populate the new tissue with mesenchymal cells and fibroblasts. The location and functional differences between the distal and proximal blastema and patterning zone are thought to be generated and maintained through epithelial–mesenchymal interactions (Poss et al., 2003; Lee et al., 2009). The final stage, regenerative termination, is not well understood.

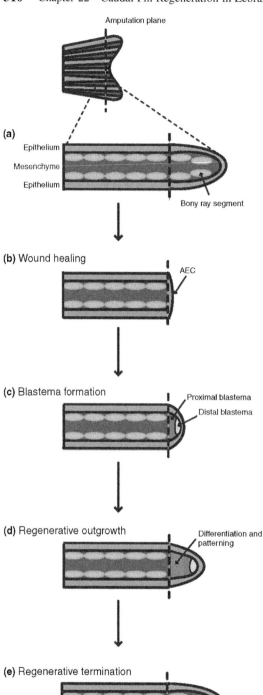

Figure 22.1 Major stages of epimorphic fin regeneration in adult zebrafish caudal fins. A series of cartoons depicting the longitudinal section of a single adult fin ray during the major stages of regeneration are shown in (a)–(d). (a) Mature fin ray covered by epithelium containing a series of bony segments that protect a mesenchymal core. (b) Following amputation, epithelial cells (blue) migrate laterally over the wound to form the AEC. (c) Following formation of the AEC and immature blastema (not pictured), blastemal cells segregate into the distal blastema (purple) and highly proliferative proximal blastema (red). (d) Following formation of the proximal and distal blastemas, a period of regenerative outgrowth is characterized by intense proliferation, cell migration, differentiation, and patterning. (e) Upon completion of regenerative outgrowth and restoration of preamputation size and form of the fin, regenerative signaling is terminated. (See the color version of this figure in Color Plates section.)

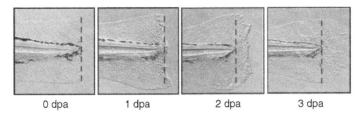

| 0 dpa | 1 dpa | 2 dpa | 3 dpa |

Figure 22.2   Larval caudal fin regeneration. Larval fin tissue was amputated at 2dpf and regenerative growth was observed at 0, 1, 2, and 3 dpa.

Fin regeneration proceeds rapidly until the preamputation fin length is reached, at which point it switches to an ontogenetic growth mechanism. It is speculated that termination occurs either by an unknown active termination mechanism or by cessation of regenerative signaling (Iovine, 2007). In an adult fin model, outgrowth occurs from 2 days post amputation (dpa) until the regenerative event is complete by about 14 dpa (at 28°C). Regeneration occurs more quickly at higher temperatures.

22.5 METHODOLOGY

22.5.1 Adult Fin Amputation

Fish between 4 and 12 months are anesthetized with 0.1% tricaine (3-amino benzoic acidethylester) adjusted to a pH between 7.0–7.5 for approximately 3 min or until gill movements have sufficiently slowed. Anesthetized animals are placed on an inverted sterile glass petri dish and visualized under a stereoscope. Using a sterile beveled razor blade, the caudal fin is partially amputated directly anterior to the fin bifurcation. Fin regeneration takes approximately 12–14 days at 28°C.

22.5.2 Larval Fin Amputation

Two-day-old embryos are manually or chemically (pronase) dechorionated or 3-day-old hatched larvae are anesthetized with 0.1% tricaine. Using a transfer pipette, anesthetized animals are placed on agar plates. Using a microinjection rig mounted with a diamond surgical blade, larvae are amputated just posterior to the notochord (Fig. 22.3). In comparison to the approach described here, a recent larval fin regeneration study describes a method that involves fin amputation anterior to end of the notochord (Rojas-Munoz et al., 2009). While this renders the procedure less technically challenging, the regenerative model then includes both notochord and fin regeneration mechanisms, thereby increasing the complexity of cell types and structures that are replaced.

22.5.3 Pronase Treatment to Dechorionate Embryos

Pronase treatment is used to remove embryonic chorions in order to amputate larval fin tissue at <3 days post fertilization (dpf). To dechorionate embryos, place 50–800

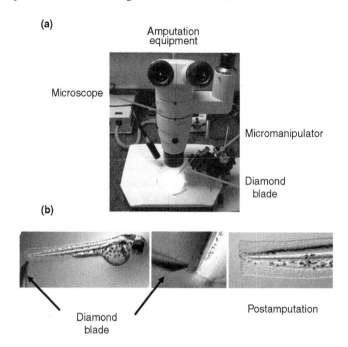

Figure 22.3 Larval caudal fin amputation rig. (a) Photograph depicting the microamputation rig. (b) Embryos at 2 or 3dpf are placed on an agar plate under a microscope and caudal fin tissue is amputated distal to the notochord with a diamond blade.

embryos (2.5–4hpf) into a 60 mm glass Petri dish with 25 mL of fish water. Add 50 μL of 50 mg/mL pronase to the center of the dish and gently and continuously swirl the solution for approximately 7 min. During this period, routinely observe the embryos under a microscope to identify embryos without chorions, chorion remnants in the solution, or deflated chorions. When the above conditions are observed or 7 min have passed, remove the pronase solution by diluting the solution with fresh fish water, swirling, and slowly decanting. Repeat this step for 10 min (approximately 10 washes) in order to dilute the pronase solution and remove displaced chorions. Allow embryos to recover at 28°C until a minimum of 6hpf. Manual dechorionation is an appropriate alternative to chemical dechorionation when using lower numbers of embryos.

22.6 STRATEGIES USED TO MANIPULATE GENE FUNCTION DURING FIN REGENERATION

22.6.1 Mutations

Chemical mutagenesis using *N*-ethyl-*N*-nitrosourea (ENU) is a commonly used chemical mutagen in zebrafish. Mutagenesis of spermatogonia introduces point mutations into the genome that result in both gain-of-function and loss-of-function mutants. Mutagenized populations are generally screened for developmental mutants and identified mutants are subsequently outcrossed and progeny are shotgun sequenced to identify heterozygous germline mutations. This approach has been successfully applied fin regeneration research. As a result, several regeneration mutants have been identified including Fgf20 and Sly1 (Nechiporuk et al., 2003; Whitehead et al., 2005). This approach was modified to allow for the identification of targeted lesions by TILLING, whereby following chemical mutagenesis specific exons are amplified and sequenced to determine whether a mutation has occurred (Wienholds et al., 2003). Insertional mutagenesis was also developed as an alternative to basic chemical mutagenesis in order to obviate the need for positional cloning. Insertional mutagenesis exploits the ability of retroviruses and transposons to insert themselves into a host genome (Gaiano et al., 1996). Following mutagenesis of males by injection of retrovirus or transposon, genomic DNA is isolated from sperm. The sequences flanking the insertions are cloned and sequenced to identify genes that were mutagenized. More recently, zinc finger nuclease (ZFN) technology has enabled targeted gene knockout in zebrafish by exploiting the inherent ability of Cys_2His_2zinc finger protein (ZFP) transcription factors to target specific DNA sequences (Doyon et al., 2008; Meng et al., 2008). ZFNs are chimeric fusions between ZFPs and the nonspecific cleavage domain of Fok1 endonuclease. The ZFPs confer exquisite binding specificity to allow the introduction of double stranded breaks by endonuclease activity. Double strand breaks are repaired by error-prone nonhomologous end joining resulting in the generation of a pool of mutants that contain both insertions and deletions at the target site. Mutants are identified by amplification and sequencing of the targeted genomic regions. ZFN technology is advantageous because in addition to allowing the introduction of specific mutations, germline transmission efficiency is high where >10% of progeny carry mutations in genes of interest (Doyon et al., 2008; Meng et al., 2008).

22.6.2 Transgenics

Transgenesis or the ability to insert foreign DNA into a host genome is used extensively in fin regeneration studies. This approach is utilized to express particular gene products in a highly controlled manner. Tissue-specific or inducible promoters are used to control the spatiotemporal expression of the transgene. Transgenesis typically employs fluorescently tagged gene products to enable efficient determination of trans-expressing progeny. Injection of plasmid DNA or bacterial artificial chromosomes (BACs) into the cytoplasm of a one-cell stage embryo has historically been the most commonly used method for creating transgenic lines (Amsterdam and Becker, 2005). Subsequently, site-specific recombination systems such as Cre/loxP and Tol2 transposon-mediated transgenesis have been successfully used to create transgenic zebrafish lines (Fisher et al., 2006; Yoshikawa et al., 2008). More recently, a temporally controlled site-specific Cre-Tol2 transposon method was developed that significantly increases transgenesis efficiency (Hans et al., 2009). In addition, next-generation ZFNs exploiting homologous end-joining repair mechanisms will soon be developed to allow for an additional method for introducing targeted gene knock-ins into the zebrafish genome.

Heat shock inducible transgenics expressing dominant negative gene products have been extremely successful at unraveling the molecular signaling pathways that dictate fin regeneration. Typically, heat shock promoters (Hsp) are used to control expression of dominant negative gene products. This strategy is particularly useful when examining the role of a gene that is required for development. Following development, fish are heat shocked at 35–38°C to express the transgene fused to a fluorescent reporter such as enhanced green fluorescent protein (EGFP). In the context of fin regeneration, Hsp transgenics have supported a role for FGF signaling through the overexpression of dominant negative Fgfr1 (Lee et al., 2005) and for Wnt–β-catenin signaling through overexpression of the negative regulator Dikkopf1, dominant negative Tcf3, wnt8a, or wnt5b (Stoick-Cooper et al., 2007b).

22.6.3 Antisense-Mediated Gene Knockdown

Morpholinos are modified antisense oligos that knockdown gene function by inhibiting mRNA splicing or translation initiation. Translation inhibition morpholinos display sequence complementarity to the 5′ untranslated region or translation start site while splice site blockers have sequence complementarity to intron–exon boundaries. Morpholino efficacy must be validated by immunohistochemistry or *in situ* hybridization dependent upon antibody availability and researchers should be cognizant of potential off-target effects (Amsterdam and Hopkins, 2006).

Morpholinos are widely used in fin regeneration studies in both adult and larval models. The zebrafish larval fin regeneration model is particularly well suited for efficient gene knockdown by morpholino injection. Morpholinos are injected into either the yolk stream or cytoplasm at the one-cell stage resulting in transient knockdown of gene expression. We have reported efficient gene knockdown in the larval fin regeneration model (Mathew et al., 2007, 2008a, 2008b). It is important to note that as embryogenesis proceeds, the morpholino concentration is exponentially reduced. While morpholino-mediated gene knockdown is possible in adult fins, it requires numerous injections per fish followed by electroporation to enable morpholinos to gain intracellular access. This is a result of more complex fin architecture present in adult fins (18 lepidotrichia and 18 blastemas following amputation). In addition, temporal restraints provide a significant limitation of morpholino injections in the adult fin model. Following amputation, researchers typically wait ~24 h before microinjection and electroporation to allow the wound to heal rendering this approach less well suited for the discovery of the signaling events that initiate blastema formation (Thummel et al., 2006; Thatcher et al., 2008).

22.6.4 Ectopic Gene Expression and Promoter Analysis

Ectopic expression of gene products suspected to impact regeneration has been used to investigate regenerative mechanisms. In the larval model, injection of plasmid DNA fused with a reporter has supported a role for sonic hedgehog signaling in bone formation and patterning during fin regeneration (Quint et al., 2002). Ectopic expression of promoter activity has recently been applied to the larval zebrafish model. Embryos are microinjected with constructs bearing a promoter of interest fused to a firefly luciferase reporter gene (Alcaraz-Perez et al., 2008).

22.6.5 Chemical Genetics

Zebrafish embryos are well suited for high-throughput chemical screening. Embryos can be loaded into 96-well plates and remain viable for 5 days without added nutrients. As such, zebrafish are commonly used in a tiered toxicology screening approach to identify classes of compounds that affect biomarkers of specific tissues' and systems' function. Chemical genetic approaches have additionally been employed to elucidate the role of Wnt signaling in epimorphic regeneration (Mathew et al., 2007, 2008b; Chen et al., 2009). Through the use of a diverse synthetic chemical library, a new class of Wnt pathway inhibitors was revealed to negatively impact fin regeneration. This class of compounds prevents destruction of Axin proteins, thereby preventing liberation of β-catenin from the APC–Axin–GSK3β inhibitory complex and increasing β-catenin degradation. We have reported that inappropriate activation of aryl hydrocarbon receptor (AHR) signaling following 2,3,7,8-tetrachlorodibenzo-p-dioxin (TCDD) exposure impairs tissue regeneration in larval and adult zebrafish (Zodrow and Tanguay, 2003; Mathew et al., 2006). Comparative toxicogenomic analysis revealed misexpression of Wnt signaling pathway genes in TCDD-exposed fin regenerates (Mathew et al., 2008a). R-Spondin1, a potent activator of Wnt/β-catenin signaling, was highly expressed in regenerating fin tissue of TCDD-exposed fish. In support of this, we have shown that transient knockdown of R-Spondin1 or LRP5/6 by morpholino injection blocked TCDD-induced inhibition of fin regeneration suggesting that TCDD impairs regeneration through inappropriate activation of canonical Wnt signaling (Mathew et al., 2008a). Together, these studies support the concept that canonical Wnt signaling plays a primary role in epimorphic regeneration and, in addition, highlight the effectiveness of chemical genetics in revealing novel molecular targets involved in epimorphic tissue regeneration.

Our group conducted a large-scale screen of a 2000-member chemical library to identify modulators of larval tissue regeneration (Mathew et al., 2007). We found that five members of the glucocorticoid family similarly impaired fin regeneration (Mathew et al., 2007). These data were confirmed using a model GR agonist beclomethasone dipropionate (BDP) (Fig. 22.4). Importantly, transient knockdown of GR activity by morpholino injection blocked BDP-induced impairment of fin regeneration (Mathew et al., 2007) (Fig. 22.4). These data indicate that the inhibitory effects of BDP exposure on tissue regeneration are GR-dependent. These data further support the relevance of using chemical genetics to unravel the signaling pathways that dictate tissue regeneration.

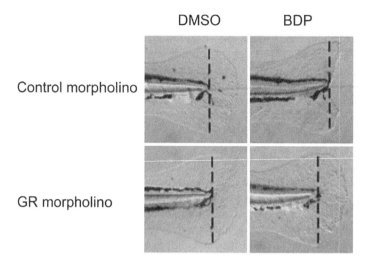

Figure 22.4 Beclomethasone-dependent inhibition of fin regeneration is GR dependent. Zebrafish embryos microinjected at the one-cell stage with GR-directed antisense oligonucleotide morpholino or control morpholino were amputated at 2dpf and exposed to beclomethasone at 2dpf. Regeneration was visually assessed at 3dpf.

22.7 THE LARVAL FIN REGENERATION MODEL

The adult fin regeneration model has unique advantages, but its usefulness as a regenerative model is limited by technical barriers such as the length of time necessary to raise adult fish (2–4 months) and allow for complete fin regeneration (12–14 days) as well as difficulty achieving genetic manipulation in the fin. It is particularly challenging to study early signaling events that promote blastema formation because wound healing must occur prior to microinjection of antisense oligonucleotides that are commonly used in loss-of-function experiments (Thatcher et al., 2008). Recently, it was reported that the fin primordia is capable of complete regeneration (2–5 dpf) similar to the adult zebrafish (Kawakami et al., 2004). Morphologically, even in the absence of complete cellular differentiation, the larval fin regenerates similarly to the adult by the development of an apical wound epithelium followed by blastema formation, which later proliferates and differentiates into the required cell types (Kawakami et al., 2004). Also, similar to the adult zebrafish, chemical inhibition of fgfr1 by SU5402, AHR activation by TCDD, and GR activation by beclomethasone abrogated larval fin regeneration (Poss et al., 2000; Zodrow and Tanguay, 2003; Mathew et al., 2006, 2007) suggesting that there are similarities at the cellular and molecular level between adult and larval regeneration. Mainly through marker analysis, Yoshinari et al. (2009) showed that a number of genes including dlx5a, msxe, junbl, ilf2, mvp, phlda2, mmp9, hspa9, junb, and smarca4 are expressed in conserved domains in both the larval and adult models. In support of this, we conducted a comparative analysis of genes similarly regulated during larval fin and adult fin and heart regeneration. 64% of the genes that are misregulated during the early stages of regeneration (1 dpa) are conserved between nonamputated control and regenerating fin tissue in the adult and larval fin regeneration models (Mathew et al., 2009). Taken together, these data suggest that there is a high probability of common hierarchical molecular signaling pathways across larval and adult regeneration systems.

22.8 SUMMARY

Regeneration is a research area that has captivated the imagination of biologists for decades, yet there are currently numerous gaps in our understanding of the complex pathways and interactions that dictate regeneration. The methods described here will facilitate the molecular dissection of the signaling events that control wound healing, blastema formation, regenerative outgrowth, and termination of regenerative signaling. Importantly, studies designed to probe the signaling events that choreograph regeneration will provide new avenues for comparative studies in mammalian models in the hope of developing novel therapeutics to help slow and/or prevent tissue loss from injury, aging, and disease, thereby improving the quality and duration of life.

ACKNOWLEDGMENTS

We thank Jill Franzosa and Margaret Corvi for helpful discussions and the critical review of this manuscript. This research was supported by the NIEHS Environmental Health Sciences Center Grant No. ES00210 and NSF Grant No. 0641409.

REFERENCES

AKIMENKO MA, MARI-BEFFA M, BECERRA J, and GERAUDIE J (2003). Old questions, new tools, and some answers to the mystery of fin regeneration. Dev Dyn 226(2): 190–201.

ALCARAZ-PEREZ F, MULERO V, and CAYUELA ML (2008). Application of the dual-luciferase reporter assay to the analysis of promoter activity in zebrafish embryos. BMC Biotechnol 8: 81.

AMSTERDAM A and BECKER TS (2005). Transgenes as screening tools to probe and manipulate the zebrafish genome. Dev Dyn 234(2): 255–268.

AMSTERDAM A and HOPKINS N (2006). Mutagenesis strategies in zebrafish for identifying genes involved in development and disease. Trends Genet 22(9): 473–478.

BECKER CG and BECKER T (2002). Repellent guidance of regenerating optic axons by chondroitin sulfate glycosaminoglycans in zebrafish. J Neurosci 22(3): 842–853.

BECKER CG, LIEBEROTH BC, MORELLINI F, FELDNER J, BECKER T, and SCHACHNER M (2004). L1.1 is involved in spinal cord regeneration in adult zebrafish. J Neurosci 24(36): 7837–7842.

CHEN B, DODGE ME, TANG W, LU J, MA Z, FAN CW, WEI S, HAO W, KILGORE J, WILLIAMS NS, ROTH MG, AMATRUDA JF, CHEN C, and LUM L (2009). Small molecule-mediated disruption of Wnt-dependent signaling in tissue regeneration and cancer. Nat Chem Biol 5(2): 100–107.

DEL RIO-TSONIS K, JUNG JC, CHIU IM, and TSONIS PA (1997). Conservation of fibroblast growth factor function in lens regeneration. Proc Natl Acad Sci USA 94(25): 13701–13706.

DOYON Y, MCCAMMON JM, MILLER JC, FARAJI F, NGO C, KATIBAH GE, AMORA R, HOCKING TD, ZHANG L, REBAR EJ, GREGORY PD, URNOV FD, and AMACHER SL (2008). Heritable targeted gene disruption in zebrafish using designed zinc-finger nucleases. Nat Biotechnol 26(6): 702–708.

FERRETTI P and GERAUDIE J. (1995). Retinoic acid-induced cell death in the wound epidermis of regenerating zebrafish fins. Dev Dyn 202(3): 271–283.

FISHER S, GRICE EA, VINTON RM, BESSLING SL, URASAKI A, KAWAKAMI K, and McCALLION AS (2006). Evaluating the biological relevance of putative enhancers using Tol2 transposon-mediated transgenesis in zebrafish. Nat Protoc 1(3): 1297–1305.

FLOSS T, ARNOLD HH, and BRAUN T (1997). A role for FGF-6 in skeletal muscle regeneration. Genes Dev 11(16): 2040–2051.

GAIANO N, ALLENDE M, AMSTERDAM A, KAWAKAMI K, and HOPKINS N (1996). Highly efficient germ-line transmission of proviral insertions in zebrafish. Proc Natl Acad Sci USA 93(15): 7777–7782.

GOESSLING W, NORTH TE, LOEWER S, LORD AM, LEE S, STOICK-COOPER CL, WEIDINGER G, PUDER M, DALEY GQ, MOON RT, and ZON LI (2009). Genetic interaction of PGE2 and Wnt signaling regulates developmental specification of stem cells and regeneration. Cell 136(6): 1136–1147.

HANS S, KASLIN J, FREUDENREICH D, and BRAND M (2009). Temporally-controlled site-specific recombination in zebrafish. PLoS One 4(2): e4640.

IOVINE MK (2007). Conserved mechanisms regulate outgrowth in zebrafish fins. Nat Chem Biol 3(10): 613–618.

KAWAKAMI A, FUKAZAWA T, and TAKEDA H (2004). Early fin primordia of zebrafish larvae regenerate by a similar growth control mechanism with adult regeneration. Dev Dyn 231(4): 693–699.

LEE Y, GRILL S, SANCHEZ A, MURPHY-RYAN M, and POSS KD (2005). Fgf signaling instructs position-dependent growth rate during zebrafish fin regeneration. Development 132(23): 5173–5183.

LEE Y, HAMI D, DE VAL S, KAGERMEIER-SCHENK B, WILLS AA, BLACK BL, WEIDINGER G, and POSS KD (2009). Maintenance of blastemal proliferation by functionally diverse epidermis in regenerating zebrafish fins. Dev Biol 331(2): 270–280.

LEPILINA A, COON AN, KIKUCHI K, HOLDWAY JE, ROBERTS RW, BURNS CG, and POSS KD (2006). A dynamic epicardial injury response supports progenitor cell activity during zebrafish heart regeneration. Cell 127(3): 607–619.

LOPEZ-SCHIER H and HUDSPETH AJ (2006). A two-step mechanism underlies the planar polarization of regenerating sensory hair cells. Proc Natl Acad Sci USA 103(49): 18615–18620.

LOVINE MK (2007). Conserved mechanisms regulate outgrowth in zebrafish fins. Nat Chem Biol 3(10): 613–618.

MATHEW LK, ANDREASEN EA, and TANGUAY RL (2006). Aryl hydrocarbon receptor activation inhibits regenerative growth. Mol Pharmacol 69(1): 257–265.

MATHEW LK, SENGUPTA S, KAWAKAMI A, ANDREASEN EA, LOHR CV, LOYNES CA, RENSHAW SA, PETERSON RT, and TANGUAY RL (2007). Unraveling tissue regeneration pathways using chemical genetics. J Biol Chem 282(48): 35202–35210.

MATHEW LK, SENGUPTA SS, LADU J, ANDREASEN EA, and TANGUAY RL (2008a). Crosstalk between AHR and Wnt signaling through R-Spondin1 impairs tissue regeneration in zebrafish. FASEB J 22(8): 3087–3096.

MATHEW LK, SIMONICH MT, and TANGUAY RL (2008b). AHR-dependent misregulation of Wnt signaling disrupts tissue regeneration. Biochem Pharmacol 77(4): 498–507.

MATHEW LK, SENGUPTA S, FRANZOSA JA, PERRY J, LA DU J, ANDREASEN EA, and TANGUAY RL (2009). Comparative expression profiling reveals an essential role for Raldh2 in epimorphic regeneration. J Biol Chem doi: 10.1074/jbc.M109.011668.

MENG X, NOYES MB, ZHU LJ, LAWSON ND, and WOLFE SA (2008). Targeted gene inactivation in zebrafish using engineered zinc-finger nucleases. Nat Biotechnol 26(6): 695–701.

MONGA SP, PEDIADITAKIS P, MULE K, STOLZ DB, and MICHALOPOULOS GK (2001). Changes in WNT/beta-catenin pathway during regulated growth in rat liver regeneration. Hepatology 33(5): 1098–1109.

NECHIPORUK A and KEATING MT (2002). A proliferation gradient between proximal and msxb-expressing distal blastema directs zebrafish fin regeneration. Development 129(11): 2607–2617.

NECHIPORUK A, POSS KD, JOHNSON SL, and KEATING MT (2003). Positional cloning of a temperature-sensitive mutant emmental reveals a role for sly1 during cell proliferation in zebrafish fin regeneration. Dev Biol 258(2): 291–306.

OSAKADA F, OOTO S, AKAGI T, MANDAI M, AKAIKE A, and TAKAHASHI M (2007). Wnt signaling promotes regeneration in the retina of adult mammals. J Neurosci 27(15): 4210–4219.

OTTO A, SCHMIDT C, LUKE G, ALLEN S, VALASEK P, MUNTONI F, LAWRENCE-WATT D, and PATEL K (2008). Canonical Wnt signalling induces satellite-cell proliferation during adult skeletal muscle regeneration. J Cell Sci 121(Pt 17): 2939–2950.

PETERSEN CP and REDDIEN PW (2008). Smed-betacatenin-1 is required for anteroposterior blastema polarity in planarian regeneration. Science 319(5861): 327–330.

POSS KD, KEATING MT, and NECHIPORUK A (2003). Tales of regeneration in zebrafish. Dev Dyn 226(2): 202–210.

POSS KD, SHEN J, NECHIPORUK A, McMAHON G, THISSE B, THISSE C, and KEATING MT (2000). Roles for Fgf signaling during zebrafish fin regeneration. Dev Biol 222(2): 347–358.

POSS KD, WILSON LG, and KEATING MT (2002). Heart regeneration in zebrafish. Science 298(5601): 2188–2190.

QUINT E, SMITH A, AVARON F, LAFOREST L, MILES J, GAFFIELD W, and AKIMENKO MA (2002). Bone patterning is altered in the regenerating zebrafish caudal fin after ectopic expression of sonic hedgehog and bmp2b or exposure to cyclopamine. Proc Natl Acad Sci USA 99(13): 8713–8718.

RAYA A, CONSIGLIO A, KAWAKAMI Y, RODRIGUEZ-ESTEBAN C, and IZPISUA-BELMONTE JC (2004). The zebrafish as a model of heart regeneration. Cloning Stem Cells 6(4): 345–351.

ROJAS-MUNOZ A, RAJADHYKSHA S, GILMOUR D, VAN BEBBER F, ANTOS C, RODRIGUEZ ESTEBAN C, NUSSLEIN-VOLHARD C, and IZPISUA BELMONTE JC (2009). ErbB2 and ErbB3 regulate amputation-induced proliferation and migration during vertebrate regeneration. Dev Biol 327(1): 177–190.

SADLER KC, KRAHN KN, GAUR NA, and UKOMADU C (2007). Liver growth in the embryo and during liver regeneration in zebrafish requires the cell cycle regulator, uhrf1. Proc Natl Acad Sci USA 104(5): 1570–1575.

STOICK-COOPER CL, MOON RT, and WEIDINGER G (2007a). Advances in signaling in vertebrate regeneration as a prelude to regenerative medicine. Genes Dev 21(11): 1292–1315.

STOICK-COOPER CL, WEIDINGER G, RIEHLE KJ, HUBBERT C, MAJOR MB, FAUSTO N, and MOON RT (2007b). Distinct Wnt signaling pathways have opposing roles in appendage regeneration. Development 134(3): 479–489.

THATCHER EJ, PAYDAR I, ANDERSON KK, and PATTON JG (2008). Regulation of zebrafish fin regeneration by microRNAs. Proc Natl Acad Sci USA 105(47): 18384–18389.

THUMMEL R, BAI S, SARRAS MP JR., SONG P, McDERMOTT J, BREWER J, PERRY M, ZHANG X, HYDE DR, and GODWIN AR (2006). Inhibition of zebrafish fin regeneration using *in vivo* electroporation of morpholinos against fgfr1 and msxb. Dev Dyn 235(2): 336–346.

WHITEHEAD GG, MAKINO S, LIEN CL, and KEATING MT (2005). fgf20 is essential for initiating zebrafish fin regeneration. Science 310(5756): 1957–1960.

WIENHOLDS E, VAN EEDEN F, KOSTERS M, MUDDE J, PLASTERK RH, and CUPPEN E (2003). Efficient target-selected mutagenesis in zebrafish. Genome Res 13(12): 2700–2707.

YOSHIKAWA S, KAWAKAMI K, and ZHAO XC (2008). G2R Cre reporter transgenic zebrafish. Dev Dyn 237(9): 2460–2465.

YOSHINARI N, ISHIDA T, KUDO A, and KAWAKAMI A (2009). Gene expression and functional analysis of zebrafish larval fin fold regeneration. Dev Biol 325(1): 71–81.

ZODROW JM and TANGUAY RL (2003). 2,3,7,8-Tetrachlorodibenzo-*p*-dioxin inhibits zebrafish caudal fin regeneration. Toxicol Sci 76(1): 151–161.

Index

Zebrafish: Methods for Assessing Drug Safety and Toxicity, Edited by Patricia McGrath.
© 2012 John Wiley & Sons, Inc. Published 2012 by John Wiley & Sons, Inc.

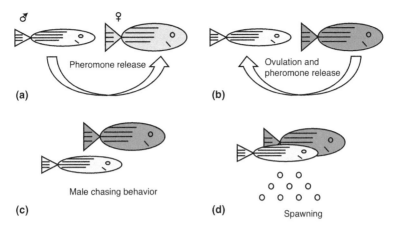

Figure 1.1 Simplified model of olfactory control of zebrafish reproduction. (*See text for full caption.*)

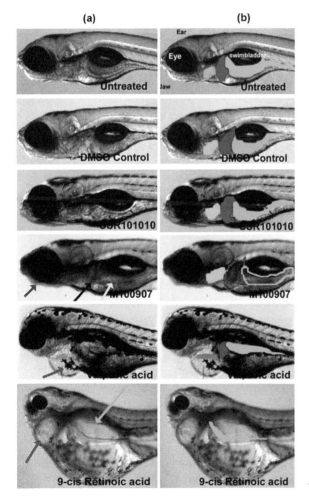

Figure 2.2 Organ structure of treated zebrafish at high magnification (8×). (*See text for full caption.*)

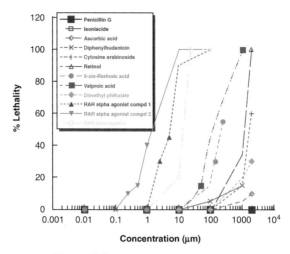

Figure 2.3 Lethality curves in zebrafish.

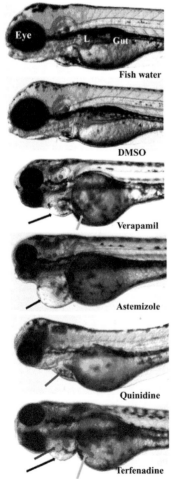

Figure 4.2 Cardiovascular morphological defects in drug-treated zebrafish. Two dpf zebrafish were treated with compounds for 24 h. Fish water and 1% DMSO-treated zebrafish exhibited normal morphology. L: liver. Black arrows: pericardial edema in verapamil-, astemizole-, and terfenadine-treated animals. Red arrow: ventricle swelling due to blood pooling in quinidine-treated animals. Blue arrow: ventricle compression due to atrial swelling in terfenadine-treated animals. Green arrow: stagnant blood flow in the cardinal vein in verapamil- and terfenadine-treated animals.

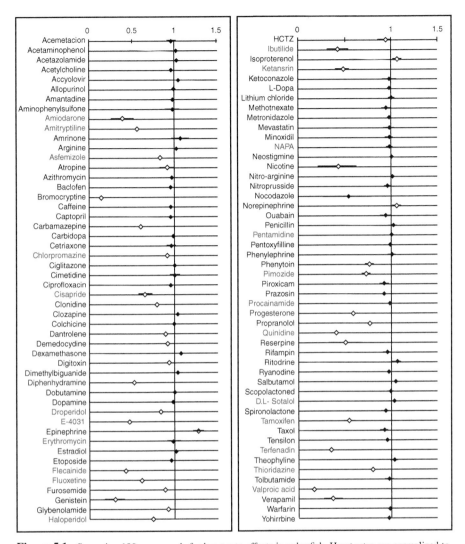

Figure 5.1 Screening 100 compounds for heart rate effects in zebrafish. Heart rates are normalized to vehicle controls. Mean heart rate and standard deviation are shown. Compounds listed in red prolong the QT interval in humans and have been associated with TdP. Blue color indicates association with QT prolongation, but not TdP in humans. Open diamonds signify a statistically significant difference from control ($p < 0.05$). (Figure updated from Milan et al., 2003)

ECG recording electrode placement

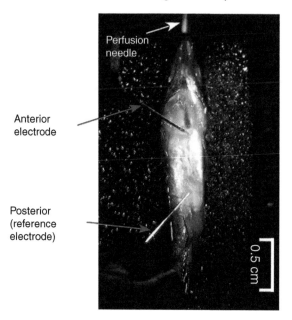

Perfusion needle

Anterior electrode

Posterior (reference electrode)

0.5 cm

Figure 6.7 Photograph of the experimental setup. The ventral surface of the adult zebrafish is visible. There is a perfusion needle resting in the mouth of the fish (white arrow) and the recording electrodes are seen in their standard locations (red arrows). (Figure adapted from Milan et al., 2006)

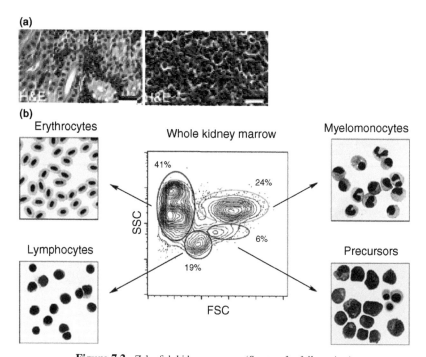

(a)

H&E   H&E

(b)

Erythrocytes   Whole kidney marrow   Myelomonocytes

41%   24%

SSC   6%

Lymphocytes   Precursors

19%

FSC

Figure 7.2 Zebrafish kidney marrow. (*See text for full caption.*)

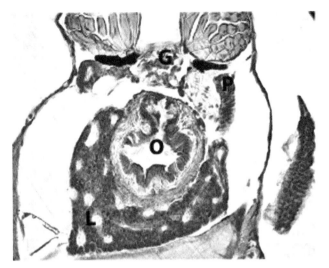

Figure 8.1 Transverse section of a wild-type Tu zebrafish larvae at 120 hpf. (*See text for full caption.*)

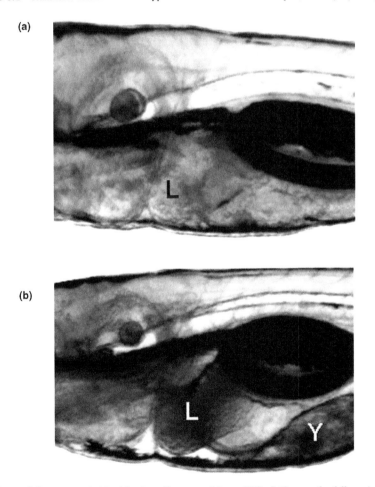

(a)

(b)

Figure 8.2 Phenotypic identification of hepatotoxicity at 120 hpf. (*See text for full caption.*)

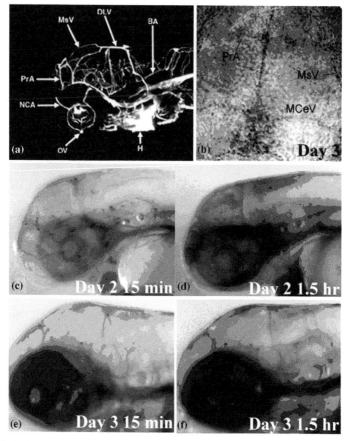

Figure 10.6 Confirmation of BBB formation in zebrafish by microangiography, antibody staining, and dye injection. (*See text for full caption.*)

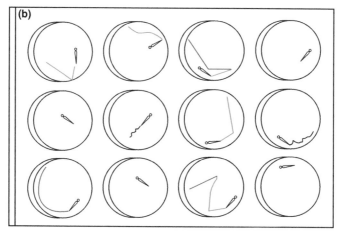

Figure 10.7 (b) Diagram of zebrafish motion tracking results. Diagram shows a portion of a 96-well microplate, one zebrafish per well. VideoTrack software analyzes length, speed, and direction of movement. Speed of movement is color coded: black, slow; green, medium; red, fast.

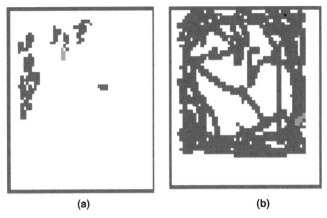

(a)   (b)

Figure 10.8 Graphical representation of movement of a single zebrafish (6 dpf). (*See text for full caption.*)

2 dpf   4 dpf

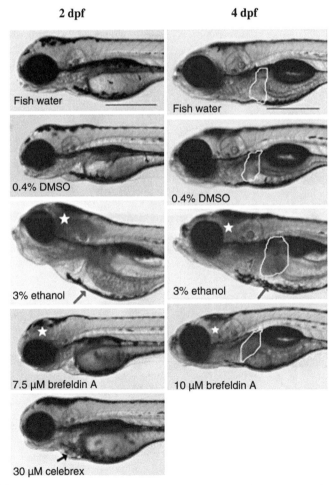

Fish water

0.4% DMSO

3% ethanol

7.5 μM brefeldin A

30 μM celebrex

Fish water

0.4% DMSO

3% ethanol

10 μM brefeldin A

Figure 11.1 Effects of carrier and positive control compounds on zebrafish organs. (*See text for full caption.*)

Heart and CNS　　　**Liver and kidney**

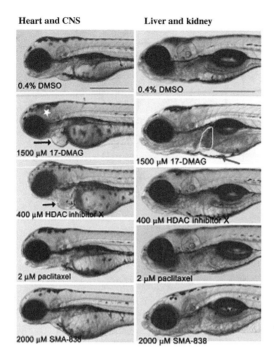

Figure 11.2 Compound-induced toxicity in zebrafish heart, CNS, liver, and kidney after 24 h treatment. (*See text for full caption.*)

Heart　　　**CNS**　　　**Liver and kidney**

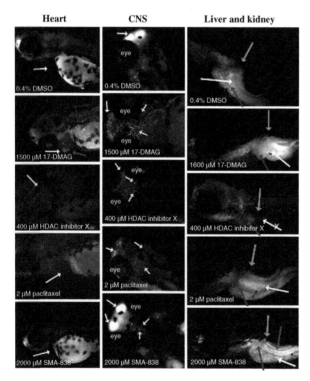

Figure 11.3 Cell death assessed in heart and CNS and liver and kidney in 3 and 5 dpf zebrafish, respectively. (*See text for full caption.*)

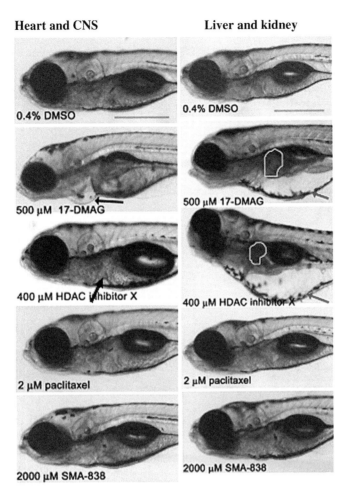

Figure 11.4 Toxicity in zebrafish heart, CNS, liver, and kidney 48 h after compound removal. (*See text for full caption.*)

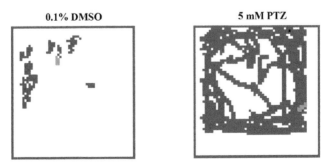

Figure 13.1 Graphical representation of movement of single larval zebrafish (left, DMSO control; right, 5 mM PTZ) during 60 min. (*See text for full caption.*)

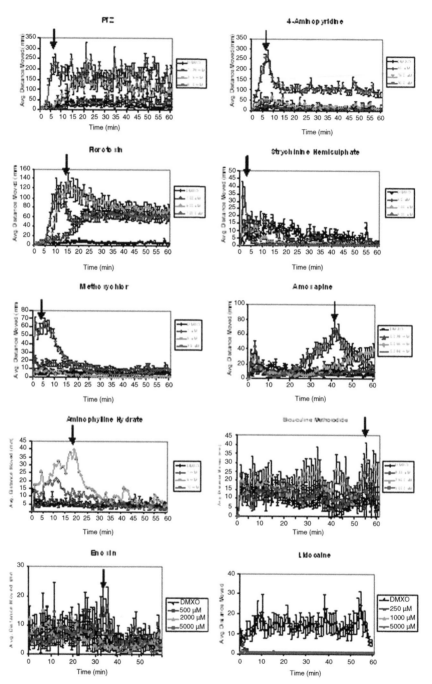

Figure 13.2 Kinetics of *D* traveled in 1 min intervals for 60 min. *D* was measured for 60 min for each compound concentration. (*See text for full caption.*)

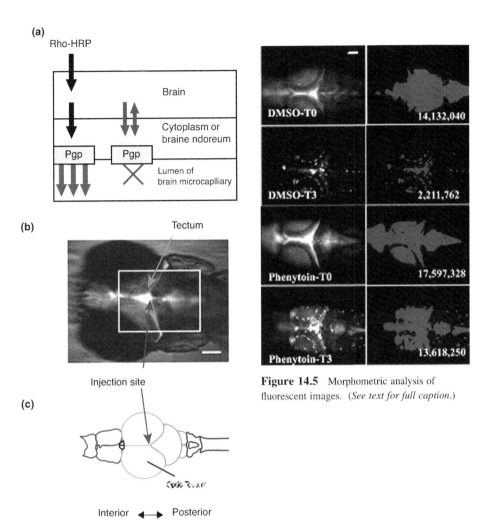

(a)

Rho-HRP

Brain

Cytoplasm or braine ndoreum

Pgp

Pgp

Lumen of brain microcaplliary

(b)

Tectum

Injection site

(c)

Interior ⟷ Posterior

Figure 14.2 Rho-HRP efflux from the brain. (*See text for full caption.*)

DMSO-T0 — 14,132,040

DMSO-T3 — 2,211,762

Phenytoin-T0 — 17,597,328

Phenytoin-T3 — 13,618,250

Figure 14.5 Morphometric analysis of fluorescent images. (*See text for full caption.*)

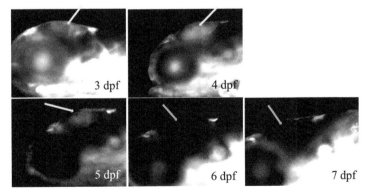

3 dpf

4 dpf

5 dpf

6 dpf

7 dpf

Figure 14.3 Optimal stage to assess Pgp efflux. (*See text for full caption.*)

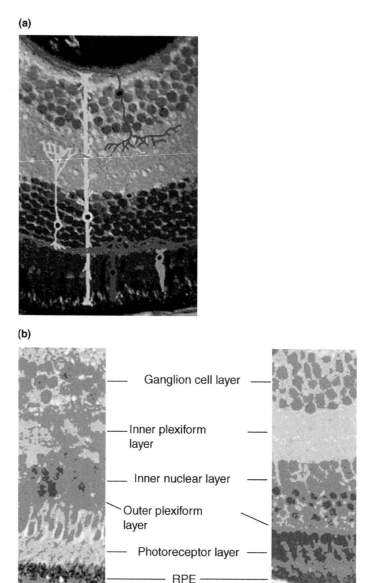

Figure 15.1 (a) The cell layout in the retina is highly stereotypical. This plastic section through an embryonic zebrafish retina has been overlaid for illustration. The lens is at the top and retinal pigment epithelium at the bottom. Red: ganglion cell; light blue: Mueller cell; green: amacrine cell; dark blue: horizontal cell; yellow: bipolar cell; pink: rod photoreceptor; orange: cone photoreceptor. (b) Plastic sections through an adult human (left) and embryonic zebrafish (right) retina. The relative positions of the various cell types are the same in both. (Reprinted from Goldsmith and Harris (2003) with permission from Elsevier.)

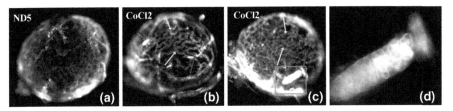

Figure 16.1 Full view of intact CVP in the back of isolated eyes. (*See text for full caption.*)

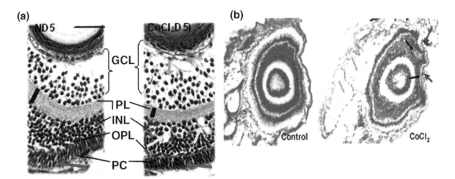

Figure 16.2 Histological assessment of CoCl$_2$-treated zebrafish eyes. (*See text for full caption.*)

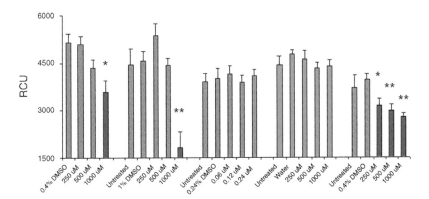

Figure 17.3 Drug effects on Xt Colo320 cells. (*See text for full caption.*)

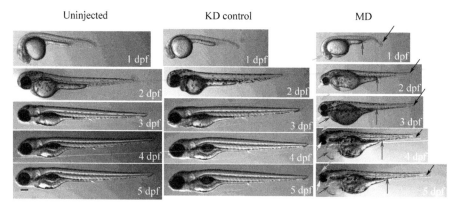

Figure 18.1 Comparison of phenotypes for uninjected, AS siRNA (KD control), and SMP siRNA (MD) injected zebrafish. (*See text for full caption.*)

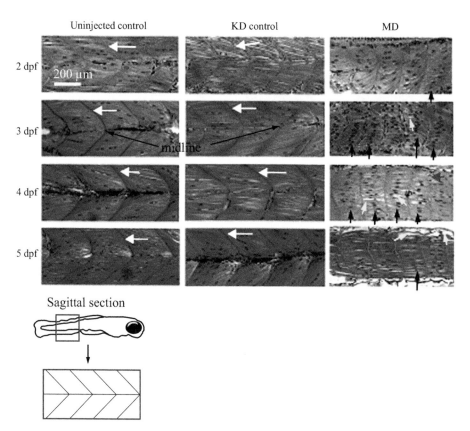

Figure 18.7 Histological assessment of muscle structure in uninjected, KD control, and MD zebrafish. (*See text for full caption.*)

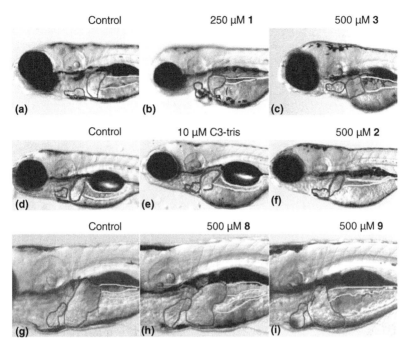

Figure 19.5 Morphological organ abnormalities in the development of liver, intestine, and heart caused by doses of fullerenes C3 (**1**), C3-tris, **2**, **3**, **8**, and **9** near LC_{50}. (*See text for full caption.*)

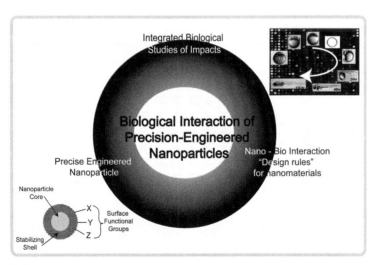

Figure 20.2 Using biological interactions of engineered nanoparticles to understand design rules for nanoparticles with enhanced biological properties.

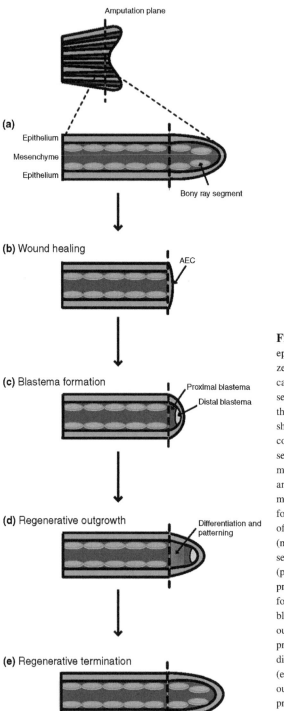

(a)

Amputation plane

Epithelium

Mesenchyme

Epithelium

Bony ray segment

(b) Wound healing

AEC

(c) Blastema formation

Proximal blastema

Distal blastema

(d) Regenerative outgrowth

Differentiation and patterning

(e) Regenerative termination

Figure 22.1 Major stages of epimorphic fin regeneration in adult zebrafish caudal fins. A series of cartoons depicting the longitudinal section of a single adult fin ray during the major stages of regeneration are shown in (a)–(d). (a) Mature fin ray covered by epithelium containing a series of bony segments that protect a mesenchymal core. (b) Following amputation, epithelial cells (blue) migrate laterally over the wound to form the AEC. (c) Following formation of the AEC and immature blastema (not pictured), blastemal cells segregate into the distal blastema (purple) and highly proliferative proximal blastema (red). (d) Following formation of the proximal and distal blastemas, a period of regenerative outgrowth is characterized by intense proliferation, cell migration, differentiation, and patterning. (e) Upon completion of regenerative outgrowth and restoration of preamputation size and form of the fin, regenerative signaling is terminated.

Printed and bound by CPI Group (UK) Ltd, Croydon, CR0 4YY

16/04/2025

14658518-0003